T0332114

# Modern Aspects *of* Superconductivity

## Theory of Superconductivity

### Second Edition

# Modern Aspects *of* Superconductivity

## Theory of Superconductivity

### Second Edition

**Sergei Kruchinin**

*Bogolyubov Institute for Theoretical Physics, Kyiv, Ukraine*

 **World Scientific**

NEW JERSEY · LONDON · SINGAPORE · BEIJING · SHANGHAI · HONG KONG · TAIPEI · CHENNAI · TOKYO

*Published by*

World Scientific Publishing Co. Pte. Ltd.

5 Toh Tuck Link, Singapore 596224

*USA office:* 27 Warren Street, Suite 401-402, Hackensack, NJ 07601

*UK office:* 57 Shelton Street, Covent Garden, London WC2H 9HE

**British Library Cataloguing-in-Publication Data**
A catalogue record for this book is available from the British Library.

**MODERN ASPECTS OF SUPERCONDUCTIVITY**
**Theory of Superconductivity**
**Second Edition**

ISBN 978-981-123-451-4 (hardcover)
ISBN 978-981-123-452-1 (ebook for institutions)
ISBN 978-981-123-453-8 (ebook for individuals)

For any available supplementary material, please visit
https://www.worldscientific.com/worldscibooks/10.1142/12215#t=suppl

Desk Editor: Rhaimie Wahap

Typeset by Stallion Press
Email: enquiries@stallionpress.com

Printed in Singapore

"The single reason for our inability to treat the problems of superconductivity consists in the absence of a sufficient imagination."

Richard P. Feynman

# Preface

The study in the field of superconductivity theory is one of the most bright, fruitful, and promising trends in the theoretical physics of condensed matter, since superconductivity remains to be one of the most interesting research areas in physics.

The goal of the book is to give representation of certain modern aspects of superconductivity. We discuss such important aspects of the theory of superconductivity as the nature of high-$T_c$ superconductivity, two-gap superconductivity, room-temperature superconductivity, mesoscopic superconductivity, the pairing state and the mechanism of cuprate high-$T_c$ superconductivity.

We will consider also the up-to-date state of the problem of real construction of a quantum computer on the base of superconducting units.

In Chapter 1, we consider the field-theoretic method of superconductivity and discuss the basic idea of superconductivity and the elaboration of the Ginzburg–Landau and BCS theories in the frame of many-particle quantum field theory.

In Chapter 2, we consider such topics of high-$T_c$ superconductivity as the structures of high-$T_c$ superconductors, phase diagrams, and the problem of pseudogaps and analyze the mechanisms of superconductivity. We present general arguments as for the pairing symmetry in cuprate superconductors and investigate their thermodynamical properties within the spin-fluctuation mechanism of superconductivity, by using the method of functional integrals.

In Chapter 3, we consider the new class of high-temperature superconductors such as Fe-based laminar compounds. We analyze the physical properties and electron models of Fe-based high-temperature superconductors. The particular interest in them is explained be perspectives of their practical application. In the chapter, we give the complete pattern of the formation of their physical properties within theoretical models on the basis of their electron structure.

Chapter 4 concentrates on the two-band and multiband superconductivity. We consider the physical properties of superconductor $MgB_2$ and use our two-band model to explain the two coupled superconductor's gaps of $MgB_2$. To study the effect of the increasing $T_c$ in $MgB_2$, we used the renormalization group approach and phase diagrams. In the field of superconductivity, we meet the problem-maximum — it consists in the creation of room-temperature superconductors. We consider this problem in our book, and give some recommendations on the search for these superconductors.

We consider the problem of high-temperature superconductivity at high pressures in hydrides and present the Éliashberg–Migdal theory explaining the room-temperature superconductivity in this case.

Chapter 5 deals with the problem of nanoscale superconductivity. We consider the two-band superconductivity in ultrasmall grains, by extending the Richardson exact solution to two-band systems, and develop the theory of interactions between nano-scale ferromagnetic particles and superconductors. The properties of nano-sized two-gap superconductors and the Kondo effect in superconducting ultrasmall grains are investigated as well.

At the present time, popular is the topic of the nano-engineering with superconducting states. In this connection, we deal with the physics of magnetic dots on the superconductors and the problem of superconducting wires. We analyze ideas of the quantum information and quantum calculations with the use of superconducting nano-units. We describe the theory of the Josephson effect and its applications in quantum calculations, as well as the idea of qubits. We consider a quantum computer on the basis of superconducting qubits and give a short review of the modern state of the problem of physical realization of quantum computers. We present the idea of a topological quantum computer with qubits in the basis of Majorana fermion quasiparticles.

This book deals with a wide scope of theoretical and experimental topics in superconductivity and is written for advanced students and researchers in the field of superconductors.

Kyiv,
November 2020                                          *S. Kruchinin*

# Contents

# CHAPTER 1

# Theory of Superconductivity

## 1.1. Introduction

In the preface of the big folio "superconductivity" edited by Parks [1], one of the editors has told, "During the preparation of this treatise one of the authors commented that it would be "the last nail in the coffin [of superconductivity]." Some specialist of superconductivity around me said that we are hardly to find articles useful for our future investigation, except for Anderson's comments at the end of this book. Further we have found that Anderson was pessimistic for the further advance of superconductivity, for example, the high-temperature or room-temperature superconductivity was almost unlikely. However, against his expectation, the discovery of high-temperature superconductivity due to Bednorz and Müller [2] is astonishing. The traditional BCS (Bardeen–Cooper–Schrieffer) theory [3] has failed to explain the mechanism of this high-temperature superconductivity. Anderson [5] proposed a new idea called the resonating valence bond (RVB) theory or the $t - J$ model. The term $t$ implies the transfer integral, and $J$ does the electron correlation. We have never heard that his theory successful.

Here is a comment of Feynman, found in his book [4], that it takes almost 50 years for the problem of superconductivity to be reduced to that of explaining the gap. Following the theory of Bardeen, Cooper, and Schrieffer, we will explain the gap, and the theory is essentially correct, but I believe it needs to be made obviously correct. As it stands now, there are a few seemingly loose ends to be clearly up.

The theory of superconductivity seems to be founded on the London postulate [7]. Associated with the gauge transformation, the conserved current is

$$\mathbf{j} = -\frac{i}{2}(\phi^* \nabla \phi - \phi \nabla \phi^*) - e|\phi|^2 \mathbf{A}. \tag{1.1}$$

1

The current due to the first term is called the paramagnetic current, while that due to the second, the diamagnetic current. In the superconducting state, the first term on the right-hand side changes very slightly, sometimes the wave function is called rigid, so that only the diamagnetic current survives. In this respect, the superconductor is the perfect diamagnetic substance. The current is dominated by

$$\mathbf{j} = -k^2 \mathbf{A}, \tag{1.2}$$

where $k$ is a properly chosen positive constant. The Meissner effect is easily derived from the Ampére's equation

$$\nabla \times \mathbf{B} = \mathbf{j}. \tag{1.3}$$

Taking rotation gives

$$\nabla^2 \mathbf{B} = k^2 \mathbf{B} \tag{1.4}$$

or

$$B_x = B_0 e^{-kx}. \tag{1.5}$$

It is important to note that $|\phi|^2$ in Eq. (1.1) is very large, the classic scale quantity, so that the magnetic field in Eq. (1.5) damps very rapidly. This is the Meissner effect. We know similar phenomena, the screening of the Coulomb interaction, or the quark confinement.

The boson model of the Cooper pairs is considerably successful. Equation (1.3) yields $\nabla^2 \mathbf{A} = k^2 \mathbf{A}$, which is

$$\partial^\mu A_\mu = -k^2 A_\mu \tag{1.6}$$

in the covariant form, suggesting that a photon is massive, which is a fundamental aspect to superconductivity [8].

The essentially same treatment has been presented by Ginzburg and Landau (GL) [9]. The superconducting state is the macroscopic state, in other words, a thermodynamic phase. They characterize this phase by introducing the order parameter, $\Psi$. This looks like the Schrödinger function $\Psi$, and then the primitive quantum theorist confused, saying that, behind the iron curtain, the quantum theory was different in feature from that of the western countries.

The theory is carried out as the phase transition. The Lagrangian for the superconducting state is postulated as

$$F_s = F_0 + a|\Psi|^2 + \frac{1}{2}b|\Psi|^4 + \frac{1}{2m^*}\left|\left(-i\hbar\nabla + \frac{e^*\mathbf{A}}{c}\right)\right|^2 + \frac{h^2}{8\pi}, \qquad (1.7)$$

where * indicates quantities in question referred to the superconductor ($\times 2$). Note that there are $|\Psi|^2$ and $|\Psi|^4$ terms in the potential parts. These drive the system to the spontaneous symmetry breaking and lead to a phase transition for the suitable choice of constants $a$ and $b$. This is called now the Higgs mechanism [8]. Certainly the Ginzburg–Landau treatment is a few years ahead the Nambu–Goldstone suggestion.

The Ginzburg–Landau theory is called the macroscopic quantum mechanism, and, in this sense, the big $\Psi$ is called sometimes amusingly cat's $\Psi$.

The very instructive presentation of the macroscopic quantum theory is found in the "Feynman Lectures on Physics, Vol. III, Chapter 21." Various topics there, for example the Josephson junction, are quite readable.

The microscopic theory is prepared by Bardeen, Cooper, and Schrieffer [3]. However, as a preliminary discussion, we present the Bogoliubov treatment. The superconductivity is a kind of many-electron problems. The most general Hamiltonian should be

$$H = \int dx\psi^+(x)h(x)\psi(x) + \frac{1}{2}\int dxdx'\psi^+(x)\psi(x)v(x,x')\psi^+(x')\psi(x').$$
$$(1.8)$$

Since the algebra of electrons is spinor, the terms other than the above identically vanish. In other words, the 3- or 4-body interactions are useless. First, we specify the spin indices, and next the plane wave representation for the spatial parts. Equation (1.8) is simply written as

$$H = \epsilon_k(a_k^+ + a_k) + v_{k,-k}a_k^*a_{-k}^*a_{-k}a_k. \qquad (1.9)$$

The simplification or the mean-field approximation for the two-body part is twofold, say,

$$\Delta_{k,-k}a_{k\alpha}a_{k\alpha}\langle a_{-k\beta}^+ a_{-k\beta}^+\rangle,$$
$$\Delta'_{k,-k}a_{k\alpha}^+ a_{k\beta}^+\langle a_{k\alpha}a_{-k\beta}\rangle + \text{c.c.} \qquad (1.10)$$

The latter looks curious, since such expectation value $\langle a_{k\alpha}a_{-k\beta}\rangle$ vanishes identically. Against this common sense, Bogoliubov put the Hamiltonian

$$H = \epsilon_k(a_k^+ + a_k) + \Delta_k a_k^* a_{-k}^* + \Delta_k^* a_{-k}a_k. \qquad (1.11)$$

We understand that Bogoliubov presumed the Cooper pair, and provided the effective Hamiltonian for pairs. His theory may be a short-hand treatment of the Bardeen–Cooper–Schrieffer (BCS) theory. This Hamiltonian is diagonalized by the so-called Bogoliubov transformation defining the quasiparticle responsible for the superconductivity as

$$
\begin{pmatrix} \gamma_{k\uparrow} \\ \gamma^{+}_{-k\downarrow} \end{pmatrix} = \begin{pmatrix} u_k & -v_k \\ v_k & u_k \end{pmatrix} \begin{pmatrix} c_{k\uparrow} \\ c^{+}_{-k\downarrow} \end{pmatrix},
\tag{1.12}
$$

with

$$
u_k^2 - v_k^2 = 1.
\tag{1.13}
$$

The spirit of the Bogoliubov transformation is to mix operators $c_{k\uparrow}$ and $c^{+}_{-k\downarrow}$ which are different in spin. The quasiparticle yields the new ground state so that the particle pair or the hole pair arises near the chemical potential. The stabilization energy thus obtained is called the gap energy $\Delta_k$ [3, 11].

We now follow the BCS microscopic treatment. The Green's function has effectively used by employing the Nambu spinor. This makes the unified treatment of normal and superconducting states possible. However, the temperature Green's function (Matsubara function) is used from the beginning.

## 1.2.  Spinors

We start with the general many-electron Hamiltonian not restricted to the BCS (Bardeen–Cooper–Schrieffer) Hamiltonian. The BCS state responsible for the superconducting state is easily recognized from the formal description. The simplest method of the quantum chemistry should be the Hückel theory. This consists of the single energy matrix,

$$
\beta_{rs} = \int dx h(x) \rho_{rs}(x),
\tag{1.14}
$$

where $h(x)$ is the single-particle quantum mechanical Hamiltonian, and the electron density $\rho_{rs}(x)$ is given by the product of the single-particle (atomic) orbitals $\chi_r$ and $\chi_s$. Note that this is the spinless theory.

We then extend the treatment into the spin space,

$$
\beta_{rs} = \begin{pmatrix} \beta_{rs}^{\uparrow\uparrow} & \beta_{rs}^{\uparrow\downarrow} \\ \beta_{rs}^{\downarrow\uparrow} & \beta_{rs}^{\downarrow\downarrow} \end{pmatrix}.
\tag{1.15}
$$

Any $2 \times 2$ matrix is expanded in the Pauli spin matrices together with the unit matrix,

$$\sigma^0 = \begin{pmatrix} 1 & 0 \\ 0 & 1 \end{pmatrix}, \quad \sigma^3 = \begin{pmatrix} 1 & 0 \\ 0 & -1 \end{pmatrix}, \quad \sigma^1 = \begin{pmatrix} 0 & 1 \\ 1 & 0 \end{pmatrix}, \quad \sigma^2 = \begin{pmatrix} 0 & -i \\ i & 0 \end{pmatrix}$$

$$(1.16)$$

However, we employ other combinations:

$$\sigma^\uparrow = \frac{1}{2}(\sigma^0 + \sigma^3) = \begin{pmatrix} 1 & 0 \\ 0 & 0 \end{pmatrix},$$

$$\sigma^\downarrow = \frac{1}{2}(\sigma^0 - \sigma^3) = \begin{pmatrix} 0 & 0 \\ 0 & 1 \end{pmatrix},$$

$$\sigma^+ = \frac{1}{2}(\sigma^1 + i\sigma^2) = \begin{pmatrix} 0 & 1 \\ 1 & 0 \end{pmatrix},$$

$$\sigma^- = \frac{1}{2i}(\sigma^1 - i\sigma^2) = \begin{pmatrix} 0 & 0 \\ 1 & 0 \end{pmatrix}. \quad (1.17)$$

In Eq. (1.24), we then have

$$\beta_{rs} = \beta_{rs}^\mu \sigma^\mu,$$

$$\beta_{rs}^\mu = \text{Tr}(\sigma^\mu \beta_{rs}), \quad (1.18)$$

in details:

$$\beta_{rs}^\uparrow = \beta_{rs}^{\uparrow\uparrow} \quad \beta_{rs}^\downarrow = \beta_{rs}^{\downarrow\downarrow},$$

$$\beta_{rs}^+ = \beta_{rs}^{\downarrow\uparrow} \quad \beta_{rs}^- = \beta_{rs}^{\uparrow\downarrow}. \quad (1.19)$$

As to the Pauli matrices, the ordinary commutators are

$$\left. \begin{array}{l} [\sigma^i, \sigma^j] = 2i\epsilon_{ijk}\sigma^k \\ [\sigma^3, \sigma^+] = 2\sigma^+ \\ [\sigma^3, \sigma^-] = -2\sigma^- \\ [\sigma^+, \sigma^-] = \sigma^3 \end{array} \right\}, \quad (1.20)$$

$$\left.\begin{array}{l} [\sigma^{\uparrow}, \sigma^{\downarrow}] = 0 \\ [\sigma^{\uparrow}, \sigma^{+}] = \sigma^{+} \\ [\sigma^{\uparrow}, \sigma^{-}] = -\sigma- \\ [\sigma^{\downarrow}, \sigma^{+}] = -\sigma^{+} \\ [\sigma^{\downarrow}, \sigma^{-}] = \sigma- \\ [\sigma^{+}, \sigma^{-}] = \sigma^{3} \end{array}\right\}, \tag{1.21}$$

and the anticommutators are

$$\left.\begin{array}{l} [\sigma^{i}, \sigma^{j}]_{+} = 2i\delta_{ij} \\ [\sigma^{+}, \sigma^{-}]_{+} = 1 \end{array}\right\}. \tag{1.22}$$

We then express the matrix in Eq. (1.15) as

$$\beta_{rs} = \beta_{rs}^{\mu}\sigma^{\mu},$$
$$\beta_{rs}^{\mu} = \text{Tr}(\sigma^{\mu}\beta_{rs}), \tag{1.23}$$

in details:

$$\beta_{rs}^{\uparrow} = \beta_{rs}^{\uparrow\uparrow}, \quad \beta_{rs}^{\downarrow} = \beta_{rs}^{\downarrow\downarrow},$$
$$\beta_{rs}^{+} = \beta_{rs}^{\downarrow\uparrow}, \quad \beta_{rs}^{-} = \beta_{rs}^{\uparrow\downarrow}. \tag{1.24}$$

Here, if the quantum mechanical Hamiltonian has the single-particle character without the external field causing a rotation in the spin space, the off-diagonal elements are meaningless. The Hückel theory involves the spin diagonal terms.

However, if we taken the electron–electron interaction into account, even in the mean-field approximation, the off-diagonal elements becomes meaningful and responsible for the superconductivity. This is what we investigate here.

### 1.2.1. *Spinor*

The algebra representing electrons is spinor. The Dirac relativistic (special relativity) function well describes this property. However, the relativity seems not so important for the present problem. We now concentrate on the spinor character of electron. The field operator has two components in the spin space.

$$\phi(x) = \begin{pmatrix} \phi_\uparrow(x) \\ \phi_\downarrow^+(x) \end{pmatrix},$$

$$\bar{\phi}(x) = \phi^+(x)\sigma^3 = (\phi_\uparrow^+(x) \quad \phi_\downarrow(x)) \begin{pmatrix} 1 & 0 \\ 0 & -1 \end{pmatrix}$$

$$= (\phi_\uparrow^+(x) \quad -\phi_\downarrow(x)). \tag{1.25}$$

This is called the Nambu representation [12]. The negative sign in front of $\phi_\downarrow$ is seen in the Dirac conjugate $\phi^+ \to \bar{\phi}$.

The field operators satisfy, of course, the anticommutators

$$[\phi_\alpha(x), \phi_\beta^+(s')]_+ = \delta_{\alpha,\beta}\delta(x - x'), \tag{1.26}$$

where $(\alpha, \beta) = (\uparrow, \downarrow)$, and $x = (\mathbf{r}, t)$. Then, for spinors (1.25), the matrix commutator holds:

$$\begin{aligned}
[\phi(x), \bar{\phi}(x')]_+ &= \left[ \begin{pmatrix} \phi_\uparrow(x) \\ \phi_\downarrow^+(x) \end{pmatrix}, (\phi_\uparrow^+(x') \quad -\phi_\downarrow(x')) \right]_+ \\
&= \begin{pmatrix} [\phi_\uparrow(x), \phi_\uparrow^+(x')]_+ & [\phi_\downarrow(x'), \phi_\uparrow(x)]_+ \\ [\phi_\downarrow^+(x), \phi_\uparrow^+(x')]_+ & [\phi_\downarrow(x'), \phi_\downarrow^+(x)]_+ \end{pmatrix} \\
&= \begin{pmatrix} \delta(x - x') & 0 \\ 0 & \delta(x - x') \end{pmatrix}.
\end{aligned} \tag{1.27}$$

### 1.2.2. *Noether theorem and Nambu–Goldstone theorem*

We seek for the meaning of the Nambu spinor [13]. Consider a global transformation of fields with the constant $\Lambda$,

$$\begin{aligned}
\phi_\alpha(x) &\to \phi_\alpha(x)e^{i\Lambda}, \\
\phi_\alpha^+(x) &\to \phi_\alpha(x)e^{-i\Lambda}.
\end{aligned} \tag{1.28}$$

It is recognized that the Hamiltonian and the equation of motion in coming are invariant under this transformation. We can observe that this transformation is a rotation around the $\sigma^3$-axis with $\Lambda$:

$$\phi(x) \to e^{i\sigma^3\Lambda}\phi(x) \tag{1.29}$$

since

$$e^{i\sigma^3 \Lambda} = \cos(\sigma^3 \Lambda) + i\sin(\sigma^3 \Lambda)$$

$$= 1 - \frac{(\sigma^3 \Lambda)^2}{2!} + \frac{(\sigma^3 \Lambda)^4}{4!} + \cdots + i\left(1 - \frac{(\sigma^3 \Lambda)^3}{3!} + \frac{(\sigma^3 \Lambda)^5}{5!} + \cdots\right)$$

$$= 1 - \frac{\Lambda^2}{2!} + \frac{\Lambda^4}{4!} + \cdots + i\sigma^3\left(1 - \frac{\Lambda^3}{3!} + \frac{\Lambda^5}{5!} + \cdots\right)$$

$$= \sigma^0 \cos \Lambda + i\sigma^3 \sin \Lambda$$

$$= \begin{pmatrix} e^{i\Lambda} & 0 \\ 0 & e^{-i\Lambda} \end{pmatrix}.$$

Here, we discuss briefly the Noether theorem and the Nambu–Goldstone theorem. The latter makes a profound investigation possible. If the Lagrangian of the system in question is invariant under some transformation which is just the present case, we have the continuity relation. Notice that the density and the current are, in terms of the Nambu spinor,

$$j^0 = \phi^+(x)\sigma^3\phi(x),$$

$$\mathbf{j} = -i\frac{\hbar}{2m}\left(\phi^+(x)\nabla\phi(x) + \nabla\phi^+(x)\phi(x)\right). \tag{1.30}$$

Then, the continuity relation is

$$\partial_t j_0(x) + \nabla \cdot \mathbf{j} = 0. \tag{1.31}$$

If the system is static, the density must be conserved,

$$\partial_t j^0 = 0. \tag{1.32}$$

Put

$$G = \int dx j^0(x), \tag{1.33}$$

and if it is found that

$$[G, \Psi'(x)] = \Psi(x), \tag{1.34}$$

and the expectation value of $\Psi(x)$ over the ground state does not vanish,

$$\langle 0|\Psi|0\rangle \neq 0, \tag{1.35}$$

that is to say, if the ground state satisfying relation (1.33) does not vanish, we can expect the appearance of a boson $\Psi(x)$, whose mass is zero. This boson is called a Goldstone boson, and the symmetry breaking takes place in the system. This is what the Goldstone theorem insists. The details are referred to the standard book of the field theory [8].

Now we apply this theorem to the superconductivity. The invariant charge is

$$Q = \int d^3x \phi_\alpha^+(x)\phi_\alpha(x) = \int d^3x \phi^+(x)\sigma^3\phi(x). \tag{1.36}$$

In terms of the Nambu spinor, here $\sigma^3$ is crucial. In commutator (2.43), we seek for the spin operators which do not commute with $\sigma^3$ and find, say, $\sigma^\pm$. We then have the Goldstone commutator as

$$\int d^3x' \langle [\phi^+(x')\sigma^3\phi(x'), \phi^+(x)\sigma^\pm\phi(x)] \rangle_{t=t'}$$

$$= \int d^3x (\pm 2)\langle 0|\phi^+(x)\sigma^\pm\phi(x')|0\rangle \delta(\mathbf{r} - \mathbf{r}'). \tag{1.37}$$

In details,

$$\phi^+(x)\sigma^+\phi(x) = (\phi_\uparrow^+(x) \quad \phi_\downarrow(x)) \begin{pmatrix} 0 & 1 \\ 0 & 0 \end{pmatrix} \begin{pmatrix} \phi_\uparrow(x) \\ \phi_\downarrow^+(x) \end{pmatrix},$$

$$= \phi_\uparrow^+(x)\phi_\downarrow^+(x) \tag{1.38}$$

$$\phi^+(x)\sigma^-\phi(x) = \phi_\uparrow(x)\phi_\downarrow(x).$$

Notice that here the same symbol $\phi$ is used for the ordinary field with spin and the Nambu spinor. The aboves are nothing but the Cooper pairs, and we now find the Goldstone bosons $\Psi^*$, $\Psi$. In the literature, it is noted that

$$\langle 0|\phi^+(x)\sigma^\pm\phi(x)|0\rangle = \pm\frac{\Delta_\pm}{g} \neq 0, \tag{1.39}$$

where $\Delta$ and $g$ are the gap and the coupling parameter, respectively. We expect the estimate

$$\Delta_+ \sim \Delta_- = \Delta$$

to be reasonable.

The Cooper pairs are now the Goldstone bosons. A comment about the Goldstone boson or the massless elementary excitation with $k = 0$ is required. Using Eq. (1.33), we write the the Goldstone commutator (1.34) as

$$\int d^3y [\langle 0|j_0(y)|n\rangle\langle n|\phi'(x)|0\rangle - \langle 0|\phi'(x)|n\rangle\langle n|j_0(y)|0\rangle]_{x_0=y_0} \neq 0, \tag{1.40}$$

where $|n\rangle$ is the intermediate state, and $x_0 = y_0$ implies that this is the equal time commutator.

Since

$$j_0(y) = e^{-ipy} j_0(0) e^{ipy}, \tag{1.41}$$

it is seen that

$$\int d^3 y [\langle 0|J_0(0)|n\rangle\langle n|\phi'(x)|0\rangle e^{ip_n y}$$

$$- \langle 0|\phi'(x)|n\rangle\langle n|J_0(0)|0\rangle e^{-ip_n y}]_{x_0=y_0}, \quad (p|n\rangle = p_n|n\rangle)$$

$$= \delta(\mathbf{p}_n)[\langle 0|J_0(0)|n\rangle\langle n|\phi'(0)|0\rangle e^{ip_{n0} y}$$

$$- \langle 0|\phi'(x)|n\rangle\langle n|J_0(0)|0\rangle e^{-ip_{n0} y}]_{x_0=y_0}]$$

$$= \delta(\mathbf{p}_n)[\langle 0|J_0(0)|n\rangle\langle n|\phi'(0)|0\rangle e^{iM_n y_0}$$

$$- \langle 0|\phi'(0)|n\rangle\langle n|J_0(0)|0\rangle e^{-iM_n y_0}]_{x_0=y_0}]$$

$$\neq 0. \tag{1.42}$$

In order to obtain the first equality, the spatial integration is carried out. Then considering $p^\mu = (\mathbf{p}, M)$, we retain the fourth component. In the last equation, when $M_n \neq 0$, cancellations will arise for the summation over $n$. We thus obtain, only for $M_n = 0$, the finite result

$$\langle 0|j_0(0)|n\rangle\langle n|\phi'(0)|0\rangle - \langle 0|\phi'(0)|n\rangle\langle n|j_0(0)|0\rangle$$

$$= \mathrm{Im}\langle 0|j_0(0)|n\rangle\langle n|\phi'(x)|0\rangle, \tag{1.43}$$

which is met with requirement (1.34). The excitation with $M_n = 0$ needs no excitation energy, suggesting the Goldstone boson.

We further note that the mass-zero excitation is the imaginary quantity. This suggests the current be the phase current, as is seen in the Josephson effect.

Before closing the preliminary discussion, we want to make a few remarks. At the beginning, we have mentioned the London's postulate that the superconducting state is characterized by a statement that the wave function is rigid, so that the current is entirely the diamagnetic current due to only the vector potential **A**. "Rigid" is not really rigid, but it is understood that the spatial derivative is vanishing, or the current flows along the entirely flat path which is described as, in a textbook, the path going around the the top of a Mexican hat. Boldly speaking, the electron in the superconducting state is massless. Also we have pointed out that the vector potential **A**, which leads to the Meissner effect, satisfies the covariant relation (1.6)

$$\partial_\mu A_\mu = -k^2 \tag{1.44}$$

so that a photon is massive in the superconductor.

In the following chapters, we develop the substantial microscopic explanation of the above assertions.

## 1.3. Propagator

In the previous section, we have found that, as is seen in Eq. (1.41), the superconducting state strongly concerns with the gap function or the anomalous Green's function. We want to deal with the solid-state substances. However, the infinite crystals described by the single band are already fully investigated in the literature, and the recent investigations are carried on objects with multiband structure [14, 15]. The infinite system with many bands is constructed from the unit cell which is really a chemical molecule. The atoms in this molecule give the band index of the real crystal. In this respect, we investigate firstly the Green's function of a unit cell. The Green's function is now presented in the site representation.

Corresponding to spinors (1.24), we define the spinor in the site representation as

$$\mathbf{a}_r = \begin{pmatrix} a_{r\uparrow} \\ a_{r\downarrow}^+ \end{pmatrix}, \quad \bar{\mathbf{a}}_r = (a_{r\uparrow}^+ \quad -a_{r\downarrow}). \tag{1.45}$$

Due to this definition, it is unnecessary to insert $\sigma^3$ in the matrix $\mathbf{G}$, as is seen in the Schrieffer's book [3]. The commutator is

$$[\mathbf{a}_r, \bar{\mathbf{a}}_s]_+ = \begin{pmatrix} [a_{r\uparrow}, a_{s\uparrow}^+]_+ & [a_{s\downarrow}, a_{r\uparrow}]_+ \\ [a_{r\downarrow}^+, a_{s\uparrow}^+]_+ & [a_{s\downarrow}, a_{r\downarrow}^+]_+ \end{pmatrix}$$

$$= \begin{pmatrix} \delta_{rs} & 0 \\ 0 & \delta_{rs} \end{pmatrix} = \delta_{rs} \mathbf{1}_{2\times2}. \tag{1.46}$$

The matrix propagator is defined by

$$\mathbf{G}_{rs}(\tau) = -\langle\langle \mathbf{a}_r(\tau_1), \bar{\mathbf{a}}_s(\tau_2)\rangle\rangle$$

$$= -\begin{pmatrix} \langle\langle a_{r\uparrow}(\tau_1), a_{s\uparrow}^+(\tau_2)\rangle\rangle & -\langle\langle a_{r\uparrow}(\tau_1), a_{s\downarrow}(\tau_2)\rangle\rangle \\ \langle\langle a_{r\downarrow}^+(\tau_1), a_{s\uparrow}^+(\tau_2)\rangle\rangle & -\langle\langle a_{r\downarrow}^+(\tau_1), a_{s\downarrow}(\tau_2)\rangle\rangle \end{pmatrix}$$

$$= -\langle \theta(\tau_1 - \tau_2)\mathbf{a}_r(\tau_1)\bar{\mathbf{a}}_s(\tau_2) + \theta(\tau_2 - \tau_1)\bar{\mathbf{a}}_s(\tau_2)\mathbf{a}_r(\tau_1)\rangle, \tag{1.47}$$

where $\tau$ is the imaginary time, so that the propagator is the temperature Green's function or the Matsubara function. In what follows, we put

$$\tau = \tau_1 - \tau_2, \tag{1.48}$$

and the system depends on $(\tau_1 - \tau_2)$. If we want to obtain the gap function, we consider

$$\mathrm{Tr}(\sigma^+ \mathbf{G}_{rs}(\tau)) = \langle\langle a_{r\downarrow}^+(\tau_1), a_{s\uparrow}^+(\tau_2)\rangle\rangle, \tag{1.49}$$

and the standard procedure will be followed.

### 1.3.1.  *Hamiltonian*

Various Hamiltonians can be written by using the charge density matrix,

$$\rho_{rs}(x) = \begin{pmatrix} \rho_{r\uparrow s\uparrow}(x) & \rho_{r\uparrow s\downarrow}(x) \\ \rho_{r\downarrow s\uparrow}(x) & \rho_{r\downarrow s\downarrow}(x) \end{pmatrix}$$

$$= \sigma^\mu \rho_{rs}^\mu, \tag{1.50}$$

where the basis orbitals are put to be real, so that we not need the conjugation procedure for field operators.

The Hückel Hamiltonian or the single-particle Hamiltonian has the structure

$$H^0 = h_{rs}^\mu \bar{\mathbf{a}}_r \sigma^\mu \mathbf{a}_r, \tag{1.51}$$

where $h$ includes the chemical potential and can be explicitly written as

$$h_{rs} = \int \rho_{rs} h(x) = \int dx h(x) \begin{pmatrix} \rho_{r\uparrow s\uparrow}(x) & \rho_{r\uparrow s\downarrow}(x) \\ \rho_{r\downarrow s\uparrow}(x) & \rho_{r\downarrow s\downarrow}(x) \end{pmatrix}$$

$$= \sigma^\mu h_{rs}^\mu. \tag{1.52}$$

Hereafter, we accept the summation convention that repeated indices imply the summation is carried out to facilitate manipulations. Other two-particle Hamiltonians are given in a similar way. Noting that

$$v_{rs;tu}^{\mu\nu} = \int dx dx' \rho_{rs}^\mu(x) v(x - x') \rho_{tu}^\nu(x'), \tag{1.53}$$

we have [18]

$$H^{\mathrm{dir}} = \frac{1}{2}(\bar{\mathbf{a}}_r \sigma^a \mathbf{a}_s) v_{rs;tu}^{ab}(\bar{\mathbf{a}}_t \sigma^b \mathbf{a}_u), \quad (a, b = \uparrow, \downarrow)$$

$$H^{\mathrm{ex}} = -\frac{1}{2}(\bar{\mathbf{a}}_r \sigma^a \mathbf{a}_s) v_{rs;ut}^{aa}(\bar{\mathbf{a}}_t \sigma^a \mathbf{a}_u), \quad (a = \uparrow, \downarrow)$$

$$H^{\text{super}} = \frac{1}{2}\{(\bar{\mathbf{a}}_r \sigma^+ \mathbf{a}_s) v_{rs;tu}^{+-}(\bar{\mathbf{a}}_t \sigma^- \mathbf{a}_u) + (\bar{\mathbf{a}}_r \sigma^- \mathbf{a}_s) v_{rs;tu}^{-+}(\bar{\mathbf{a}}_t \sigma^+ \mathbf{a}_u)\}. \quad (1.54)$$

For the direct interaction, the quantum mechanical Hamiltonian is

$$H^{\text{dir}} = \int dx dx' \chi_{r\uparrow}^*(r) \chi_{s\uparrow}(r)(x) v(x - x') \chi_{\downarrow}^*(x') \chi_{u\downarrow}^*(x').$$

For the field theoretical Hamiltonian, the wave functions are replaced by the creation–annihilation operators $a_{r\uparrow}^+$, $a_{r\uparrow}$, and so on. These are written in the spinor notation, for example, as

$$a_{r\uparrow}^+ a_{s\uparrow} = \begin{pmatrix} a_{r\uparrow}^+ & a_{r\downarrow} \end{pmatrix} \begin{pmatrix} 1 \\ 0 \end{pmatrix} \begin{pmatrix} 1 & 0 \end{pmatrix} \begin{pmatrix} a_{s\uparrow} \\ a_{s\downarrow}^+ \end{pmatrix}$$

$$= \mathbf{a}_r^+ \begin{pmatrix} 1 & 0 \\ 0 & 0 \end{pmatrix} \mathbf{a}_s = \mathbf{a}_r^+ \sigma^\uparrow \mathbf{a}_s.$$

Note that $H^{\text{ex}}$ is obtained by reversing indices $(u \leftrightarrow t)$ in $v_{rs;tu}$.

## 1.4. Non-interacting

'Non-interacting' implies that the Hamiltonian is bilinear with respective to operators so that the diagonalization is always possible. It should be instructive to begin with the single-particle case, since even if we manipulate the complicated two-particle case, the procedures are almost the same when the mean-field approximation is employed.

The energy for Hamiltonian (1.52) is

$$E^0 = \text{Tr}(\hat{\rho} H^0) = h_{rs} \langle \bar{\mathbf{a}}_r \mathbf{a}_s \rangle, \quad (1.55)$$

where $\hat{\rho}$ is the statistical operator,

$$\hat{\rho} = e^{\beta(H^0 - \Omega)}, \quad (1.56)$$

with the normalization factor $\Omega$. We now define the temperature Green's function, in which $\tau$ is the imaginary time, $\tau = it$, [16, 17],

$$G_{rs}(\tau) = -\text{Tr}\hat{\rho} \left[ (\theta(\tau_1) \mathbf{a}_r(\tau_1) \mathbf{a}_s^+(\tau_2) - \theta(-\tau) \mathbf{a}_s^+(\tau_2) \mathbf{a}_r(\tau_1) \right]$$

$$= \langle\langle \mathbf{a}_r(\tau_1) \mathbf{a}_s^+(\tau_2) \rangle\rangle, \quad (1.57)$$

where

$$\tau = \tau_1 - \tau_2, \quad (1.58)$$

and it is assumed that the system is only dependent of the relative time $\tau$. Then we can written $E^0$ in terms of the temperature Green's function,

$$E^0 = \text{Tr}[h_{rs} G_{sr}(\tau = 0^-)]. \tag{1.59}$$

Note that $h_{rs}$ and $G_{sr}$ are matrices.

The equation of motion (1.57) for the non-interacting Hamiltonian (1.51) can be read as

$$
\begin{aligned}
\partial_{\tau_1} \langle\langle \mathbf{a}_s(\tau_1)\bar{\mathbf{a}}_r(\tau_2)\rangle\rangle &= \delta(\tau_1 - \tau_2)\langle[\mathbf{a}_s(\tau_1), \bar{\mathbf{a}}_r(\tau_2)]_+\rangle \\
&\quad + \langle[\mathbf{a}_s(\tau_1), \bar{\mathbf{a}}_{s'}\mathbf{a}_{r'} h_{s'r'}(\tau_1)]_-, \bar{\mathbf{a}}_r(\tau_2)\rangle \\
&= \delta(\tau_1 - \tau_2)\delta_{sr} + \langle\delta_{ss'} h_{r's'}(\tau_1)\mathbf{a}_{r'}(\tau_1), \bar{\mathbf{a}}_r(\tau_2)\rangle \\
&= \delta(\tau_1 - \tau_2)\delta_{sr} + \{h_{sr'}\}\langle\langle \mathbf{a}_{r'}(\tau_1), \bar{\mathbf{a}}_r(\tau_2)\rangle\rangle. \tag{1.60}
\end{aligned}
$$

The equation of motion can be solved by the Fourier transformation [16]

$$\langle\langle \mathbf{a}_s(\tau_1)\bar{\mathbf{a}}_r(\tau_2)\rangle\rangle = \frac{1}{\beta}\sum_n e^{i\omega_n \tau}\langle\langle \mathbf{a}_s\bar{\mathbf{a}}_r; \omega_n\rangle\rangle \tag{1.61}$$

with

$$\omega_n = (2n+1)\pi/\beta, \tag{1.62}$$

where the odd number indicates that particles are fermions. Then Eq. (1.50) becomes

$$(i\omega_n\delta_{sr'} - h_{sr'})\langle\langle \mathbf{a}_{r'}\bar{\mathbf{a}}_r; \omega_n\rangle\rangle = \delta_{sr}. \tag{1.63}$$

At this step, the matrix structure of the above should be carefully investigated. Let us assume the single-particle Hamiltonian is spin-diagonal,

$$h_{sr} = \begin{pmatrix} h_{sr}^{\uparrow} & 0 \\ 0 & h_{sr}^{\downarrow} \end{pmatrix}. \tag{1.64}$$

It is preferable to introduce the flame diagonalizing each element:

$$h_{sr}^{\uparrow} = (\langle s|i\rangle\langle i|h|i\rangle|i\rangle\langle ir\rangle)^{\uparrow} = \langle s|i\rangle^{\uparrow}\epsilon_{i\uparrow}\langle i|r\rangle^{\uparrow}. \tag{1.65}$$

Then we have

$$\langle\langle \mathbf{a}_s\bar{\mathbf{a}}_r\rangle\rangle = \frac{1}{\beta}\sum_n \frac{1}{\begin{pmatrix} \frac{(\langle s|i\rangle\langle i|r\rangle)^{\uparrow}}{i\omega_n - \epsilon_{i\uparrow}} & 0 \\ 0 & \frac{(\langle s|i\rangle\langle i|r\rangle)^{\downarrow}}{i\omega_n - \epsilon_{i\downarrow}} \end{pmatrix}}$$

$$= \frac{1}{\beta} \sum_n \begin{pmatrix} \frac{(\langle s|i\rangle\langle i|r\rangle)^\uparrow}{i\omega_n - \epsilon_{i\uparrow}} & 0 \\ 0 & \frac{(\langle s|i\rangle\langle i|r\rangle)^\downarrow}{i\omega_n - \epsilon_{i\downarrow}} \end{pmatrix}$$

$$= \begin{pmatrix} (\langle s|i\rangle n(\epsilon_i)\langle i|r\rangle)^\uparrow & 0 \\ 0 & (\langle s|i\rangle n(\epsilon_i)\langle i|r\rangle)^\downarrow \end{pmatrix}, \tag{1.66}$$

where

$$n(\epsilon_{i\uparrow}) = \frac{1}{1 + e^{\epsilon_{i\uparrow}/k_B T}}. \tag{1.67}$$

Another sophisticated way starts from the decomposed Hamiltonian

$$H^0 = \bar{\mathbf{a}}_r \sigma^\mu h^\mu_{rs} \mathbf{a}_s. \tag{1.68}$$

The commutator is evaluated as

$$[\mathbf{a}_s, \bar{\mathbf{a}}_{r'} \sigma^\mu h^\mu_{r's'} \mathbf{a}_{s'}] = \delta_{sr'} \sigma^0 \sigma^\mu h^\mu_{r's'} \mathbf{a}_{s'} = \sigma^\mu h^\mu_{ss'} \mathbf{a}_{s'}. \tag{1.69}$$

The equation of motion

$$(i\omega_n \delta_{ss'} - \sigma^\mu h^\mu_{ss'})\langle\langle \mathbf{a}_{s'}, \bar{\mathbf{a}}_r; \omega_n \rangle\rangle = \sigma^0 \delta_{sr}. \tag{1.70}$$

In the matrix notation,

$$(i\omega_n - \sigma^\mu h^\mu)\langle\langle \mathbf{a}, \bar{\mathbf{a}}; \omega_n \rangle\rangle = \sigma^0$$

or

$$\langle\langle \mathbf{a}, \bar{\mathbf{a}}; \omega_n \rangle\rangle = \frac{\sigma^0}{i\omega_n - \sigma^\mu h^\mu}. \tag{1.71}$$

In the representation where $h$ is diagonal,

$$\langle r|h|s\rangle = \langle r|i\rangle\langle i|h|i\rangle\langle i|s\rangle = \epsilon_i \langle r|i\rangle\langle i|s\rangle, \tag{1.72}$$

relation (1.71) becomes

$$\langle\langle \mathbf{a}, \bar{\mathbf{a}}; \omega_n \rangle\rangle = \frac{(|i\rangle\langle i|)(i\omega_n + \sigma^\mu \epsilon^\mu_i)}{(i\omega_n - \sigma^\mu \epsilon^\mu_i)(i\omega_n + \sigma^\mu \epsilon^\mu_i)}$$

$$= \frac{(|i\rangle\langle i|)(i\omega_n + \sigma^\mu \epsilon^\mu_i)}{(i\omega_n)^2 - (\epsilon^\mu_i)^2}. \tag{1.73}$$

Now

$$\frac{i\omega_n + \sigma^\mu \epsilon_i^\mu}{(i\omega_n)^2 - (\epsilon_i^\mu)^2} = \frac{i\omega_n}{(i\omega_n)^2 - (\epsilon_i^\mu)^2} + \frac{\sigma^\mu \epsilon_i^\mu}{(i\omega_n)^2 - (\epsilon_i^\mu)^2}$$

$$= \frac{1}{2}\left(\frac{1}{i\omega_n - \epsilon_i^\mu} + \frac{1}{i\omega_n + \epsilon_i^\mu}\right) + \frac{\sigma^\mu}{2}\left(\frac{1}{i\omega_n - \epsilon_i^\mu} - \frac{1}{i\omega_n + \epsilon_i^\mu}\right)$$

$$= \frac{1}{2}\left(\frac{1}{i\omega_n + \epsilon_i^\mu} + \frac{1}{i\omega_n + \epsilon_i^\mu}\right) \quad \text{(summing } \omega_n)$$

$$\rightarrow \frac{1}{2}\left(n(\epsilon_i^\mu) + n(-\epsilon_i^\mu)\right) + \frac{\sigma^\mu}{2}\left(n(\epsilon_i^\mu) - n(-\epsilon_i^\mu)\right). \tag{1.74}$$

We evaluate

$$G_{rs}^\uparrow = \text{Tr}\sigma^\uparrow \left\{ <r|i><i|s> \frac{\sigma^0}{2}\left(n(\epsilon_i^\mu) + n(-\epsilon_i^\mu)\right) + \frac{\sigma^\mu}{2}\left(n(\epsilon_i^\mu - n(-\epsilon_i^\mu)\right) \right\}$$

$$= \langle r|i\rangle\langle i|s\rangle]^\uparrow n(\epsilon_i^\uparrow). \tag{1.75}$$

We then have

$$\frac{1}{\beta}\sum_n G_{rs}^\uparrow(0^-) = \frac{1}{\beta}\sum_n \frac{\langle r|i\rangle\langle i|s\rangle}{i\omega_n - \epsilon_i^\uparrow} = \langle r|i\rangle n(\epsilon_i^\uparrow)\langle i|s\rangle. \tag{1.76}$$

This relation holds for both ↑ and ↑, and we recover result (1.66).

By the way, we give the energy expression for Eq. (1.51):

$$E^0 = \text{Tr}(h_{rs}G_{sr})$$

$$= \text{Tr}\begin{pmatrix} h_{rs}^\uparrow & 0 \\ 0 & h_{rs}^\downarrow \end{pmatrix}\begin{pmatrix} \langle s|i\rangle n(\epsilon_i^\uparrow)\langle i|r\rangle & 0 \\ 0 & \langle s|i\rangle n(\epsilon_i^\downarrow)\langle i|r\rangle \end{pmatrix}. \tag{1.77}$$

It may be needless to present another illustration:

$$E^0 = \text{Tr}(h_{rs}G_{sr}) = \text{Tr}(\sigma^\mu h_{rs}^\mu)(\sigma^\nu G_{sr}^\nu). \tag{1.78}$$

The result is meaningful if $\sigma^\mu \sigma^\nu = \sigma^0$, which leads to, in the present case,

$$\sigma^\mu = \sigma^\nu = \sigma^\uparrow, \text{ or } \sigma^\mu = \sigma^\nu = \sigma^\downarrow.$$

Such manipulations will be used in the later investigation.

## 1.5.  Interacting

In this chapter, the electron-electron interactions are taken into account, and we will discuss how these interactions lead to the superconducting state. The

Hamiltonians given in Eq. (1.54) are

$$H^{\text{dir}} = \frac{1}{2}(\bar{\mathbf{a}}_r \sigma^a \mathbf{a}_s) v^{ab}_{rs;tu}(\bar{\mathbf{a}}_t \sigma^b \mathbf{a}_u), \quad (a, b = \uparrow, \downarrow)$$

$$H^{\text{ex}} = -\frac{1}{2}(\bar{\mathbf{a}}_r \sigma^a \mathbf{a}_s) v^{aa}_{rs;ut}(\bar{\mathbf{a}}_t \sigma^a \mathbf{a}_u), \quad (a = \uparrow, \downarrow)$$

$$H^{\text{sup}} = \frac{1}{2}\{(\bar{\mathbf{a}}_r \sigma^+ \mathbf{a}_s) v^{+-}_{rs;tu}(\bar{\mathbf{a}}_t \sigma^- \mathbf{a}_u) + (\bar{\mathbf{a}}_r \sigma^- \mathbf{a}_s) v^{-+}_{rs;tu}(\bar{\mathbf{a}}_t \sigma^+ \mathbf{a}_u)\}.$$

These are written in the mean-field approximation. The estimate beyond this approximation is not the case of the present consideration. We have

$$H^0 = h^a_{rs} \bar{\mathbf{a}}_r \sigma^a \mathbf{a}_r \quad (a = \uparrow, \downarrow)$$

$$H^{\text{dir}} = (\bar{\mathbf{a}}_r \sigma^a \mathbf{a}_s) v^{ab}_{rs;tu} \langle \bar{\mathbf{a}}_t \sigma^b \mathbf{a}_u \rangle = (\bar{\mathbf{a}}_r \sigma^a \mathbf{a}_s) D^{a:\text{dir}}_{rs} \quad (a, b = \uparrow, \downarrow)$$

$$H^{\text{ex}} = -(\bar{\mathbf{a}}_r \sigma^a \mathbf{a}_s) v^{aa}_{rs;ut} \langle \bar{\mathbf{a}}_t \sigma^b \mathbf{a}_u \rangle = (\bar{\mathbf{a}}_r \sigma^a \mathbf{a}_s) D^{a:\text{ex}}_{rs} \quad (a = \uparrow, \downarrow)$$

$$H^{\text{sup}} = (\bar{\mathbf{a}}_r \sigma^+ \mathbf{a}_s) v^{+-}_{rs;ut} \langle \bar{\mathbf{a}}_t \sigma^- \mathbf{a}_u \rangle + (\bar{\mathbf{a}}_r \sigma^- \mathbf{a}_s) v^{-+}_{rs;ut} \langle \bar{\mathbf{a}}_t \sigma^+ \mathbf{a}_u \rangle$$

$$= (\bar{\mathbf{a}}_r \sigma^+ \mathbf{a}_s) D^{-:\text{sup}}_{rs} + \bar{\mathbf{a}}_r \sigma^- \mathbf{a}_s) D^{+:\text{sup}}_{rs}, \tag{1.79}$$

where

$$D^{a:\text{dir}}_{rs} = v^{ab}_{rs;tu} \langle \bar{\mathbf{a}}_t \sigma^b \mathbf{a}_u \rangle,$$

$$D^{a:\text{ex}}_{rs} = -v^{ab}_{rs;ut} \langle \bar{\mathbf{a}}_t \sigma^b \mathbf{a}_u \rangle \delta_{ab},$$

$$D^{+:\text{sup}}_{rs} = v^{+-}_{rs;tu} \langle \bar{\mathbf{a}}_t \sigma^- \mathbf{a}_u \rangle, \tag{1.80}$$

where $\langle \cdots \rangle$ implies the ground-state average, which is actually obtained with wave functions in the previous step during the SCF calculations.

These Hamiltonians are classified into two kinds called modes, the normal many-electron problem and that for the superconductivity:

$$H = H^{\text{norm}} + H^{\text{sup}}, \tag{1.81}$$

where

$$H^{\text{norm}} = (\bar{\mathbf{a}}_r \sigma^a \mathbf{a}_s)(h^a_{rs} \delta_{rs} + D^{a:\text{dir}}_{rs} + D^{a:\text{ex}}_{rs}),$$

$$H^{\text{sup}} = \bar{\mathbf{a}}_r \sigma^\pm \mathbf{a}_s D^{\mp:\text{sup}}_{rs}. \tag{1.82}$$

The main difference between two is that, in the former, we have the single-particle Hamiltonian, while in the latter not. Various $D_{rs}$ are complicated, but they are merely the *c*-numbers in this treatment. The propagator in question in Eq. (1.57) is presented here again:

$$G_{rs}(\tau) = -\mathrm{Tr}\hat{\rho}\left[(\theta(\tau)\mathbf{a}_r(\tau_1)\mathbf{a}_s^+(\tau_2) - \theta(-\tau)\mathbf{a}_s^+(\tau_2)\mathbf{a}_r(\tau_1)\right]$$

$$= \langle\langle\mathbf{a}_r(\tau_1)\mathbf{a}_s^+(\tau_2)\rangle\rangle.$$

The equation of motion can be read as

$$\partial_{\tau_1}\langle\langle\mathbf{a}_s(\tau_1)\bar{\mathbf{a}}_r(\tau_2)\rangle\rangle = \delta(\tau_1 - \tau_2)\langle[\mathbf{a}_s(\tau_1), \bar{\mathbf{a}}_r(\tau_2)]_+\rangle$$

$$+\langle[\mathbf{a}_s(\tau_1), \bar{\mathbf{a}}_{s'}\mathbf{a}_{r'}H_{s'r'}^a(\tau_1)]_-, \bar{\mathbf{a}}_r(\tau_2)\rangle$$

$$(a; \text{ norm, sup}) = \delta(\tau_1 - \tau_2)\delta_{sr} + D_{sr'}^a\sigma^a\langle\langle\mathbf{a}_{r'}(\tau_1), \bar{\mathbf{a}}_r(\tau_2)\rangle\rangle. \quad (1.83)$$

Making the Fourier transformation with respect to $\tau = \tau_2 - \tau_2$ gives

$$(i\omega_n - D_{rs'}^a\sigma^a)G_{s'r} = \delta_{sr} \qquad (1.84)$$

or, in the matrix form,

$$G(\omega_n) = \frac{1}{i\omega_n - D^a\sigma^a} = \frac{i\omega_n + D^a\sigma^a}{(i\omega_n)^2 - (D^a)^2}. \qquad (1.85)$$

Note that $D^a$ consists of the single-electron part and the two-electron interaction term which involves another mate $\rho_{tu}^a$ combined with the propagator $\langle\bar{\mathbf{a}}_t\sigma^a\mathbf{a}_u\rangle$:

$$D^a(x) = h^a(x) + \int dx' v(x - x')\rho_{tu}^b(x')\sigma^b\langle\bar{\mathbf{a}}_t\mathbf{a}_t\rangle. \qquad (1.86)$$

However, the mean-field approximation makes this as if it were the single-electron interaction. A few comments will be given about the matrix character of $G$. This is a big matrix with site indices, and each element is a $2 \times 2$ matrix in the spin space. The index $a$ characterizes the mode of the mean-field potential. Every mode is independent of each other and is individually diagonalized. We now introduce a flame, in which these are diagonal,

$$\langle i^a|D^a|i^a\rangle = \eta_i^a. \qquad (1.87)$$

Look at the right-hand side of Eq. (1.85) and remember the Einstein convention that repeated indices imply summation,

$$(D^a)^2 = (\eta^\uparrow)^2 + (\eta^\downarrow)^2 + (\eta^+)^2 + (\eta^-)^2$$

$$= (\eta^{\mathrm{norm}})^2 + (\eta^{\mathrm{sup}})^2, \qquad (1.88)$$

where the second line is in a simple notation. Then we have

$$G(\omega_n) = \frac{|i^a\rangle(i\omega_n + \langle i^a|D^a|i^a\rangle\sigma^a)\langle i^a|}{\{i\omega_n - (\eta_i^{\mathrm{norm}} + \eta_i^{\mathrm{sup}})\}\{i\omega_n + (\eta_i^{\mathrm{norm}} + \eta_i^{\mathrm{sup}})\}}$$

$$= |i^a\rangle\frac{1}{2}\left(\frac{1}{i\omega_n - (\eta_i^{\mathrm{norm}} + \eta_i^{\mathrm{sup}})} + \frac{1}{i\omega_n + (\eta_i^{\mathrm{norm}} + \eta_i^{\mathrm{sup}})}\right)\langle i^a|$$

$$+ |i^a\rangle\frac{D_{ii}^a\sigma^a}{2(\eta_i^{\mathrm{norm}} + \eta_i^{\mathrm{sup}})}$$

$$\times\left(\frac{1}{i\omega_n - (\eta_i^{\mathrm{norm}} + \eta_i^{\mathrm{sup}})} - \frac{1}{i\omega_n - (\eta_i^{\mathrm{norm}} + \eta_i^{\mathrm{sup}})}\right)\langle i^a|.$$

Taking the $(r, s)$-matrix elements and the mode $c$ which is achieved by the operation,

$$\mathrm{Tr}\ \sigma^c G_{rs} = G^c rs \tag{1.89}$$

we select the terms on the right-hand side with the mode $c$ satisfying

$$\mathrm{Tr}\ \sigma^c\sigma^a = 1. \tag{1.90}$$

Otherwise, the Tr operation leads to the vanishing result.

Carrying out the summation over $\omega_n$, we get

$$G_{rs}^c(0^-) = [2\mathrm{mm}]\frac{1}{\beta}\sum_n G_{rs}^c(\omega_n)$$

$$= \frac{1}{\beta}\sum_n\langle r^c|i^a\rangle\frac{1}{2}\left(\frac{1}{i\omega_n - (\eta_i^{\mathrm{norm}} + \eta_i^{\mathrm{sup}})} + \frac{1}{i\omega_n + (\eta_i^{\mathrm{norm}} + \eta_i^{\mathrm{sup}})}\right)$$

$$\times\langle i^a|s^c\rangle + \frac{1}{\beta}\sum_n\langle r^c|i^a\rangle\frac{D_{ii}^a}{2(\eta_i^{\mathrm{norm}} + \eta_i^{\mathrm{sup}})}$$

$$\times\left(\frac{1}{i\omega_n - (\eta_i^{\mathrm{norm}} + \eta_i^{\mathrm{sup}})} - \frac{1}{i\omega_n + (\eta_i^{\mathrm{norm}} + \eta_i^{\mathrm{sup}})}\right)\langle i^a|s^c\rangle$$

$$= \langle r^c|i^a\rangle\frac{1}{2}(n(\eta_i^{\mathrm{norm}} + \eta_i^{\mathrm{sup}}) + n(-\eta_i^{\mathrm{norm}} - \eta_i^{\mathrm{sup}}))\langle i^a|s^c\rangle$$

$$+ \langle r^c|i^a\rangle\frac{D_{ii}^a}{2(\eta_i^{\mathrm{norm}} + \eta_i^{\mathrm{sup}})}$$

$$\times(n(\eta_i^{\mathrm{norm}} + \eta_i^{\mathrm{sup}}) - n(-\eta_i^{\mathrm{norm}} - \eta_i^{\mathrm{sup}}))\langle i^a|s^c\rangle$$

$$= \frac{1}{2}\delta_{rs} - \langle r^c|i^a\rangle\frac{D_{ii}^a}{2(\eta_i^{\mathrm{norm}} + \eta_i^{\mathrm{sup}})}\tanh(\eta_i/2k_BT)\langle i^a|s^c\rangle, \tag{1.91}$$

where

$$n(\eta_i^a) = \frac{1}{1 + e^{\eta_i^a/kBT}} \qquad (1.92)$$

and

$$n(\eta) + n(-\eta) = 1,$$

$$n(\eta) - n(-\eta) = -\tanh\left(\frac{\eta}{2k_BT}\right).$$

In the estimation of matrix elements, the chemical potential which was disregarded up to now is taken into account. Namely, the Hamiltonian has an additive term $-\mu\bar{a}_r \mathbf{a}_r$ which causes

$$\eta_i^a \to \eta_i^a - \mu < 0, \text{ while } - \eta_i^a \to -(\eta_i^a - \mu) > 0.$$

The mean-field potentials are carefully treated. In modes with $\uparrow$ and $\downarrow$, we have the non-vanishing single particle parts, $h_{rs}^{\uparrow}\delta_{rs}$ and $h_{rs}^{\downarrow}\delta_{rs}$ which are usually negative. However, for superconducting modes, $h_{rs}^{\pm} = 0$, and the chemical potential is lost for the same reason. The latter may be closely related to the fact that the number of particles is not conserved in a superconductor. These circumstances are crucial for the superconducting mode.

### 1.5.1.  *Unrestricted HF*

Let us review the SCF (self-consistent field) procedure. We now discuss the ordinary many-electron system. As an example, the propagator with the up spin, Eq. (1.89), is

$$G_{rs}^{\uparrow}(0^-) = \frac{1}{2}\langle r^{\uparrow}|s^{\uparrow}\rangle - \langle r^{\uparrow}|i^a\rangle \frac{D_{ii}^a}{2(\eta_i^{\text{norm}} + \eta_i^{\text{sup}})}\tanh(\eta_i^a/2k_BT)\langle i^a|s^{\uparrow}\rangle, \quad (1.93)$$

where $\langle r^{\uparrow}|s^{\uparrow}\rangle$ is the overlap integral between sites $r$ and $s$ and approximately vanishes, and

$$D_{ii}^a = h_{ii}^{\uparrow} + v_{ii;tu}^{\uparrow a}G_{tu}^a - v_{ii;tu}^{\uparrow\uparrow}G_{tu}^{\uparrow}.$$

$$(a :\uparrow,\downarrow) \qquad (1.94)$$

Note that

$$\tanh(\eta_i^a/2k_BT) < 0, \text{ since } \eta_i^a < 0. \qquad (1.95)$$

Look at the potential of the mode ↑,

$$D^\uparrow(x) = h^\uparrow(x) + \int dx' v(x - x') \rho_{tu}^b(x') \sigma^b \langle \bar{a}_t a_t \rangle. \tag{1.96}$$

In this case, there is $h^\uparrow(x)$, whose matrix element should be negative, so that even if the matrix elements of the second terms are positive, the $(r, s)$ matrix element of $D^\uparrow(x)$ is probably negative. This is usually the cases in atoms, molecules, and solids. In evaluating $D^\uparrow(x)$, the propagators (wave functions) of all other modes are required. In this respect, Eq. (1.93) is the self-consistent relation between propagators. Usually, the self-consistent relation between wave functions is given in a such a way that, at beginning, the total energy is given by the potentials given in terms of tentatively approximated wave functions, and the new approximate wave functions in the next step are obtained by optimizing the total energy. This procedure is lacking in the present consideration.

### 1.5.2. *Gap equation for superconductivity*

In the case of superconductivity, since $\sigma^+(\sigma^-)$ is traceless, the first term of Eq. (1.91) vanishes, and also $h^+ = 0$. Now Eq. (1.91) can be read as

$$G_{rs}^+(0^-) = -\left\{ \langle r^+ | i^+ \rangle \frac{v_{ii;tu}^{+-} \langle \bar{a}_t a_u \rangle^-)}{2(\eta_i^{\text{norm}} + \eta_i^{\text{sup}})} \langle i^+ | s^+ \rangle \right\} \tanh(\eta_i^+ / 2k_B T).$$

$$= -\left\{ \langle r^+ | i^+ \rangle \frac{v_{ii;tu}^{+-} G_{tu}^-}{2(\eta_i^{\text{norm}} + \eta_i^{\text{sup}})} \langle i^+ | s^+ \rangle \right\} \tanh(\eta_i^+ / 2k_B T). \tag{1.97}$$

This complicated equation gives the relation between $G_{rs}^+$ and $G_{rs}^-$ both referring to the superconductivity and is called the gap equation. A few points should be presented. While selecting the superconducting mode, it is used that

$$\text{Tr}(\sigma^+ \sigma^-) = \text{Tr} \begin{pmatrix} 0 & 1 \\ 0 & 0 \end{pmatrix} \begin{pmatrix} 0 & 0 \\ 1 & 0 \end{pmatrix} = \text{Tr} \begin{pmatrix} 1 & 0 \\ 0 & 0 \end{pmatrix} = 1, \tag{1.98}$$

so that

$$\text{Tr}(\sigma^+ v^{+a} \sigma^a) = v_{ii;tu}^{+-}. \tag{1.99}$$

Relation (1.97) yields

$$-\left\{ \langle r^+ | i^+ \rangle \frac{v_{ii;tu}^{+-}}{2(\eta_i^{\text{norm}} + \eta_i^{\text{sup}})} \langle i^+ | s^+ \rangle \right\} > 0. \tag{1.100}$$

As has been mentioned previously, we have no chemical potential in the superconducting state, so that $\eta_i^+$ is positive, as same as $v_{ii;tu}^{+-}$ is. Therefore, it is required that $\langle r^+|i^+\rangle\langle i^+|s^+\rangle$ be negative in order that Eq. (1.100) be hold. This is really possible, as will be mentioned below. Thus, we have nothing to do with the electron–phonon coupling. Let us perform the successive approximations as

$$\langle r^+|i^+\rangle v_{ii;tu}^{+-}\langle i^+|s^+\rangle\tanh(\eta_i^+/2k_BT)\approx\sum_i^{occ}\langle r^+|i^+\rangle v_{ii;tu}^{+-}\langle i^+|s^+\rangle \qquad (T\to 0)$$

$$= q_{rs}v_{ii;tu}^{+-}(<0). \qquad (1.101)$$

Here, $v_{ii;tu}^{+-}$ is the electron–electron interaction between two electron densities and certainly positive (repulsive). However, the bond order is not necessarily, but $q_{14}<0$ in the following example. The last relation was usually assumed at the beginning of the superconductivity theory.

We thus obtain the condition for the superconducting state to appear which is pure electronic and is apart from the electron–phonon coupling mechanism. Actually, for the chain molecule of four carbon atoms, called butadiene, the matrix $q_{rs}$, $(r,s)=(1-4)$ is

$$\{q_{rs}\}=\begin{pmatrix} 1.000 & 0.894 & 0.000 & -0.447 \\ & 1.000 & 0.447 & 0.000 \\ & & 1.000 & 0.894 \\ & & & 1.000 \end{pmatrix}. \qquad (1.102)$$

We can clearly observe

$$v_{11;14}^{+-}q_{14}<0. \qquad (1.103)$$

## 1.6.   Illustrative Example, Critical Temperature

The gap equation (1.97) is, in appearance, considerably different from the usual one. We rewrite this equation in a form similar to the usual one. To this end, we adopt, as an example, a polyacene high polymer. Benzene, naphthalene, anthracene, ... are series of polyacene shown in Fig. 1.1. Here, the unit cell which is the butadiene molecule is in the dotted rectangle numbered by $n$. The interactions $t_1$ and $t_2$ are given for the corresponding bonds.

polyacene

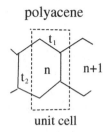

unit cell

Fig. 1.1.   Polyacene.

## 1.6.1.   *Bond alternation*

At beginning, we discuss the bond alternation or the Peierls instability of these molecules. The infinite chain of acetylene, the so-called polyacetylene, has the bond alternation, that is to say, the long and short bonds do not loose their memories in the limit that an infinite chain has been formed. This is popular for chemists [20], but physicists call it Peierls distortion [21]. The bond alternation causes the gap between the conduction and valence bands. It has been said that this discontinuity prevents the superconducting phase from arising. However, this is an old-fashioned assertion and now seems suspicious.

In the Hückel theory, the interaction matrix elements are put $t_1$ for the shorter bond and $t_2$ for the longer bond:

$$H^0 = -t_1(a_{2n}^+ a_{1n} + a_{1n}^+ a_{2n}) - t_2(a_{1n+1}^+ a_{2n} + a_{2n}^+ a_{n+1}), \tag{1.104}$$

where we consider the neighboring unit cells numbered $n$ and $n+1$, and each cell has two kinds of bonds. The transfer integrals are parametrized as

$$t_1 = t - \delta, \quad t_2 = t + \delta, \quad (\delta > 0, \quad \delta/t \ll 1), \tag{1.105}$$

and then the Hamiltonian is easily diagonalized as

$$\epsilon_k = \pm [t^2(1 + \cos k) + \delta^2(1 - \cos k)]^{1/2}, \tag{1.106}$$

where $+$ and $-$ correspond to the conduction and valence bands, respectively. We are interested in the features at the zone boundary, $k = \pi$,

$$\epsilon_\pi^c = 2\delta, \quad \epsilon_\pi^v = -2\delta. \tag{1.107}$$

Here, the superscripts $c$ and $v$ indicate the conduction and valence bands, respectively. When $\delta \neq 0$, certainly we have the gap, and if $\delta = 0$, both bands continuously join to a single band called the half-filled band,

Next, we turn to the polyacene, whose unit cell is the butadiene molecule. In this case, we obtain four bands

$$\epsilon_k^{c'} = \frac{1}{2}[t_3 + (t_3^2 + 4|\tilde{t}_k|^2)^{1/2}] = -\epsilon_k^{v'},$$

$$\epsilon_k^{c} = \frac{1}{2}[-t_3 + (t_3^2 + 4|\tilde{t}_k|^2)^{1/2}] = -\epsilon_k^{v}, \qquad (1.108)$$

where

$$\tilde{t}_k = t_1 + t_2 e^{ik}. \qquad (1.109)$$

The usual pairing property in alternant hydrocarbons is also seen in this case. Employing parameterization (1.105) gives, at $k = \pi$,

$$\epsilon_\pi^{v} = \frac{1}{2}[t_3 + (t_3^2 + 16\delta^2)^{1/2}] \approx 4t_3(\delta/t_1)^2. \qquad (1.110)$$

When $\delta = 0$ (without bond alternation), we have

$$\epsilon_k^{v} = \frac{1}{2}\{t_3 - [t_3^2 + 8t^2(1 + \cos k)]^{1/2}\}. \qquad (1.111)$$

Consider single-particle states. If $\delta = 0$ and $k = \pi$, we have, from Eq. (1.109), $\tilde{t} = 0$, so that the amplitudes at sites 2 and 5 vanish. Therefore, when $\delta = 0$, we get, at $k = \pi$,

$$|\psi_k^{v}\rangle = \frac{1}{2}(a_{1k}^{+} - a_{4k}^{+})|0\rangle,$$

$$|\psi_k^{c}\rangle = \frac{1}{2}(a_{1k}^{+} + a_{4k}^{+})|0\rangle. \qquad (1.112)$$

It is seen that $|\psi_k^{v}\rangle$ is antisymmetric about the $C_{2v}$ symmetry axis, and $|\psi_k^{v}\rangle$ is symmetric. Thus, $v$ and $c$ bands are not continuous at $k = \pi$.

## 1.6.2.  *Deformation energy*

For these systems, let us study whether the bond alternation is energetically favorable or not. We assume that the energy gain due to the bond alternation mainly contributes to the highest valence band energy, $\epsilon_k^{v}$.

The case of polyacetylene. The energy gain $\Delta E$ is

$$\Delta E = \int_0^{2\pi} \frac{dk}{2\pi}(\epsilon_k^{v} - \epsilon_k^{v}(0)), \qquad (1.113)$$

where the second term refers to the case without bond alternation ($\delta = 0$). In Eq. (1.106), we shift the integration origin from 0 to $\pi$, then approximating

$$\cos k = \cos(\pi + p) \approx -1 + p^2/2.$$

For small $p$, we obtain

$$\Delta E = -\int_0^{2\pi} \frac{dk}{2\pi} \left\{ \sqrt{(tp)^2 + 4\delta^2} - tp \right\}$$

$$= \frac{2t}{\pi} \left( \frac{\delta}{t} \right)^2 \ln \left( \frac{\delta}{\pi t} \right), \quad (<0). \tag{1.114}$$

For polyacene, the similar treatment of $\epsilon_k^v$ of Eq. (1.108) leads to

$$\epsilon_k^v = \frac{1}{2} \left( 1 - \left\{ 1 + 16 \left( \frac{\delta}{t} \right)^2 + 4p^2 \left[ 1 - \left( \frac{\delta}{t} \right)^2 \right] \right\}^{1/2} \right), \tag{1.115}$$

and then the deformation energy becomes

$$\Delta E = t \left( \frac{\delta}{t} \right)^2 \left( \frac{\pi}{4} - \frac{2}{\pi} \ln 4\pi \right) \approx -0.83 \, t \left( \frac{\delta}{t} \right)^2. \tag{1.116}$$

The bond alternation looks favorable for both cases. However, when the effect of the $\sigma$ bond is taken into account, this almost cancels out the stabilization energy of the $\pi$ system in the case of polyacene. On the other hand, this is not the case for polyacetylene due to the singular term in relation (1.116).

Therefore in what follows, by concentrating on polyacene, we are free from the bond alternation.

### 1.6.3. *Polyacene, gap equation, critical temperature*

The unit cell of polyacene is a butadiene molecule composed of four $2p\pi$ carbon atoms. The Hamiltonian in the tight-binding approximation is given as

$$H^0 = t(a_{n1}^+ a_{n2} + a_{n2}^+ a_{n3} + a_{n4}^+ a_{n3} + \text{h.c.})$$

$$+ t(a_{n+1,1}^+ a_{n2} + a_{n+1,4}^+ a_{n3} + \text{h.c.}), \tag{1.117}$$

where the second line connects the unit cells $n$ and $n+1$ [22].

The band structure of levels and the LCAO (linear combination of atomic orbitals) coefficients $U$ are

$$\epsilon_1(k) = \frac{t}{2}\{1 + s(k)\},$$

$$\epsilon_2(k) = -\frac{t}{2}\{1 - s(k)\},$$

$$\epsilon_3(k) = \frac{t}{2}\{1 - s(k)\},$$

$$\epsilon_4(k) = -\frac{t}{2}\{1 + s(k)\} \tag{1.118}$$

with

$$s(k) = \sqrt{9 + 8\cos k}, \tag{1.119}$$

$$U = \begin{pmatrix} N_4 & N_3 & N_2 & N_1 \\ -N_4\epsilon_4/\tilde{t}_k & -N_3\epsilon_3/\tilde{t}_k & -N_2\epsilon_2/\tilde{t}_k & -N_1\epsilon_1/\tilde{t}_k \\ N_4\epsilon_4/\tilde{t}_k & -N_3\epsilon_3/\tilde{t}_k & N_2\epsilon_2/\tilde{t}_k & -N_1\epsilon_1/\tilde{t}_k \\ -N_4 & N_3 & -N_2 & N_1 \end{pmatrix}, \tag{1.120}$$

where, for example,

$$N_1^2 = \frac{|\tilde{t}_k|^2}{2(|\tilde{t}_k|^2 + \epsilon_1^2)}, \qquad 1/\tilde{t}_k = \frac{e^{ik/2}}{2t\cos(k/2)}. \tag{1.121}$$

At this stage, we have finished, in principle, the usual many-electron problem. The mean-field approximation makes the interaction problem to be a one-particle problem even though the SCF (self-consistent field) treatment is required at each step. In other words, from the viewpoint of the Hückel theory, the spin-diagonal parts provide the answer. On the other hand, the spin-off-diagonal part, which means less in the case without electron–electron interactions, is responsible for the superconductivity.

Up to the previous chapter, the problem has been investigated in the site representation. That is, the system is considered to be composed of $N$ sites. However, the real substance is formed from unit cells, so that the system is a repetition of the unit cell. The usual band theory of polyacene thus has been finished at this stage.

We turn to the onset of superconductivity. In this case, the single-particle approach is almost meaningless, but the pair state, say, the wave function of a Cooper pair should be investigated. For this purpose, the Green's function of a Cooper pair is most preferable. The gap equation (1.97) is nothing but the SCF equation for a Cooper pair.

The electronic structure of the single butadiene molecule referring to $k = 0$ is suggestive. Let the total number of sites of high polymer polyacene be $N$. The number of sites in the unit cell is four. Then $N = 4n$ with the number of unit cells, $n$. The numbers $1 - 4$ are the band indices; then we have the chemical potential between the $2n$ level and $3n$ levels. The Cooper pair should be the hole pair of the $2n$ level indicated by mode $(-)$ or the particle pair of the $3n$ level. The discussion is confined to these levels in solid-state physics, even if the interaction with other bands is taken into account.

The $2n$ and $2n + 1$ levels are called HOMO (highest occupied molecular orbital) and LUMO (lowest unoccupied molecular orbital). These features and behaviors are not so far or qualitatively the same as those for $k = 0$.

The electronic structure of a single butadiene molecule is suggestive. The levels and the bond orders are (the unit $= t$):

$$
\begin{array}{c|c}
\epsilon_1 & -1.618 \\
\epsilon_2 & -0.618 \\
\epsilon_3 & 0.618 \\
\epsilon_4 & 1.618
\end{array}
\tag{1.122}
$$

Here, $\epsilon_1$ and $\epsilon_2$ are occupied (valence) levels, while $\epsilon_3$ and $\epsilon_4$ are unoccupied (conduction) ones. Let the probability amplitude, with which the electron on the level $i$ is found at the site $r$, be $\langle r|i \rangle$, then the bond order $q_{rs}$ is defined as (at the zero temperature)

$$
q_{rs} = \sum_i^{occ} \langle r|i \rangle \langle i|s \rangle,
\tag{1.123}
$$

where the summation includes the spin state. In the determination of the attractive electron–electron interaction, the bond order is of crucial importance.

The bond orders are

$$
\begin{array}{c|cccc}
\text{site} & & & & \\
\hline
1 & 1.000 & 0.894 & 0.000 & -0.447 \\
2 & 0.894 & 1.000 & 0.447 & 0.000 \\
3 & 0.000 & 0.447 & 1.000 & 0.894 \\
4 & -0.447 & 0.000 & 0.894 & 1.000
\end{array}
\tag{1.124}
$$

We indicate the negative value of $q_{14}$. In solid-state physics, the discussion is concentrated on the highest valence band or the lowest conduction band. In the quantum-chemical language, the partial bond orders referring only to level 2, $q_{rs}^H$, which seem not so far from those with $k$, are listed as

$$
\begin{array}{c|cccc}
\text{site} & & & & \\
\hline
1 & 0.362 & 0.224 & -0.224 & -0.362 \\
2 & 0.224 & 0.138 & -0.138 & -0.224 \\
3 & -0.224 & -0.138 & 0.138 & 0.224 \\
4 & -0.362 & -0.224 & 0.224 & 0.362
\end{array}
\qquad (1.125)
$$

We indicate the negative value of $q_{14}^H$.

   If we want to look at the band structure, for example $q_{rs}^H$, it is multiplied by the third column of Eq. (1.120).

   Basing on these results, we may perform successive approximations or simplifications.

(1) The denominator of Eq. (1.97) is the sum of the ordinary Hartree–Fock energy and that of the superconducting state, and the latter is

$$
\eta_i^{\text{sup}} = -v_{ii;tu}^{+-} G_{tu}^{-}. \qquad (1.126)
$$

The total energy is obtained by summing

$$
\eta_i = \eta_i^{\text{norm}} + \eta_i^{\text{sup}} \qquad (1.127)
$$

relative to the chemical potential.

(2) Each level with $\eta_i$ has really a band structure and then is written as $\eta_{i;k}$, by stressing the band structure by $k$ with the band index $i$. Then $i$ is put to be the highest occupied level, and the integration over $k$ is carried out. The bond orders thus obtained are approximated by the partial bond orders $q_{rs}^H$.

(3) The electron–electron interaction which is effectively negative due to the chemical structure of species can be simply written for $g < 0$ as

$$
g = q_{rs} v_{rs;tu}^{-+}. \qquad (1.128)
$$

(4) The energy interval establishing the superconductivity in the BCS theory is related to the electron–phonon interaction, $\hbar\omega_D$ (Debye frequency). In the present theory, it is replaced by the band width nearly equal to $|g|$.

(5) Assuming

$$
G_{s\uparrow,r\downarrow}^{+}(0^-) = G_{t\uparrow,u\downarrow}^{+}(0^-) \qquad (1.129)
$$

in Eq. (1.97), we have

$$1 = gN(0) \int_0^{|g|} \frac{d\xi}{\xi + \eta^{\text{sup}}} \tanh \left( \frac{\xi + \eta^{\text{sup}}}{2k_B T} \right), \qquad (1.130)$$

where $\xi = \eta^{\text{norm}}$. The critical temperature $T_C$ is determined by the condition that $\eta^{\text{super}}$ vanishes at this temperature. The integration in Eq. (1.130) is carried out as usual. Approximating

$$\int \frac{d^3 k}{(2\pi)^3} = N(0) \int d\xi$$

gives

$$\int_0^{|g|} \frac{d\xi}{\xi} \tanh \left( \frac{\xi}{2k_B T} \right) = \int_0^Z \frac{dz}{z} \tanh z, \quad z = |g|/2k_B T$$

$$= [\ln z \tanh z]_0^Z - \int_0^\infty dz \ln z \, \text{sech}^2 z, \quad (1.131)$$

where the upper limit in the second integration is replaced by $\infty$ which makes it integrable [16]:

$$\int_0^\infty dz \ln z \, \text{sech}^2 z = -\ln \frac{4e^\gamma}{\pi}, \quad \gamma \text{ is the Euler constant}$$

The simple rearrangement of the result gives

$$k_B T_C = \frac{2e^\gamma}{\pi} |g| e^{-1/N(0)g} \sim 1.13 |g| e^{-/N(0)g}. \qquad (1.132)$$

The result is entirely same as the current one. However, since it is probable that

$$|g|/k_B T \sim 100, \qquad (1.133)$$

the critical temperature is, at most, enhanced by this value, even though it is considerably reduced by the factor $e^{-1/N(0)g}$.

### 1.6.4.  *Conclusion*

As has been presented, is not a too complicated phenomenon. If we employ the spinor representation, superconductivity is described in parallel to the normal electronic processes. If we find, in the copper oxide complex, the four-site unit as a butadiene molecule, it might be the origin of superconductivity of this material. We think that it is not a so difficult problem for the quantum chemists.

## 1.7. Linear Response Magnetic Resonance in Normal and Superconducting Species: Spin-Lattice Relaxation Time

### 1.7.1. *Introduction*

The theory of linear response is one of the main topics in solid-state physics, and its application to superconductivity is also a fundamental problem. Perhaps, the most important problem must be the Meissner effect. However, we are now interested in the magnetic resonance, whose main theme should concern with the relaxation time. Let us discuss the spin-lattice relaxation time $T_1$ in the nuclear magnetic resonance. The elegant theory has been provided by Kubo and Tomita [23] and revised by us with the temperature Green's function [24].

The spin-lattice relaxation time $T_1$ remarkably increases in a superconductor just below the critical temperature. This is explained by the BCS pairing theory and is said to be its brilliant triumph [3,25,26]. The external perturbation acting on the electron is written as

$$H' = B_{k\sigma,k'\sigma'} c^+_{k\sigma} c_{k'\sigma'}, \tag{1.134}$$

where $c^+_{k\sigma}$, $c_{k\sigma}$, etc. are the creation and annihilation operators of an electron in the normal phase, and $B_{k\sigma,k'\sigma'}$ is the matrix element of the perturbation operator between the ordinary one-electron states in the normal phase. The problem is as follows: If we rewrite it in terms of operators of a quasiparticle in the superconducting phase, what will arise?

The time reversal to the above, $B_{-k'-\sigma',-k-\sigma}$, has the same absolute value, but the phase is the same or the reverse.

It is possible to classify as:

$1_\pm$. The spin flip-flop does not arise,

$$B_{k\sigma,k'\sigma}(c^+_{k\sigma} c_{k'\sigma} \pm c^+_{-k'-\sigma} c_{-k-\sigma}).$$

$2_\pm$. The spins flip-flop does arise,

$$B_{k\sigma,k'-\sigma}(c^+_{k\sigma} c_{k'-\sigma} \pm c^+_{-k'\sigma} c_{-k-\sigma}).$$

As is seen above, the theory implicitly assumes that the system is a perfect crystal and described by the single wave vector $k$. The positive and negative signs of $k$ indicate waves propagating from the vertex or off the vertex. Before entering the discussion about the relaxation time of the magnetic resonance, we briefly review the Bogoliubov theory [3, 11, 16].

The Bogoliubov transformation defines a quasiparticle responsible to the superconductivity as

$$
\begin{pmatrix} \gamma_{k\uparrow} \\ \gamma_{-k\downarrow}^{+} \end{pmatrix} = \begin{pmatrix} u_k & -v_k \\ v_k & u_k \end{pmatrix} \begin{pmatrix} c_{k\uparrow} \\ c_{-k\downarrow}^{+} \end{pmatrix}
\tag{1.135}
$$

with

$$
u_k^2 - v_k^2 = 1.
$$

The spirit of the Bogoliubov transformation is to mix operators $c_{k\uparrow}$ and $c_{-k\downarrow}^{+}$ which are different in spin (as to the wave number, this mixing is not so serious) and not mixed in the normal situation. The quasiparticle yields the new ground state near the chemical potential. The stabilization energy thus obtained is called the gap energy, $\Delta_k$. What we have done in Sec. 1.3 is a substantial understanding of this reason. However, if we want to make the $\pm k$ distinction meaningful, it is natural to adopt the four-component spinor, say, the extended Nambu spinor (perhaps, spurious) [27]. For ordinary states,

$$
\mathbf{c}_k^{+} = (c_{k\uparrow}^{+} \quad c_{-k\downarrow} \quad c_{-k\uparrow} \quad c_{k\downarrow}^{+}), \quad \mathbf{c}_k = \begin{pmatrix} c_{k\uparrow} \\ c_{-k\downarrow}^{+} \\ c_{-k\uparrow}^{+} \\ c_{k\downarrow} \end{pmatrix},
\tag{1.136}
$$

and, for the superconducting state,

$$
\gamma_k^{+} = (\gamma_{k\uparrow}^{+} \quad \gamma_{-k\downarrow} \quad \gamma_{-k\uparrow} \quad \gamma_{k\downarrow}^{+}), \quad \gamma_k = \begin{pmatrix} \gamma_{k\uparrow} \\ \gamma_{-k\downarrow}^{+} \\ \gamma_{-k\uparrow}^{+} \\ \gamma_{k\downarrow} \end{pmatrix}.
\tag{1.137}
$$

These are connected by the Bogoliubov transformation with each other as

$$
\gamma_k = U_k \mathbf{c}_k, \quad \gamma_k^{+} = \mathbf{c}_k U_k^{+},
\tag{1.138}
$$

where

$$
U_k = \begin{pmatrix} \mathbf{u}_k & 0 \\ 0 & \mathbf{u}_k^{+} \end{pmatrix}, \quad \text{with} \quad \mathbf{u}_k = \begin{pmatrix} u_k & v_k \\ -v_k & u_k^{+} \end{pmatrix}.
\tag{1.139}
$$

It is instructive to manipulate carefully.

Case $1_+$:

$$(c_{k\sigma}^+ c_{k'\sigma} + c_{-k'-sigma}^+ c_{-k-\sigma}) = \mathbf{c}_k^+ \Sigma^3 \mathbf{c}_{k'} = \gamma_k^+ U_k \Sigma^3 U_{k'}^+ \gamma_{k'}, \tag{1.140}$$

where

$$\Sigma^3 = \begin{pmatrix} \sigma^3 & \\ & -\sigma^3 \end{pmatrix}. \tag{1.141}$$

This is the $4 \times 4$ matrix manipulation; however, it is enough to note the upper half of the result.

The upper half of Eq. (1.138) is

$$(c_{k\uparrow}^+ \quad c_{-k\downarrow}) \begin{pmatrix} 1 & 0 \\ 0 & -1 \end{pmatrix} \begin{pmatrix} c_{k'\uparrow} \\ c_{-k'\downarrow}^+ \end{pmatrix}$$

$$= (\gamma_{k\uparrow}^+ \quad \gamma_{-k\downarrow}) \begin{pmatrix} u & v \\ -v & u \end{pmatrix} \begin{pmatrix} 1 & 0 \\ 0 & -1 \end{pmatrix} \begin{pmatrix} u' & -v' \\ v' & u' \end{pmatrix} \begin{pmatrix} \gamma_{k'\uparrow} \\ \gamma_{-k'\downarrow}^+ \end{pmatrix}$$

$$= (\gamma_{k\uparrow}^+ \quad \gamma_{-k\downarrow}) \begin{pmatrix} uu' - vv' & -uv' - vu' \\ -vu' - uv' & vv' - uu' \end{pmatrix} \begin{pmatrix} \gamma_{k'\uparrow} \\ \gamma_{-k'\downarrow}^+ \end{pmatrix}, \tag{1.142}$$

where $u$ and $v$ are the abbreviations of $u_k$ and $v_k$, respectively, while $u'$ and $v'$ are those of $u_{k'}$ and $v_{k'}$.

Case $1_-$:

$$\sum_\sigma (c_{k\sigma}^+ c_{k'\sigma} - c_{-k'-\sigma}^+ c_{-k-\sigma}) = \mathbf{c}_k^+ \mathbf{1} \mathbf{c}_{k'} = \gamma_k^+ U_k \mathbf{1} U_{k'}^+ \gamma_{k'}. \tag{1.143}$$

The upper half of the above is

$$(\gamma_{k\uparrow}^+ \quad \gamma_{-k\downarrow}) \begin{pmatrix} uu' + vv' & -uv' + vu' \\ -vu' + uv' & vv' + uu' \end{pmatrix} \begin{pmatrix} \gamma_{k'\uparrow} \\ \gamma_{-k'\downarrow}^+ \end{pmatrix}. \tag{1.144}$$

As is shown clearly, the original term is transformed in the quasiparticle representation to a combination of the scattering term with the diagonal element in the $\mathbf{u}_k$ matrix and the creation or annihilation of a pair with the off-diagonal element of $\mathbf{u}$. These matrix elements are called the coherent factors.

Let us turn to the case where the spin flip-flop is allowed.

Case $2_+$:

$$c^+_{k\sigma}c_{k'-\sigma} + c^+_{-k'\sigma}c_{-k-\sigma} = \mathbf{c}^+_k\Sigma^J\mathbf{c}_{k'} = \gamma^+_k U_k\Sigma^J U^+_{k'}\gamma_{k'} \tag{1.145}$$

with

$$\Sigma^J = \begin{pmatrix} & & & 1 \\ & -1 & & \\ & & -1 & \\ 1 & & & \end{pmatrix}. \tag{1.146}$$

This relation connects the left half of $\gamma^+_k$ and the lower half of $\gamma_{k'}$, by giving

$$\text{Eq. (1.146)} = (\gamma^+_{k\uparrow} \quad \gamma_{-k\downarrow}) \begin{pmatrix} uu' + vv' & uv' - vu' \\ -vu' + uv' & -vv' - uu' \end{pmatrix} \begin{pmatrix} \gamma_{-k'\uparrow} \\ \gamma^+_{k'\downarrow} \end{pmatrix}. \tag{1.147}$$

Case $2_-$:

$$c^+_{k\sigma}c_{k'-\sigma} - c^+_{-k'\sigma}c_{-k-\sigma} = \mathbf{c}^+_k\mathbf{J}\mathbf{c}_{k'} = \gamma^+_k U_k\mathbf{J}U^+_{k'}\gamma_{k'} \tag{1.148}$$

with

$$\mathbf{J} = \begin{pmatrix} & & & 1 \\ & & 1 & \\ & 1 & & \\ 1 & & & \end{pmatrix}. \tag{1.149}$$

This relation also connects the left half of $\gamma^+_k$ and the lower half of $\gamma_{k'}$, and we have

$$\text{Eq. (1.149)} = (\gamma^+_{k\uparrow} \quad \gamma_{-k\downarrow}) \begin{pmatrix} uu' - vv' & uv' + vu' \\ -vu' - uv' & -vv' + uu' \end{pmatrix} \begin{pmatrix} \gamma_{-k'\uparrow} \\ \gamma^+_{k'\downarrow} \end{pmatrix}. \tag{1.150}$$

Note that, at the present, the scattering terms are off-diagonal, and the creation and annihilation terms are diagonal.

By the use of the relations [3, 11, 16]

$$u^2_k = \frac{1}{2}\left(1 + \frac{\epsilon_k}{E_k}\right), \quad v^2_k = \frac{1}{2}\left(1 - \frac{\epsilon_k}{E_k}\right),$$

$$E^2_k = \epsilon^2_k + \triangle^2_k \tag{1.151}$$

($\triangle_k$ is the gap energy), the coherent factors are expressed substantially as

$$1_{\pm}. \quad (uu' \mp vv')^2 = \frac{1}{2}\left(1 + \frac{\epsilon_k \epsilon_{k'}}{E_k E_{k'}} \mp \frac{\triangle_k \triangle_{k'}}{E_k E_{k'}}\right),$$

$$2_{\pm}. \quad (uv' \mp vu')^2 = \frac{1}{2}\left(1 \mp \frac{\triangle_k \triangle_{k'}}{E_k E_{k'}}\right). \tag{1.152}$$

In case $1_+$, we have the ultrasonic attenuation, while the electromagnetic interaction is in case $2_+$, and the magnetic resonances are in case $2_-$.

### 1.7.2.  $T_1$ in NMR

The detailed analysis of the spin-lattice relaxation time $T_1$ in the nuclear magnetic resonance will be present in the next section. Here, the results are given briefly:

$$T_1 \sim \sum_{kk'} |B_{kk'}|^2 \frac{1}{2}\left(1 + \frac{\triangle_k \triangle_{k'}}{E_k E_{k'}}\right) n_k (1 - n_{k'})\delta(E_k - E_{k'} - w), \tag{1.153}$$

where $w$ is the applied radio frequency. By converting the summation (explicitly shown) to the integration, and further by using the relation of state densities,

$$N(E)dE = N(\epsilon)d\epsilon, \quad \text{then}$$

$$\frac{N(E)}{N(\epsilon)} = \frac{d\epsilon}{dE} = \begin{cases} \dfrac{E}{(E^2 - \triangle^2)^{1/2}} & (E > \triangle), \\ 0 & (E < \triangle), \end{cases} \tag{1.154}$$

we rewrite Eq. (1.151) for $w \ll \triangle$ as

$$T_1 \sim |B|^2 N^2(0) \int_{\triangle}^{\infty} \frac{1}{2}\left(1 + \frac{\triangle^2}{E(E + w)}\right)$$

$$\times \frac{E(E + w)k_B T(-\partial n/\partial E)dE}{(E^2 - \triangle^2)^{1/2}[(E + w)^2 - \triangle^2]^{1/2}}, \tag{1.155}$$

where the coupling constant and the state density are replaced by their suitable averages. This integral is divergent. Therefore, $T_1$ of a superconductor is strongly enhanced just below the critical temperature. This phenomenon was observed and explained by Slichter *et al.* [25, 26]. This was told one of the brilliant victories of the BCS theory. However, it is found that, in the recent experiments of the high-temperature superconductors or the copper-oxide superconductors, the $T_1$ enhancement is lost. This phenomenon is considered deeply connected with the mechanism of the high-temperature superconductivity of these species and attracted the interests of many investigators [28, 30]. However, so long as we know, the theory of magnetic resonance of a superconductor has been done almost all in the scheme mentioned in this introduction. Then it will be preferable to develop the theory of magnetic resonance in accordance with the sophisticated recent theory of superconductivity.

### 1.7.3. *Theory with Green's function*

Our idea is as follows: the algebra of electrons is related to their field operators. In the same way, we assume the field for nuclei. For example, the creation operator for a nucleus $a_{KM}^+$ yields the nuclear motion with $K$ and $M$ which are spatial and spin quantum numbers, respectively. The energy spectrum of the propagator $\mathcal{G}_{KM}(\tau) = \langle\langle a_{KM}(\tau) a_{KM}^+ \rangle\rangle$ gives the line shape of the magnetic resonance.

The nuclear propagator $\mathcal{G}_{KM}(\tau)$ sees the electron sea, followed by the electron excitation in the spin space. This gives the additional line width of the nuclear magnetic resonance. The phenomenon looks like the vacuum polarization in quantum electrodynamics. The self-energy part thus arisen in the nuclear energy is the source of the line shape of the nuclear magnetic resonance [24].

The spin-lattice relaxation time of the nuclear spin $I^z$ is given by the imaginary part of the magnetic susceptibility $\chi_{zz}$ which is equal to $(\chi_{+-} + \chi_{-+})/2$ in the spatially homogeneous system. Here, $\pm$ correspond to $(I^x \pm iI^y)/2$, respectively. The ensemble average of a change, $\delta\langle I^+(t)\rangle$, is given by the linear response theory as

$$\delta\langle I^+(t)\rangle = i \int_{-\infty}^{t} dt' \text{Tr}\{\rho_G [H^{\text{ex}}(t'), I^+(t)]_-\}, \tag{1.156}$$

where $\rho_G$ is the grand canonical statistical operator. However, the chemical potential is not given explicitly, otherwise stated. The rotating magnetic field causing the magnetic transition is, assuming a single mode for simplicity,

$$H^{\text{ex}}(t) = H_R(I^+(t)e^{i\omega t} + I^-(t)e^{-i\omega t}). \tag{1.157}$$

As has been said, the spin-lattice relaxation arises from the interaction between the nuclear spin and the electron spin. In other words, the electron spins play the role of a lattice system.

$$H' = \hbar\gamma g\beta_B F(R, r)\mathbf{I} \cdot \mathbf{S}, \tag{1.158}$$

where $F(R, r)$ is a function of spatial coordinates of the nucleus $R$ and that of the electron, $r$. The term $\gamma$ is the gyromagnetic ratio of the nucleus, and $g$ and $\beta_B$ are the $g$-factor and the Bohr magneton of electron, respectively.

Now the second quantization of the aboves are carried out. First of all, the orthonormalized wave function describing the nuclear behavior, $|\xi_K(R)M\rangle$, is introduced as

$$(H_N + H_M)|\xi_K(R)M\rangle = (\epsilon_K + M)|\xi_K(R), M\rangle$$
$$= \epsilon_{KM}|\xi_K(R,)M\rangle \qquad (1.159)$$

with

$$\epsilon_{KM} = \epsilon_K + M,$$

where $H_N$ is the spatial part, and $H_M$ is the Zeeman part. Then we have

$$I^\alpha \rightarrow \langle \xi_K(R)M|I^\alpha|\xi_{K'}(R)M'\rangle a^+_{KM} a_{K'M'}$$
$$= \langle M|I^\alpha|M'\rangle a^+_{KM} a_{K'M'} \delta_{KK'}. \qquad (1.160)$$

The similar equation is given for electrons

$$(H_S + H_m)\phi_k(r)|m\rangle = (\epsilon_k + m)|\phi_k(r), m\rangle$$
$$= \epsilon_{km}|\phi_k(r), m\rangle, \qquad (1.161)$$

where

$$\epsilon_{km} = \epsilon_k + m,$$

so that

$$H' \rightarrow h\gamma g\beta_B \langle M|I^\beta|M'\rangle \langle m|S^\alpha|m'\rangle$$
$$\times \langle \xi_K(R)\phi_k(r)|F(R,r)|\xi_{K'}(R)\phi_{k'}(r)\rangle (a^+_{KM} a_{K'M'})(c^+_{km} c_{k'm'}). \qquad (1.162)$$

If the nuclear motion is assumed as that of a harmonic oscillator, $a^+_{KM}$ and $a_{KM}$ are the creation and annihilation operators of vibrational excitations. When the nuclei carry a non-integer spins, these are considered to obey the Fermi statistics or satisfy the anticommutation relation

$$[a^+_{KM}, a_{K'M'}]_+ = \delta_{KK'}\delta_{MM'}. \qquad (1.163)$$

However, as will be seen in the following, this selection of the statistics is not fatal for the theory. It is needless to say that the operators for electrons satisfy the anticommutation relations.

The change of $I^+$ in Eq. (1.156) can be written, in the interaction representation, as (we retain the $I^-$ term in Eq. (1.157))

$$\delta\langle I^+(t)\rangle = i\gamma H_R \int_{-\infty}^{t} dt' e^{-i\omega t'}$$

$$\times \mathrm{Tr}\{\rho_G \langle M-1|I^-|M\rangle\langle M|I^+|M-1\rangle$$

$$\times [a_{K,M-1}^+(t')a_{K,M}(t'), a_{K,M}^+(t)a_{K,M-1}(t)]_-\}$$

$$= -\gamma H_R \int_{-\infty}^{t} dt' e^{-i\omega t'} D_{K,M-1,M}(t'-t)$$

$$= -\frac{1}{2}\gamma H_R e^{-i\omega t} \int_{-\infty}^{\infty} ds e^{-i\omega s} \sum_{KM} D_{K,M-1,M}(s)$$

$$= -\frac{1}{2}\gamma H_R e^{-i\omega t} D_{K,M-1,M}(\omega). \tag{1.164}$$

From this result, the magnetic susceptibility of the present system is

$$\chi_{+-}(\omega) = -\frac{\gamma}{2} D_{K,M-1,M}(\omega). \tag{1.165}$$

In the course of derivation, the matrix elements of spin operators are put to be equal to 1, and then the $Tr$ operation is carried out. Here, $D_{K,M-1,M}(s)$ is a retarded Green's function,

$$D_{K,M-1,M}(s) = -i\theta(s)\mathrm{Tr}\left\{\rho_G[a_{KM-1}^+(s)a_{KM}(s), a_{KM}^+ a_{KM-1}]_-\right\}, \tag{1.166}$$

and $D_{K,M-1,M}(\omega)$ is its Fourier transform. Now our problem is to estimate this retarded function.

The retarded Green's function is easily obtained by the analytical continuation from the Matsubara function (or the temperature Green's function with imaginary time $\tau$) which is causal in $\tau$,

$$\mathcal{D}_{K,M-1,M}(\tau) = -\mathrm{Tr}\{\rho_G T\tau[a_{K,M-1}^+(\tau)a_{K,M}(\tau)a_{K,M}^+ a_{K,M-1}]\}, \tag{1.167}$$

for which the Feynman diagram analysis is available [16].

### 1.7.4. *Non-interacting*

Here, we deal with the case without the spin-lattice interaction. It might be trivial; however, it seems instructive for the later investigation. By the use of the simplified notation, $\langle \cdots \rangle = \mathrm{Tr}(\rho_G \cdots)$, the Green's function in this case is written as

$$\mathcal{D}^0_{K,M-1,M}(\tau) = \langle T\tau[a^+_{KM-1}(\tau)a_{K,M}(\tau)a^+_{KM}a_{KM-1}]\rangle$$
$$= \mathcal{G}^0_{KM}(\tau)\mathcal{G}^0_{KM-1}(-\tau), \tag{1.168}$$

where

$$\mathcal{G}^0_{KM}(\tau) = -\langle T\tau[a_{KM}(\tau)a^+_{KM}]\rangle, \tag{1.169}$$

and the corresponding Fourier transform of Eq. (1.168) is

$$\mathcal{D}^0_{K,M-1,M}(\omega_n) = \frac{1}{\beta}\sum_{\nu_n}\mathcal{G}^0_{KM}(\nu_n)\mathcal{G}^0_{KM-1}(\nu_n - \omega_n), \tag{1.170}$$

where

$$\mathcal{G}^0_{KM}(\nu_n) = \frac{1}{i\nu_n - \epsilon_{KM}}. \tag{1.171}$$

Therefore,

$$\mathcal{D}^0_{K,M-1,M}(\omega_n) = \frac{1}{\beta}\sum_{\nu_n}\frac{1}{i\nu_n - i\omega_n - \epsilon_{KM-1}} \cdot \frac{1}{i\nu_n - \epsilon_{KM}}$$

$$= \frac{1}{\beta}\sum_{\nu_n}\left(\frac{1}{i\nu_n - i\omega_n - \epsilon_{KM-1}} - \frac{1}{i\nu_n - \epsilon_{KM}}\right)$$

$$\times \frac{1}{i\omega_n + \epsilon_{KM-1} - \epsilon_{KM}}$$

$$= [n(\epsilon_{KM-1}) - n(\epsilon_{KM})]\frac{1}{i\omega_n - \epsilon_{KI}}, \tag{1.172}$$

where

$$\epsilon_{KI} = \epsilon_{KM} - \epsilon_{KM-1}$$

and it is noted that $\omega_n$ is even, it does not matter in obtaining the particle number. The retarded Green's function is obtained simply by replacing $i\omega_n$ by $\omega + i\eta$ ($\eta$ is a positive infinitesimal). Thus, we obtain

$$\delta\langle I^+(t)\rangle = e^{-i\omega t}H_R(n(\epsilon_{K-}) - n(\epsilon_{K+}))\frac{1}{\omega - \epsilon_{KI} + i\eta}, \tag{1.173}$$

where the radio frequency arousing the resonance is rewritten by $\omega^0$. The magnetic susceptibility $\chi_{+-}$ thus becomes

$$\chi_{+-}(\omega) = \gamma(n(\epsilon_{K-}) - n(\epsilon_{K+}))\frac{1}{\omega - \epsilon_{KI} + i\eta},$$

$$\gamma = e^{-i\omega t} H_R, \tag{1.174}$$

whose imaginary part is

$$\chi''_{+-}(\omega) = -\pi\gamma(n(\epsilon_{K-}) - n(\epsilon_{K+}))\delta(\omega - \epsilon_{KI}). \tag{1.175}$$

This gives the sharp $\delta$-function type energy spectrum. We have no line width or the relaxation time, and states are stationary.

### 1.7.5. *Interacting; normal*

In the interacting system, $\mathcal{G}^0$ in Eq. (1.168) has to be replaced by $\mathcal{G}$ including the interaction which, in the present case, is the spin–spin interaction between nuclei and electrons, as has been given in Eq. (1.156),

$$\mathcal{D}'_{KM-1,M}(\omega_n) = \frac{1}{\beta}\sum_{\nu_n} \mathcal{G}_{KM}(\nu_n)\mathcal{G}_{KM-1}(\nu_n - \omega_n), \tag{1.176}$$

where, for instance,

$$\mathcal{G}_{KM}(\nu_n) = [(\mathcal{G}^0_{KM}(\nu_n))^{-1} - \Sigma_{KM}(\omega)]^{-1}, \tag{1.177}$$

where $\omega$ is the interaction energy.

Our procedure is going on as follows. The two Green's functions with the self-energy part are evaluated. Combining these gives the retarded Green's function $\mathcal{D}$ to estimate the magnetic susceptibility.

The most important (divergent) self-energy part of this self-energy is due to the ring diagram shown in Fig. 1.2. The problem is to examine how the

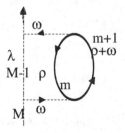

Fig. 1.2. Self-energy part $\mathcal{G}(\nu_n)$.

energy of the nucleus propagator (two-dot-interrupted line) is changed by the ring diagram of the electron (full line). That is to say,

$$\mathcal{G}(\nu_n)^{-1} = \mathcal{G}(\nu_n)^0 + \Sigma(\omega),$$

$$\Sigma(\omega) = -|K|^2 \mathcal{S}(\rho_n)\mathcal{S}(\rho_n + \omega), \tag{1.178}$$

where $\mathcal{S}$ is the electron propagator, and the minus sign is due to the fermion loop. The self-energy part includes the coupling terms, where

$$|K|^2 = \langle M|I^+|M-1\rangle\langle m-1|S^-|m\rangle|^2$$

$$\times |\langle \chi_K(R)\phi_k(r)|F(R,r)|\phi_{k'}(r)\chi_{K'}(R)\rangle|^2 \tag{1.179}$$

and the minus sign is due to a Fermion loop. The ring diagram is calculated as follows:

$$\mathcal{S}_{km}(\rho_n)\mathcal{S}_{km-1}(\rho_n + \omega) = \sum_{\rho_n} \frac{1}{i\rho_n - \epsilon_{km}} \cdot \frac{1}{i\rho_n + \omega - \epsilon_{km+1}}$$

$$= \sum_{\rho_n} \left( \frac{1}{i\rho_n - \epsilon_{km}} - \frac{1}{i\rho_n + \omega - \epsilon_{km+1}} \right)$$

$$\times \frac{1}{\omega - \epsilon_{km+1} + \epsilon_{km}}$$

$$= (n(\epsilon_{km}) - n(\epsilon_{km+1}))\frac{1}{\omega - \epsilon_{km+1} + \epsilon_{km}}. \tag{1.180}$$

The lattice (electron) gets $\omega$ from a nucleus to lift the electron spin from $m$ to $m+1$, and, at the other vertex, the inverse process occurs. This looks like the radiation process in the photochemistry. We thus have the self-energy part of $\mathcal{G}_{K;M-1,M}$,

$$\Sigma_{K;M-1,M}(\nu_n) = |K|^2(n(\epsilon_{km}) - n(\epsilon_{km+1}))\frac{1}{\omega - \epsilon_{ks}}, \tag{1.181}$$

where

$$\epsilon_{ks} = \epsilon_{km+1} - \epsilon_{km}.$$

Another propagator $\mathcal{G}_{K;M}$ has the self-energy part $\Sigma_{K;M,M-1}(\nu_n)$. This is built by replacing $\rho_n + \omega$ by $\rho_n - \omega$ in Fig. 1.2:

$$\Sigma_{K;M,M-1}(\nu_n) = |K|^2(n(\epsilon_{km}) - n(\epsilon_{km-1}))\frac{1}{\omega - \epsilon_{km-1} + \epsilon_{km}}$$

$$= |K|^2(n(\epsilon_{km}) - n(\epsilon_{km-1}))\frac{1}{\omega - \epsilon_{ks}}$$

$$\approx |K|^2(n(\epsilon_{km+1}) - n(\epsilon_{km}))\frac{1}{\omega - \epsilon_{ks}}. \tag{1.182}$$

We now turn to the evaluation of the propagator $\mathcal{D}_{K;M-1,M}(\tau)$,

$$\mathcal{D}_{K;M-1,M}(\omega_n)$$

$$= \frac{1}{\beta} \sum_{\nu_n} \mathcal{G}_{KM}(\nu_n) \mathcal{G}_{KM-1}(\nu_n - \omega_n)$$

$$= \frac{1}{\beta} \sum_{\nu_n} (i\nu_n - \epsilon_{KM} - \Sigma_{KM-})^{-1} (i\nu_n - i\omega_n - \epsilon_{KM-1} - \Sigma_{K;M,M-1})^{-1}$$

$$= \frac{1}{\beta} \sum_{\nu_n} \left\{ \frac{1}{i\nu_n - i\omega_n - \epsilon_{KM-1} - \Sigma_{KM-1}} - \frac{1}{i\nu_n - \epsilon_{KM} - \Sigma_{KM}} \right\}$$

$$\times \left\{ \frac{1}{i\omega_n - \epsilon_{KM-1} - \Sigma_{KM-1} + \epsilon_{KM} + \Sigma_{KM}} \right\}$$

$$= (N(\epsilon_{KM}) - N(\epsilon_{KM-1})) \left\{ \frac{1}{i\omega_n - \epsilon_{KI} + \Sigma_{KM} - \Sigma_{KM-1}} \right\}. \quad (1.183)$$

Here,

$$\epsilon_{KI} = \epsilon_{KM} - \epsilon_{KM-1},$$

and the self-energy parts are disregarded in obtaining the particle density. We can observe that the additional terms in the denominator modify the line shape.

If we put $i\omega_n \to \omega + i\eta$, we can obtain the retarded Green's function, whose imaginary part gives the line shape:

$$D(\omega) \sim \left\{ \frac{1}{\omega + i\eta - \epsilon_{KI} + \Sigma_{KM} - \Sigma_{KM-1}} \right\}$$

$$= \left\{ \mathrm{P} \left( \frac{1}{\omega - -\Sigma_{KM-1}\epsilon_{KI}} \right) - i\pi\delta(\omega - \epsilon_{KI} + \Sigma_{KM} - \Sigma_{KM-1}) \right\}$$

$$= \frac{i(\mathrm{Im})}{(\mathrm{Re})^2 + (\mathrm{Im})^2} \quad \text{(at the resonance point)}. \quad (1.184)$$

The line shape now changed from the $\delta$-function type to the Lorentz type, as expected.

## 1.8. Interacting; Superconductor

### 1.8.1. *The extended Nambu spinor*

We then investigate how the line shape obtained above is further modified in a superconductor. The electron propagators in the previous section are replaced by those in a superconductor. These have been studied already in Sec. 1.3 and are present here in a new fashion adequate for the following investigation. As has been done by BCS, let us consider the attractive two-body potential $g$ which is assumed constant for simplicity and further keep in mind the Cooper pair. The Hamiltonian

$$H_{el} = \left\{ \epsilon_{k\alpha} c_{k\alpha}^+ c_{k\alpha} + \frac{1}{2} g c_{k\alpha}^+ c_{-k\beta}^+ c_{-k\beta} c_{k\alpha} \right\}, \tag{1.185}$$

where $\epsilon_{k\alpha}$ is the orbital energy, includes the Zeeman energy in the present case.

Now we use the extended Nambu representation of Eq. (1.136),

$$\mathbf{c}_k = \begin{pmatrix} c_{k\alpha} \\ c_{-k\beta}^+ \\ c_{-k\alpha}^+ \\ c_{k\beta} \end{pmatrix}, \qquad \mathbf{c}_k^+ = \begin{pmatrix} c_{k\alpha}^+ & c_{-k\beta} & c_{-k\alpha} & c_{k\beta}^+ \end{pmatrix} \tag{1.186}$$

with the equal time commutator,

$$[\mathbf{c}_k, \mathbf{c}_{k'}^+]_+ = 1\delta_{kk'}. \tag{1.187}$$

In these terms, the Hamiltonian is rewritten as

$$H_{el} = \epsilon_k \mathbf{c}_k^+ \Sigma^3 \mathbf{c}_k + \frac{1}{2} g (\mathbf{c}_k^+ \Sigma^+ \mathbf{c}_k)(\mathbf{c}_{k'}^+ \Sigma^- \mathbf{c}_{k'}), \tag{1.188}$$

where $\epsilon_k$ is the diagonal matrix of $\epsilon_\gamma$:

$$\epsilon_k = \begin{pmatrix} \epsilon_{k\alpha} & & & \\ & \epsilon_{-k\beta} & & \\ & & \epsilon_{-k\alpha} & \\ & & & \epsilon_{k\beta} \end{pmatrix}. \tag{1.189}$$

Let us define

$$\Sigma^3 = \begin{pmatrix} \sigma^3 & 0 \\ 0 & -\sigma^3 \end{pmatrix}, \quad \Sigma^+ = \begin{pmatrix} \sigma^+ & 0 \\ 0 & -\sigma^- \end{pmatrix}, \quad \Sigma^- = \begin{pmatrix} \sigma^- & 0 \\ 0 & -\sigma^+ \end{pmatrix} \tag{1.190}$$

with

$$\sigma^3 = \begin{pmatrix} 1 & 0 \\ 0 & -1 \end{pmatrix}, \quad \sigma^1 = \begin{pmatrix} 0 & 1 \\ 1 & 0 \end{pmatrix}, \quad \sigma^2 = \begin{pmatrix} 0 & -i \\ i & 0 \end{pmatrix},$$

$$\sigma^+ = \frac{1}{2}(\sigma^1 + i\sigma^2), \quad \sigma^- = \frac{1}{2}(\sigma^1 - i\sigma^2). \tag{1.191}$$

As for the electron–electron interaction, those for the Cooper pairs are only selected: namely $\Sigma^+$ selects a Cooper pair in the particle state, while $\Sigma^-$ that in the hole state.

## 1.8.2. *Green's function*

Here, the discussions done in Sec. 1.4 are repeated briefly. The above Hamiltonian is invariant under the scale transformation. Then we have a current conservation; especially the charge conservation in the static state (the Noether's theorem). If we have any quantity which does not commute with this invariant charge, we can expect a phase transition (the Goldstone theorem).

The charge proportional to

$$\langle c_{k\gamma}^+ c_{k\gamma} \rangle = \sum_{k>0} \langle c_k^+ \Sigma^3 c_k \rangle \tag{1.192}$$

is invariant under the rotation about the $\Sigma^3$ axis in the space spanned by $\Sigma^3$, $\Sigma^+$, and $\Sigma^-$. Observing that

$$[\Sigma^3, \Sigma^\pm] = \pm 2\Sigma^\pm \tag{1.193}$$

suggests the phase transitions along the $\Sigma^\pm$ directions.

If we define

$$\bar{\mathbf{c}} = \mathbf{c}^+ \Sigma^3$$

as has been done in Sec. 1.43, the discussions parallel to those there will be possible in the following. However, this is not employed in this case.

The phase transition cannot be achieved by the perturbational approach, but the effective Hamiltonian giving the phase transition should be included at the beginning. Say, the modified Hamiltonian

$$H^0 = \mathbf{c}_k^+ (\epsilon_k \Sigma^3 + \rho \Sigma^+ + \eta \Sigma^-) \mathbf{c}_k,$$

$$H^{\text{int}} = \frac{1}{2} g \sum_{kk'} \{ (\mathbf{c}_k^+ \Sigma^+ \mathbf{c}_k) \cdot (\mathbf{c}_{k'}^+ \Sigma^- \mathbf{c}_{k'}) - \mathbf{c}_k^+ (\rho \Sigma^+ + \eta \Sigma^-) \mathbf{c}_k \}, \tag{1.194}$$

where $\rho$ and $\eta$ are the so-called gap energies which are assumed independent of $k$ for simplicity. Note that the Hamiltonian $H^0$ is already symmetry-broken. Here, we adopt a conventional method. In what follows, $H^{\text{int}}$ is neglected, that is to say, we start from the Green's function due to the effective Hamiltonian and then take interaction (1.157) into account. At this stage, we may expect the result that will be obtained in such a manner that the normal and superconducting states contribute additively.

The temperature Green's function with the imaginary time $\tau$ is defined as

$$G_{kk'}^0(\tau) = -\langle T_\tau [\mathbf{c}_k(\tau) \mathbf{c}_{k'}^+] \rangle, \tag{1.195}$$

where $\langle \cdots \rangle = \mathrm{Tr}\{\rho_G \cdots\}$. The equation of motion of $G^0_{kk'}$ is

$$\partial_\tau G^0_{kk'}(\tau) = -\partial_\tau[\theta(\tau)\langle c_k(\tau)c^+_{k'}(0)\rangle - \theta(-\tau)\langle c^+_{k'}(0)c_k(\tau)\rangle]$$

$$= \delta(\tau)\langle [c_k(\tau), c^+_{k'}(0)]_+\rangle + \langle T\tau[c_k(\tau), H^0]_-, c^+_k(0)\rangle$$

$$= \delta_{kk'} + (\epsilon_k \Sigma^3 + \rho\Sigma^+ + \eta\Sigma^-)G^0_{kk'}. \tag{1.196}$$

In the course of the above derivation, the commutator in Eq. (6.65) was used.

We make the Fourier transformation,

$$G^0_{kk'}(\tau) = \frac{1}{\beta}\sum_{\omega_n} e^{-i\omega_n \tau} G^0_{kk'}(\omega_n), \tag{1.197}$$

where $\beta = k_B T$ and $k_B$ is the Boltzmann constant. Then the equation of motion becomes

$$(i\omega_n + \epsilon_k \Sigma^3 + \rho\Sigma^+ + \eta\Sigma^-)G^0_{kk'}(\omega_n) = \delta_{kk'}. \tag{1.198}$$

Namely,

$$G^0_{kk'}(\omega_n) = \frac{1}{i\omega_n - \epsilon_k\Sigma^3 - \rho\Sigma^+ - \eta\Sigma^-}$$

$$= \delta_{kk'}\left\{\frac{i\omega_n + \epsilon_k\Sigma^3 + \rho\Sigma^+ + \eta\Sigma^-}{(i\omega_n)^2 - E^2_k}\right\}, \tag{1.199}$$

where

$$E^2_k = \epsilon^2_k + \rho\eta. \tag{1.200}$$

### 1.8.3.  *Spin dynamics*

The propagator in Eq. (1.199) is the $4 \times 4$ matrix. If we ignore $\rho$ and $\eta$, this is reduced to the normal propagator. We have already tried this simple case.

Now we assume the terms responsible for the superconductivity as a perturbation. Namely ($S$ is the electron propagator),

$$S(\nu_n) = \frac{i\nu_n + \epsilon_k\Sigma^3 + \rho\Sigma^+ + \eta\Sigma^-}{(i\nu_n)^2 - E^2_k}$$

$$= \frac{i\nu_n + \epsilon_k\Sigma^3}{(i\nu_n)^2 + \epsilon^2_k} + \frac{\rho\Sigma^+ + \eta\Sigma^-}{(i\nu_n)^2 + \epsilon^2_k} + \cdots . \tag{1.201}$$

It is helpful to separate this relation into components and to manipulate each one individually. Noticing

$$\sigma^0 + \sigma^3 = \sigma^\uparrow + \sigma^\downarrow,$$

we have, for example by operating $\sigma^\uparrow$ followed by Tr,

$$S^\uparrow(\nu_n) = \frac{i\nu_n + \epsilon_k \Sigma^3 + \rho \Sigma^+ + \eta \Sigma^-}{(i\nu_n)^2 - E_k^2}$$

$$= \frac{1}{i\nu_n - \epsilon_{k\uparrow}} + \frac{\rho + \eta}{(i\nu_n)^2 + \epsilon_k^2} + \cdots$$

$$= \frac{1}{i\nu_n - \epsilon_{k\uparrow}} + \frac{\rho + \eta}{2\epsilon_{k\uparrow}} \left( \frac{1}{i\nu_n - \epsilon_{k\uparrow}} - \frac{1}{i\nu_n + \epsilon_{k\uparrow}} \right) + \cdots . \quad (1.202)$$

In order to get the self-energy part of the nuclear propagator, we have to evaluate

$$\Sigma_{k;M-1,M}(\omega) = K^2 \frac{1}{\beta} \sum_{\nu_n} S(\nu_n) S(\nu_n + \omega). \quad (1.203)$$

The first-order term is already evaluated in Eq. (1.180). Then we have to carry out the complicated manipulation due to the second term of the last line in Eq. (1.202). However, the combinations other than those satisfying the resonance condition $\omega \sim \epsilon_{ks}$ (see Eq. (1.181)) should give a small effect to be neglected.

We then consider that the procedures of the previous section need not be repeated, and we are allowed to multiply the results there by the factor $(\rho + \eta)/2\epsilon_{k\uparrow}$ as the perturbing correction or the superconducting effect.

### 1.8.4. *Conclusion*

If we review the present investigations from the viewpoints of the line shape problem in the magnetic resonance, three cases are clearly distinguished: In the non-interacting case, the line shape is written in terms of the delta function. In the interacting case of the normal phase, it is presented by the Lorentz-like function, and, in the superconducting phase, it is further multiplied by a coherent factor; whereas the statistical factors referring to the nuclear states never change throughout.

For a superconductor, the present theory has almost nothing to give more than the current one has done. However, the theory is not merely to reproduce the experimental result, but to predict the mechanism hidden in observations, in such a way that the manipulations tell step-by-step what are working inside the matter. We might be satisfied with a little deeper understanding of the superconductivity.

Among various opinions we give, as an example, Scalpino's one [30]. He pointed out three possibilities that the $T_1$ enhancement is lost in the copper oxide superconductor:

(1) The $d$-wave single particle density of states has a logarithmic singularities, rather than the square-root singularity for the $s$-wave gap.
(2) The coherent factor for the quasiparticle scattering vanishes for $k \sim (\pi, \pi)$ for a $d_{x^2-y^2}$ gap.
(3) The inelastic scattering acts to suppress the peak just as for an $s$-wave.

The author had an opinion that the theory of superconductivity is already so well furnished that other fundamental ideas beyond the original BCS's are almost needless, except for a few smart equipments. As the phenomena observed in the copper oxide superconductor are rather qualitative and fairly clear cut, so the explanation of them must be simple. It is expected that a quantum chemical speculation could make this possible.

Let us address the author's opinion. The divergent character of Eq. (1.153) seems a merely mathematical problem. The difference between Eq. (1.184) and that (will be obtained) for the superconducting case is the coherent factor, (1.202). In the case of a BCS superconductor, it holds that $E_k \gg \rho$ so that the enhancement of the spin-lattice relaxation time $T_1$ is observed due to this factor, which leads to the superconductivity. However, in the case of a high-temperature superconductor, $\rho$, $\eta \sim E_k$ as has been seen in the previous chapter, then we cannot observe the sharp onset of superconductivity. Then we miss the coherent effect.

There is presumably a simpler reason why the $T_1$ enhance is not observed. In copper oxide, electrons responsible for superconductivity are probably the $d$-electrons, then we have the vanishing interaction term if it is the Fermi contact term between a nucleus and an electron, say, $F(R,r) \sim \delta(R-r)$ in Eq. (1.158).

## 1.9. Ginzburg–Landau Theory from BCS Hamiltonian

The macroscopic quantum theory of superconductivity has been given by Ginzburg and Landau (in short, GL) [9]. This looks rather phenomenological, however since a microscopic justification has been done by Gorkov and others [17, 31], this obtains a substantial foundation. We also reviewed the GL theory, in the introduction of Sec. 1.4, from the viewpoint of the Landau's general theory of phase transitions. It is crucial that the Lagrangian of the system is written as the fourth-order function of the order parameter $\Psi$ which is the electron field. If the coefficients of the second- and fourth-order terms are suitably chosen, the new ground state shapes a champagne bottle, or the Mexican hat is built. If electrons move on this flat route around the top, the derivative of the orbital vanishes or the kinetic energy vanishes, which implies that the wave function is rigid. We may say that the electron mass is effectively zero. We thus have the current only due to the vector potential, i.e., the diamagnetic current. This causes the Meissner effect.

The GL theory is quite helpful for applications, since the microscopic theory by itself is too complicated to manipulate the large-scale problems. If we can solve the GL equation for a real problem under an appropriate boundary condition, various information on this system can be obtained [11]. The macroscopic wave function or GL order parameter $\Psi$ is related to the gap function and understood as the field of Cooper pairs. The parameters in the GL equation are also written in the microscopic terms or by the experimental values. In [29], the GL function $\Psi$ is derived directly from the Bardeen–Cooper–Schrieffer Hamiltonian, not via the gap function or the anomalous Green's function related with the gap function. The electron–electron interaction composed of the four fermion operators is transformed by the Hubbard–Stratonovitch transformation to an auxiliary complex boson field $\phi$, in which electrons behave as if they were free. Using the path integral method, we can carry out the integration up to the quadratic terms of the electron operators. If we carefully analyze the resulting effective Lagrangian for the boson field, it is found that this boson field, which is described by a complex function, suggests a phase transition: the condensation arises in particles described by the real part of the boson field, while particles in the imaginary (or phase) part turn out to be massless Goldstone bosons. It is then clear that the boson field is to be the GL order parameter. In the course of analysis, the concept of supersymmetry is effectively used.

### 1.9.1. *BCS theory*

We assume the classical field of an electron is described by the Grassmann algebra or the anticommuting $c$-number, namely the creation and annihilation operators $a_k^*$ and $a_k$ are treated as anticommuting $c$-numbers. The BCS Hamiltonian is written as

$$H = H_0 + H_{\text{int}},$$

$$H_0 = \sum_{k\sigma} \epsilon_{k\sigma} a_{k\sigma}^* a_{k\sigma}, \qquad (1.204)$$

$$H_{\text{int}} = \sum_{kk'} -g_{k-k,-k'k'}(a_{k\uparrow}^* a_{-k\downarrow}^*)(a_{-k'\downarrow} a_{k'\uparrow}), \quad g_{k-k,k'-k'} > 0,$$

and

$$-g_{k-k,-k'k'} = \int d\mathbf{r}_1 d\mathbf{r}_2 \; \chi_{k\uparrow}^*(\mathbf{r}_1)\chi_{-k\downarrow}^*(\mathbf{r}_2)v(\mathbf{r}_1,\mathbf{r}_2)\chi_{-k'\downarrow}(\mathbf{r}_2)\chi_{k'\uparrow}(\mathbf{r}_1), \quad (1.205)$$

where $v(\mathbf{r}_1, \mathbf{r}_2)$ is the effective coupling giving an attractive character for the electron–electron interaction. Here, the summation convention that repeated indices imply summation is used. This is helpful to facilitate the manipulation. We would consider that the BCS Hamiltonian is an attractive interaction between Cooper pairs rather than an attractive interaction between electrons (Figs. 1.3 and 1.4). In the following, it is assumed that $\epsilon_{k\sigma}$ is independent of spin and $g_{k-k,k'-k'}$ is independent not only of spin but also, finally, of $k$ and $k'$.

Let us investigate the partition function

$$Z = \text{Tr} e^{-\beta(H_0 + H_{\text{int}})}. \tag{1.206}$$

This is expressed by the use of the path integral method [32–34]. The spirit of the Feynman path integral is sometimes written as follows: the (imaginary) time interval $0 \rightarrow \beta$ is sliced into numerous pieces, each one being labeled by $\tau_p$. Since each slice is made arbitrarily small, the quantum effect arising from the commutation relation can be neglected, so that the operators are regarded as $c$-numbers in each slice. Instead, this $c$-number function can take any value even in the small time-slice. Connecting these precise values from

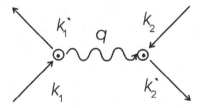

Fig. 1.3.   In the BCS model, superconductivity arises due to the attractive electron–electron interaction which appears owing to the scattering by phonons.

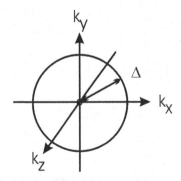

Fig. 1.4.   Excitation of the ground state of a Cooper pair demonstrates the presence of the energy gap $\Delta$ which is (very approximately) isotropic in the momentum space.

$0 \to \beta$, we can draw all of the paths in this interval. If we count the effects from all of these paths, the quantum mechanical result of the subject can be obtained. However, this statement seems rather misleading.

Feynman's path integral is the third method of quantization [35]. We begin with the classical treatment, and then if we apply the path integral procedure upon this, the quantum effect is certainly taken into account. This corresponds to the conceptual development from the geometric optics to the physical optics. The Bose system (in quantum mechanical sense) is written by an ordinary $c$-number, but the Fermi system should be described by a Grassmann number. Thus, we do not worry about the commutation relations of field operators and obtain

$$Z = \text{Tr } e^{-\beta \hat{H}}$$

$$= \lim_{N \to \infty} \int \prod_{p=1}^{N} da^*(\tau_p) da(\tau_p) \exp \sum_p [\epsilon(\dot{a}^*(\tau_p)a(\tau_p) - H(\tau_p)], \quad \epsilon = \beta/N$$

$$= \int \mathcal{D}a^*(\tau) \mathcal{D}a(\tau) \exp \int_0^\beta d\tau \{ a_{k\sigma}^*(\tau)(-\partial_\tau - \epsilon_{k\sigma})a_{k\sigma}(\tau)$$

$$+ g_{k-k,-k'k'} a_{k\uparrow}^*(\tau) a_{-k\downarrow}^*(\tau) a_{-k'\downarrow}(\tau) a_{k'\uparrow}(\tau) \}, \tag{1.207}$$

where

$$\mathcal{D}a^*(\tau)\mathcal{D}a(\tau) = \prod_{p=1}^{\infty} da_{k\sigma}^*(\tau_p) da_{k\sigma}(\tau_p). \tag{1.208}$$

### 1.9.2.  *Hubbard–Stratonovitch transformation*

Difficulty lies in the quartic term of the electron–electron interaction. Let us define

$$B_\alpha^* = a_{k\uparrow}^* a_{-k\downarrow}^* \quad \text{and} \quad B_{\alpha'} = a_{-k'\downarrow} a_{k'\uparrow} \tag{1.209}$$

to write, in each time slice,

$$g_{k-k,-k'k'} a_{k\uparrow}^* a_{-k\downarrow}^* a_{-k'\downarrow} a_{k'\uparrow} = g_{\alpha\alpha'} B_\alpha^* B_{\alpha'}. \tag{1.210}$$

This is simplified by using the identity called the Hubbard–Stratonovitch transformation [36]. We introduce a complex boson field $\phi$, since $(B_\alpha)^* \neq B_\alpha$:

$$1 = \int_{-\infty}^{\infty} d\phi^* d\phi \, e^{-\pi \phi^* \phi} = \int_{-\infty}^{\infty} d\phi^* d\phi \, e^{-\pi(\phi^* - i\sqrt{g/\pi}B^*)(\phi + i\sqrt{g/\pi}B)}$$

$$= \int_{-\infty}^{\infty} d\phi^* d\phi \, e^{-\pi \phi^* \phi + i\sqrt{g\pi}(B^*\phi - B\phi^*)} e^{-gB^* B},$$

namely

$$\int_{-\infty}^{\infty} d\phi^* d\phi e^{-\pi\phi^*\phi + i\sqrt{g\pi}(B^*\phi - B\phi^*)} = e^{gB^*B}, \tag{1.211}$$

where it is defined that

$$d\phi^* d\phi = d(\text{Re } \phi)d(\text{Im } \phi).^{\text{a}} \tag{1.212}$$

We cannot apply this identity for the quantum mechanical partition function because of the non-commutativity of the operators involved, but now we have no trouble since, in the present path-integral treatment, operators turn out to be $c$-numbers. Then Eq. (1.207) becomes

$$Z = \int\int \mathcal{D}a^*(\tau)\mathcal{D}a(\tau)d\phi^*(\tau)d\phi(\tau)$$

$$\times \exp \int_0^\beta d\tau \{-a_{k\sigma}^*(\tau)\triangle_{k\sigma}^{-1}(\tau)a_{k\sigma}(\tau) - \pi\phi_\alpha^*(\tau)\phi_{\alpha'}(\tau)$$

$$+ i\sqrt{\pi g_{k-k,\alpha'}}a_{k\uparrow}^*(\tau)a_{-k\downarrow}^*(\tau)\phi_{\alpha'}(\tau) - i\sqrt{\pi g_{\alpha,-k'k'}}a_{-k'\downarrow}(\tau)a_{k'\uparrow}(\tau)\phi_\alpha^*(\tau)\} \tag{1.213}$$

with

$$\triangle_{k\sigma}^{-1}(\tau) = -\partial_\tau - \epsilon_{k\sigma}.$$

The essential feature of this expression is as follows: the operator appearing in the exponent in Eq. (1.213) is only quadratic in $a_k$ and $a_k^*$, so that the evaluation with respect to the fermion variables is similar to the evaluation for the non-interacting system, in which particles are moving in an effective boson field $\phi_k$. Equation (1.213) shows that, inside the exponent, $Z$ is a weighted average of the second and third terms over the field $\phi_\alpha$.

### 1.9.3.  *Fourier transform*

The Fourier transformation with respect to $\tau$ is performed as

$$a_k(\tau) = \sum_{\omega_n} e^{i\omega_n\tau} a_k(\omega_n), \quad \phi_\alpha(\tau) = \sum_{\nu_n} e^{i\nu_n\tau}\phi_\alpha(\nu_n),$$

$$\omega_n = (2n+1)\pi/\beta, \quad \nu_n = 2n\pi/\beta. \tag{1.214}$$

Note that $\omega_n$ refers to a fermion, while $\nu_n$ to a boson.

---

[a]Really $d\phi^R d\phi^I = \frac{1}{2i}d\phi^* d\phi$. The constants are adsorbed in the normalization factor.

Using the above and the relation

$$\int_0^\beta d\tau e^{i(\omega_n - \omega_m)\tau} = \beta\delta_{nm},\tag{1.215}$$

we have

$$Z = \int \mathcal{D}a^*(\omega_n)\mathcal{D}a(\omega_n)\mathcal{D}\phi^*(\nu_n)\mathcal{D}\phi(\nu_n)$$

$$\times \exp \sum_{\omega_n\nu_n}\{-\beta a_{k\sigma}^*(\omega_n)\triangle_{k\sigma}(i\omega_n)^{-1}a_{k\sigma}(\omega_n) - \beta\pi\phi_\alpha^*(\nu_n)\phi_{\alpha'}(\nu_n)$$

$$+i\sqrt{\pi}g_{k-k,\alpha'}a_{k\uparrow}^*(\omega_n+\nu_n)a_{-k\downarrow}^*(\omega)\phi_{\alpha'}(\nu_n)$$

$$-i\sqrt{\pi}g_{\alpha,-k'k'}a_{k'\downarrow}(\omega_n)a_{-k'\uparrow}(\omega_n+\nu_n)\phi_\alpha^*(\nu_n)\},\tag{1.216}$$

where $\triangle_{k\sigma}(i\omega_n)$ is Green's function given as

$$\triangle_{k\sigma}(i\omega_n) = \frac{1}{i\omega_n - \epsilon_{k\sigma}}.\tag{1.217}$$

We can observe in Eq. (1.216): the third line tells that, at the vertex indicated by the coupling $g_{k-k,\alpha'}$, the boson field $\phi_{\alpha'}$ sinks, and then the particle pair, $a_{k\uparrow}^*a_{-k\downarrow}^*$ arises. The last line displays the reverse phenomenon. Note the energy conservations at vertices.

### 1.9.4. *Nambu spinor*

At this stage, we would rather use a spinor notation due to Nambu [12] very useful to investigate the superconductivity:

$$\mathbf{a}_k = \begin{pmatrix} a_{k\uparrow} \\ a_{-k\downarrow}^* \end{pmatrix}, \quad \mathbf{a}_k^* = (a_{k\uparrow}^* \quad a_{-k\downarrow}),$$

$$\mathbf{a}_{-k} = \begin{pmatrix} a_{-k\downarrow} \\ a_{k\uparrow}^* \end{pmatrix}, \quad \mathbf{a}_{-k}^* = (a_{-k\downarrow}^* \quad a_{k\uparrow}).\tag{1.218}$$

It is instructive to manipulate the complicated last line in (1.216) in the present language: it has a structure such that

$$\phi_{\alpha'}a_{k\uparrow}^*a_{-k\downarrow}^* - \phi_\alpha^* a_{-k'\downarrow}a_{k'\uparrow}$$

$$= \phi_{\alpha'} (a_{k\uparrow}^* \quad a_{-k\downarrow}) \begin{pmatrix} 1 \\ 0 \end{pmatrix} (0 \quad 1) \begin{pmatrix} a_{k\uparrow} \\ a_{-k\downarrow}^* \end{pmatrix} - \phi_\alpha^* (a_{-k'\downarrow}^* \quad a_{k'\uparrow}) \begin{pmatrix} 0 \\ 1 \end{pmatrix} (1 \quad 0) \begin{pmatrix} a_{-k'\downarrow} \\ a_{k'\uparrow}^* \end{pmatrix}$$

$$= \phi_{\alpha'} \mathbf{a}_k^*\sigma^+ \mathbf{a}_k + \phi_\alpha^* \mathbf{a}_{-k'}^*\sigma^- \mathbf{a}_{-k'}.$$

The positive sign in the second term is due to the Grassmann character of $a_{k\uparrow}$ and $a_{-k\downarrow}$. Also

$$a_{k\sigma}^* \epsilon_k a_{k\sigma} = \mathbf{a}^* \sigma_k^3 \epsilon_k \mathbf{a}_k, \tag{1.219}$$

since $\epsilon_k = \epsilon_{-k}$.

In the above, the Pauli matrices and related ones are rewritten as

$$\sigma^1 = \begin{pmatrix} 0 & 1 \\ 1 & 0 \end{pmatrix}, \quad \sigma^2 = \begin{pmatrix} 0 & -i \\ i & 0 \end{pmatrix}, \quad \sigma^3 = \begin{pmatrix} 1 & 0 \\ 0 & -1 \end{pmatrix},$$

$$\sigma^+ = \frac{1}{2}(\sigma^1 + i\sigma^2) = \begin{pmatrix} 0 & 1 \\ 0 & 0 \end{pmatrix}, \quad \sigma^- = \frac{1}{2}(\sigma^1 - i\sigma^2) = \begin{pmatrix} 0 & 0 \\ 1 & 0 \end{pmatrix}. \tag{1.220}$$

Equation (1.216) is now written as

$$Z = \int \int \mathcal{D}\mathbf{a}^* \mathcal{D}\mathbf{a} \mathcal{D}\phi^* \mathcal{D}\phi \exp \sum_{\omega_n \nu_n} \{ -\beta\pi\phi_\alpha^*(\nu)\phi_\alpha(\nu)$$

$$+ \beta\mathbf{a}_k^*(\omega_n)(i\omega_n - \epsilon_k\sigma^3)\mathbf{a}_k(\omega_n)$$

$$+ i\beta\sqrt{\pi g_{\alpha,k'-k'}}\phi_\alpha(\nu_n)\mathbf{a}_{k'}^*(\omega_n + \nu_n)\sigma^+ \mathbf{a}_{k'}(\omega_n)$$

$$+ i\beta\sqrt{\pi g_{\alpha',k-k}}\phi_{\alpha'}^*(\nu_n)\mathbf{a}_{-k}^*(\omega_n)\sigma^- \mathbf{a}_{-k}(\omega_n + \nu_n) \}. \tag{1.221}$$

We can integrate with respect to fermion variables: note that [32]

$$\int dz^* dz e^{(z^* Az)} = \det A \quad \text{for a fermion,}$$

$$\int dz^* dz e^{(z^* Az)} = (\det A)^{-1} \quad \text{for a boson.} \tag{1.222}$$

Using the upper one, we can immediately obtain

$$Z = \int \mathcal{D}\phi^* \mathcal{D}\phi \exp \sum_{\nu_n} \left\{ -\beta\pi\phi_\alpha^*(\nu)\phi_\alpha(\nu) \right.$$

$$\left. + \det \left\| \sum_{\omega_n} \beta(i\omega_n - \epsilon_k\sigma^3) + i\beta\sqrt{\pi g_{\alpha,k'-k'}}(\phi_\alpha(\nu_n)\sigma^+ + \phi_{\alpha'}^*(\nu_n)\sigma^-) \right\| \right\}, \tag{1.223}$$

where the use was made of a symmetry property, $g_{\alpha,k'-k'} = g_{\alpha',k-k}$, and $\| \cdots \|$ stands for a matrix. This can be also rewritten as

$$Z = \int \mathcal{D}\phi^* \mathcal{D}\phi \exp^{-\beta W(\phi)} \tag{1.224}$$

with the effective action for the boson field

$$W(\phi) = \pi \sum_{\nu_n} \phi_\alpha^*(\nu_n)\phi_\alpha(\nu_n) - \frac{1}{\beta} \sum_{\omega_n} \mathrm{Tr} \log \beta \tag{1.225}$$

$$\times \left\{ (i\omega_n - \epsilon_k \sigma^3) + i \sum_{\nu_n} \sqrt{u_{\alpha,k}}(\phi_\alpha(\nu_n)\sigma^+ + \phi_\alpha^*(\nu_n)\sigma^-) \right\}, \tag{1.226}$$

where the relation log·det = Tr·log and the simplified notation

$$u_{\alpha,k} = \pi g_{\alpha,k'-k'} = \pi g_{\alpha',k-k} \tag{1.227}$$

were used.

By making use of the steepest descent method, we can obtain a particular $\phi$, by optimizing $W(\phi)$, which will be denoted $\bar\phi$. For example, by differentiating $W(\phi)$ with respect to $\phi_\delta^*(\nu_n)$, we have

$$\frac{\partial W(\phi)}{\partial \phi_\delta^*} = \pi \phi_\alpha - \frac{1}{\beta} \sum_{\omega_n} \mathrm{Tr} \frac{i\sqrt{u_{\delta,k}}\sigma^-}{(i\omega_n - \epsilon_k\sigma^3) + i\sqrt{u_{\alpha,k}}(\phi_\alpha(\nu_n)\sigma^+ + \phi_\alpha^*(\nu_n)\sigma^-)}$$

$$= 0. \tag{1.228}$$

We thus obtain

$$\pi\bar\phi_\delta(\nu_n) = \frac{1}{\beta} \sum_{i\omega_n} \frac{\sqrt{(u_{\delta,k}u_{\alpha,k})} \cdot \bar\phi_\alpha}{(i\omega_n)^2 - E_k^2}, \tag{1.229}$$

where

$$E_k^2(\nu) = \epsilon_k^2 + \sqrt{(u_{\alpha,k}u_{\alpha',k})}\bar\phi_\alpha^*(\nu_n)\bar\phi_{\alpha'}(\nu_n). \tag{1.230}$$

Note that the field strength $\phi_\alpha^*(\nu_n)\phi_{\alpha'}(\nu_n)$ has the dimension of energy. In obtaining Eq. (1.229), we first rationalize the denominator, and then the Tr operation on $\sigma$'s is carried out. The numerator remained is the coefficient of Tr $\sigma^+\sigma^- = 1$. The similar equation is obtained for $\phi^*$. Equation (1.229) corresponds to the gap equation.

Now we employ an approximation in the rest of this study that, as was mentioned at the beginning, $u_{\alpha,k} = u$ (a positive constant). Thus we have, to the first approximation, that $\phi$ is nearly constant,

$$\frac{\pi}{u} = \frac{1}{\beta} \sum_{\omega_n} \frac{1}{(i\omega_n)^2 - E_k^2}. \tag{1.231}$$

Carrying out the frequency sum, we get

$$\frac{1}{\beta} \sum_{\omega_n} \frac{1}{i\omega_n - x} = \pm n(x), \qquad (1.232)$$

where $n(x)$ is the Fermi or Bose function according to $\omega_n$ is odd or even, respectively, and we obtain (for fermions)

$$\frac{\pi}{u} = \frac{n(E_k) - n(-E_k)}{2E_k} = \frac{\tanh(\beta E_k/2)}{2E_k}. \qquad (1.233)$$

From the above relation combined with Eq. (1.229), we can estimate $\bar{\phi}$ which corresponds to the gap energy.

### 1.9.5.   *Critical temperature*

It is straightforward to obtain the critical temperature, $T_c$, for the superconductivity from Eq. (1.233) [16]. Here, it is repeated for completeness. In the limit $T \to T_c$, the gap, or $\phi_\alpha$ in the present case, disappears. The summation over $k$ is approximately replaced by the integration:

$$\frac{2\pi}{uN(0)} = \int_0^{\omega_D} \frac{d\epsilon}{\epsilon} \tanh\frac{\epsilon}{2k_B T_c}, \qquad (1.234)$$

where $N(0)$ is the state density at the Fermi surface, and the cut-off parameter $\omega_D$ is, in the case of the BCS theory, the Debye frequency of phonons associated with the attractive interaction between electrons.

The integration in Eq. (1.241) proceeds as follows:

$$\int_0^{\omega_D} \frac{d\epsilon}{\epsilon} \tanh\frac{\epsilon}{2k_B T_c} = \int_0^{\beta_C \omega_D/2} \frac{dz}{z} \tanh z \quad (z = \epsilon\beta_c/2)$$

$$= \log z\tanh z\Big]_0^{\beta_c\omega_D/2} - \int_0^{\beta_c\omega_D/2} dz \log z \, \mathrm{sech}^2 z$$

$$\approx \log \frac{\beta_c\omega_D}{2} - \int_0^\infty dz \log z \, \mathrm{sech}^2 z \qquad \left(\tanh\frac{\beta_c\omega_D}{2} \approx 1\right)$$

$$= \log \frac{\beta_c\omega_D}{2} + \log \frac{4e^\gamma}{\pi}. \qquad (1.235)$$

Combined with (1.233), this yields

$$\log \frac{\beta_c\omega_D}{2} = \frac{2\pi}{uN(0)} - \log \frac{4e^\gamma}{\pi}, \qquad (1.236)$$

where $\gamma = 0.5772$ is the Euler's constant. This relation will be used in the next section. Simple rearrangements yield

$$T_c \approx 1.13\omega_D e^{-2\pi/N(0)u}. \tag{1.237}$$

### 1.9.6. *Temperature dependence of $\phi$*

For the later investigation, we derive a temperature dependence of $\phi_\alpha$ near the critical temperature [17]. Coming back to (1.231) and restoring the hidden summation with respect to $k$, we can rewrite this as

$$\frac{2\pi}{uN(0)} = \frac{1}{\beta} \sum_{\omega_n} \int_0^{\omega_D} d\epsilon \frac{1}{(i\omega_n)^2 - (\epsilon^2 + u|\phi_\alpha|^2)}. \tag{1.238}$$

Considering that, $|\phi_\alpha|^2 \ll 1$ near the critical temperature, we expand the above as

$$\frac{2\pi}{uN(0)} = \frac{1}{\beta} \sum_{\omega_n} \int_0^{\omega_D} d\epsilon$$

$$\times \left\{ \frac{1}{(i\omega_n)^2 - \epsilon^2} + \frac{u|\phi_\alpha|^2}{((i\omega_n)^2 - \epsilon^2)^2} + \frac{(u|\phi_\alpha|^2))^2}{((i\omega_n)^2 - \epsilon^2)^3} + \cdots \right\}. \tag{1.239}$$

For the convergent integrals which are the second and third terms, the integration limit is extended to infinity, and then we obtain (putting $\epsilon = \omega \tan \theta$),

$$\frac{2\pi}{uN(0)} = \int_0^{\omega_D} \frac{d\epsilon}{\epsilon} \tanh\frac{\beta\epsilon}{2} + \frac{1}{\beta}\frac{\pi}{4} u|\phi_\alpha|^2 \sum \frac{1}{\omega_n^3} + \frac{1}{\beta}\frac{3\pi}{16} (u|\phi_\alpha|^2)^2 \sum \frac{1}{\omega_n^5} + \cdots. \tag{1.240}$$

The integration of the first term on the right-hand side which is similar to (1.235) has $\beta$ instead of $\beta_c$. Then, using (1.236), we have, in the lowest approximation,

$$\log\frac{\beta_c}{\beta} = \frac{1}{\beta}\frac{\pi}{4}u|\phi_\alpha|^2 \sum \frac{1}{\omega_n^3} + \cdots. \tag{1.241}$$

Using

$$\sum_0^\infty \frac{1}{(2n+1)^p} = \frac{2^p - 1}{2^p}\zeta(p) \tag{1.242}$$

in (1.241), we have the temperature dependence of $\phi_\alpha$,

$$\sqrt{u|\phi_\alpha|^2} = \pi k_B T_c \sqrt{\frac{16}{7\zeta(3)}} \sqrt{1 - \frac{T}{T_c}} \qquad \zeta(3) = 1.202, \qquad (1.243)$$

which is the same as that of the energy gap [16].

### 1.9.7. *Dynamics of the boson field, symmetry breaking*

The boson field $\phi_\alpha(\nu_n)$ satisfying (1.227) is written as $\bar{\phi}_\alpha$, so that

$$\phi_\alpha(\nu_n) = \bar{\phi}_\alpha + (\phi_\alpha - \bar{\phi}_\alpha) = \bar{\phi}_\alpha + \Phi_\alpha(\nu_n). \qquad (1.244)$$

Now $\Phi_\alpha(\nu_n)$ is a physical or called fluctuation in solid-state physics. Our aim is to find the effective Lagrangian or Hamiltonian for $\Phi_\alpha(\nu_n)$.

In order to find the terms proportional to $\Phi_\alpha(\nu_n)$ in (1.244), we first observe that

$$\phi_\alpha(\nu_n)\tau^+ + \phi_\alpha^*(\nu_n)\tau^- \equiv \phi_\alpha(\nu_n) \cdot \tau = \phi_\alpha^{(a)}(\nu_n)\tau^{(a)}$$
$$= \bar{\phi}_\alpha(\nu_n) \cdot \tau + \Phi_\alpha(\nu_n) \cdot \tau, \quad a = 1, 2. \quad (1.245)$$

Expanding $W(\phi)$ in (1.226) in $\Phi_\alpha(\nu_n)$ and noticing that, upon the Tr operation, the odd terms with respect to $\tau$ will vanish, we have

$$W(\phi^*, \phi) = \pi \sum_{\nu_n} \phi_\alpha^* \phi_\alpha - \frac{1}{\beta} \sum_{\omega_n} \text{Tr} \log \beta(i\omega_n - \epsilon\tau^3 - i\sqrt{u}(\bar{\phi}_\alpha \cdot \tau)$$

$$- \frac{1}{\beta} \sum_{\omega_n \nu_n} \sum_n^{\text{even}} \frac{1}{n} \text{Tr}[\triangle(\omega_n) i\sqrt{u}(\Phi_\alpha(\nu_n) \cdot \tau)]^n, \qquad (1.246)$$

where

$$\triangle_k(\omega_n) = \frac{i\omega_n + \epsilon_k\tau^3 + i\sqrt{u}(\bar{\phi}_\alpha \cdot \tau)}{(i\omega_n)^2 - E_k^2}. \qquad (1.247)$$

Note that now the propagator includes the mean-field effect. What we are interested in is the second line of Eq. (1.246).

The term with $n = 2$ is precisely written as

$$\frac{-1}{2\beta} \sum_{\omega_n \nu_n} \text{Tr} \, u\triangle_k(\omega_n)(\Phi_\alpha^{(a)}(\nu_n)\tau^{(a)})\triangle_{k+\alpha}(\omega_n + \nu_n)(\Phi_\alpha^{*(a')}(\nu_n)\tau^{(a')})$$

$$= \sum_{\nu_n} u|\Phi_\alpha(\nu_n)|^2 P_\alpha(\nu_n), \qquad (1.248)$$

where careful manipulations about $\tau$'s are required. Except for the explicit ones combined with $\Phi$'s, we have $\tau$'s inside $\triangle$'s. Let us call the terms with $a = a' = 1$ and $a = a' = 2$ the direct interactions, while the terms with $a = 1$, $a' = 2$, and $a = 2$, $a' = 1$ the cross interactions. We then have

$$P_\alpha(\nu_n) = \frac{-2}{\beta} \sum_{\omega_n} \frac{i\omega_n(i\omega_n + i\nu_n) - \epsilon_k\epsilon_{k+\alpha} + u\bar{\phi}^* \cdot \bar{\phi}}{[(i\omega_n)^2 - E_k^2][(i\omega_n + i\nu_n)^2 - E_{k+\alpha}^2]}, \tag{1.249}$$

where the first two terms in the numerator arise from the direct interaction and the third from the cross one. Here the use was made that $\mathrm{Tr}\,\tau^3\tau^+\tau^3\tau^- = -1$. Using (1.231), we can carry out a lengthy but not difficult calculation of $P_\alpha(\nu_n)$. Note that an even frequency $\nu_n$ has nothing to do with obtaining the Fermi functions. The result is

$$P_\alpha(\nu_n) = -n(E_k)(1 - n(E_{k+\alpha}))$$
$$\times \frac{E_{k+\alpha} - E_k}{(i\nu)^2 - (E_{k+\alpha} - E_k)^2} \left[ \frac{E_{k+\alpha}E_k - \epsilon_{k+\alpha}\epsilon_k + u(\bar{\phi}^* \cdot \bar{\phi})}{E_{k+\alpha}E_k} \right]. \tag{1.250}$$

Here, we again note that, in the square brackets, the first two terms are obtained from the direct interaction and the third term from the cross interaction.

It might be convenient, for the later investigation, to carry out the frequency sum for $P_\alpha(\nu_n)$:

$$\sum_{\nu_n} \frac{E_{k+\alpha} - E_k}{(i\nu_n)^2 - (E_{k+\alpha} - E_k)^2}$$

$$= \beta \frac{1}{\beta} \sum_{\nu_n} \frac{1}{2} \left[ \frac{1}{i\nu_n - (E_{k+\alpha} - E_k)} - \frac{1}{i\nu_n + (E_{k+\alpha} - E_k)} \right]$$

$$= \frac{\beta}{2}[-n_B(E_{k+\alpha} - E_k) + n_B(-E_{k+\alpha} + E_k)],$$

$$= \frac{-\beta}{2}\coth\left[ \frac{\beta(E_{k+\alpha} - E_k)}{2} \right],$$

where $n_B$ is the Bose function, and the use of Eq. (1.232) was made for the even frequency $\nu_n$. Then Eq. (1.250) becomes

$$\sum_{\nu_n} P_\gamma(\nu_n) = n(E_\alpha)(1 - n(E_{\alpha+\gamma}))\frac{\beta}{2}\coth\left[ \frac{\beta(E_{\alpha+\gamma} - E_\alpha)}{2} \right]$$

$$\times \left[ \frac{E_{\alpha+\gamma}E_\alpha - \epsilon_{\alpha+\gamma}\epsilon_\alpha + u(\bar{\phi}^* \cdot \bar{\phi})}{E_{\alpha+\gamma}E_\alpha} \right]. \tag{1.251}$$

We then obtain the effective action for $\phi$ up to the second order in $\Phi$ as follows:

$$
W_\alpha(\phi^*, \phi) = \pi \sum_{\nu_n} \phi_\alpha^*(\nu_n)\phi_\alpha(\nu_n)
$$

$$
- \frac{1}{\beta} \sum_{\omega_n \nu_n} \mathrm{Tr} \, \log\{-\beta(i\omega_n - \epsilon_k \tau^3 - \sqrt{u}\bar{\phi}_\alpha(\nu_n) \cdot \tau)\}
$$

$$
- n(E_\alpha)(1 - n(E_{\alpha+\gamma})) \sum_{\nu_n} \frac{E_{k+\alpha} - E_k}{(i\nu)^2 - (E_{k+\alpha} - E_k)^2}
$$

$$
\times \left[ \frac{E_{k+\alpha}E_k - \epsilon_{k+\alpha}\epsilon_k + u(\bar{\phi}^* \cdot \bar{\phi})}{E_{k+\alpha}E_k} \right]. \tag{1.252}
$$

### 1.9.8.  *Instability*

In order to discuss the dynamics of $\Phi_\alpha(\nu_n)$, we must obtain an expression for the action of $\Phi_\alpha(\nu_n)$ similar to that for $a_\alpha(\omega_n)$ given in the exponent of Eq. (1.207). Since we are interested in the energy region much lower than that for the electronic excitation $(E_{\alpha+\gamma} - E_\alpha)$, the denominator of $P_\alpha(\nu_n)$ in Eq. (1.250) is expanded as

$$
\frac{1}{(i\nu)^2 - (E_{k+\alpha} - E_k)^2} = \frac{-1}{(E_{k+\alpha} - E_k)^2} \left[ 1 + \frac{(i\nu_n)^2}{(E_{k+\alpha} - E_k)^2} + \cdots \right]. \tag{1.253}
$$

Substituting this into $P_\alpha(\nu_n)$ in Eq. (1.250) and taking up the terms of the order of $(i\nu)^2$ in the above, we have

$$
(i\nu_n)^2 + u(\bar{\phi}^*\bar{\phi})\frac{(E_{k+\alpha} - E_\gamma)^2}{E_{k+\alpha}E_k} + \cdots . \tag{1.254}
$$

The second term is a small and complicated, however positive, quantity. We thus anyway obtain the effective Lagrangian for $\Phi$ up to the second order of $\Phi^*\Phi$ as

$$
\mathcal{L}_B \sim \Phi^*(i\nu_n)^2\Phi + \eta\Phi^*\Phi + \cdots , \quad \eta > 0. \tag{1.255}
$$

In the above, $i\nu_n$ is replaced by $\partial/\partial\tau$ in the future. At this stage, the details of $\eta$ are immaterial except that this is positive.

To put forward the problem, we need the term proportional to $\Phi^*\Phi^*\Phi\Phi$. This is obtainable from the term with $n = 4$ in Eq. (1.246). Without any detailed calculation, we may presume a positive quantity for this, noted by

$\Lambda^2$. Then the total Lagrangian of the boson field becomes

$$\mathcal{L}_B \sim \Phi^*(i\nu_n)^2\Phi + \eta\Phi^*\Phi - \Lambda^2\Phi^*\Phi^*\Phi\Phi. \tag{1.256}$$

This looks like the Lagrangian that Ginzburg and Landau have used for investigating the superconductivity. We thus expect a kind of phase transition. Let us optimize, with respect to $\Phi^*$, the potential part of Eq. (1.256),

$$\mathcal{V} = -\eta\Phi^*\Phi + \Lambda^2\Phi^*\Phi^*\Phi\Phi, \tag{1.257}$$

$$\frac{\partial\mathcal{V}}{\partial\Phi^*} = -\eta\Phi + 2\Lambda^2\Phi^*\Phi\Phi$$

$$= (-\eta + 2\Lambda^2\Phi^*\Phi)\Phi = 0. \tag{1.258}$$

If $\Phi = 0$, nothing happens. While, in the case of $\Phi \neq 0$, it is possible that

$$\overline{\Phi^*\Phi} = \frac{\eta}{2\Lambda^2} \equiv \bar{a}^2. \tag{1.259}$$

We now put

$$\Phi = (\rho + \bar{a})e^{i\theta}, \tag{1.260}$$

namely only the radial part is affected. This choice is allowed in the spirit of the unitary gauge. If we insert this into (1.257), it follows that

$$\mathcal{V} = \Lambda^2[(\rho + \bar{a})^2 - \bar{a}^2]^2 - \Lambda^2\bar{a}^4 = \Lambda^2[\Phi^*\Phi - \bar{a}^2]^2 - \Lambda^2\bar{a}^4. \tag{1.261}$$

Observing the final result, we realize that the potential becomes minimum at $|\Phi| = \bar{a}$, namely at this position apart from the origin, $|\Phi| = 0$, which corresponds to the Hartree–Fock (HF) state. The mass of the field $\Phi$ is given by the coefficient of $\Phi^*\Phi$, which is $4\Lambda^2\bar{a}^2$ for a particle of the $\rho$ field, while the term involving $\theta$ is completely lost. That is, the particle in the $\theta$ field is massless, which is called the Goldstone boson. This is due to the occurrence of the infinite degeneracies along the direction perpendicular to the $\rho$ coordinate.

### 1.9.9. *Supersymmetry*

The total Lagrangian of the system is

$$\mathcal{L}_{\text{tot}} = \mathcal{L}_e + \mathcal{L}_B + \mathcal{L}_{\text{int}}, \tag{1.262}$$

where

$$\mathcal{L}_e = \mathbf{a}_i^*(i\omega_n - \epsilon_i \tau^3)\mathbf{a}_i,$$

$$\mathcal{L}_B = -\dot{\Phi}^*\dot{\Phi} - \Lambda^2[\Phi^*\Phi - \bar{a}^2]^2, \qquad (1.263)$$

$$\mathcal{L}_{\text{int}} = -i\sqrt{u}(\Phi \mathbf{a}^* \tau^+ \mathbf{a} + \Phi^* \mathbf{a}^* \tau^- \mathbf{a}).$$

In the above, the second line is obtained from Eqs. (1.256) and (1.261), and then the third line from Eq. (1.245) supplied by necessary terms, and a compact notation is used, and the immaterial constant terms are omitted. The $\mathcal{L}$ in the above is, strictly speaking, the Lagrangian density, and the action which we are interested in is the space–time integral of the Lagrangian density, so that the partial integration with respect to $\tau$ (after replacing $i\nu_n \to \partial_\tau$) gives the first term with the minus sign. If we define the momentum conjugate to $\Phi$,

$$\Pi = \frac{\partial \mathcal{L}_{\text{tot}}}{\partial \dot{\Phi}} = -\dot{\Phi}^* \quad \Pi^* = \frac{\partial \mathcal{L}_{\text{tot}}}{\partial \dot{\Phi}^*} = -\dot{\Phi}, \qquad (1.264)$$

the total Hamiltonian for the boson field is obtained as

$$\mathcal{H}_{\text{tot}} = \Pi^*\dot{\Phi} + \Pi\dot{\Phi}^* - \mathcal{L}$$

$$= \Pi^*\Pi + i\sqrt{u}(\mathbf{a}^*\Phi\tau^+\mathbf{a} + \mathbf{a}^*\Phi^*\tau^-\mathbf{a}) + \Lambda^2[\Phi^*\Phi - \bar{a}^2]^2. \quad (1.265)$$

The momenta, $\Pi^*$ and $\Pi$ conjugate to $\Phi^*$ and $\Phi$ respectively, have properties

$$[\Pi, f(\Phi^*, \Phi)]_- = \frac{\partial}{\partial \Phi} f(\Phi^*, \Phi), \quad [\Pi^*, f(\Phi^*, \Phi)]_- = \frac{\partial}{\partial \Phi^*} f(\Phi^*, \Phi), \quad (1.266)$$

since we are now dealing with the classical operators in the framework of the path integral formalism.

Let us turn to a supersymmetric treatment [37–39]. The fermionic composite charge operators are defined as

$$Q^* = \Pi^*\mathbf{a}^* - i\sqrt{u}(\Phi^*\Phi - \bar{a}^2)\mathbf{a}^*\tau^-,$$

$$Q = \Pi\mathbf{a} + i\sqrt{u}(\Phi^*\Phi - \bar{a}^2)\tau^+\mathbf{a}. \qquad (1.267)$$

First of all, we have to manipulate the commutator

$$[Q^*, Q]_+ = [\Pi^*\mathbf{a}^* - i\sqrt{u}(\Phi^*\Phi - \bar{a}^2)\mathbf{a}^*\tau^-, \Pi\mathbf{a} + i\sqrt{u}(\Phi^*\Phi - \bar{a}^2)\tau^+\mathbf{a}]_+. \quad (1.268)$$

Straightforward calculations give

$$[\Pi^*\mathbf{a}^*, \Pi\mathbf{a}]_+ = [\Pi^*\mathbf{a}^*, \mathbf{a}]_+\Pi - \mathbf{a}[\Pi, \Pi^*\mathbf{a}^*]_+$$

$$- \Pi^*[\mathbf{a}^*, \mathbf{a}]_+\Pi - \mathbf{a}[\Pi, \Pi^*]_+\mathbf{a}^*$$

$$= \Pi^*\Pi.$$

Here, for the Grassmann numbers, the use was made of the relation

$$[a^*, a]_+ = \frac{\partial}{\partial a}a - a\frac{\partial}{\partial a} = \frac{\partial a}{\partial a} + a\frac{\partial}{\partial a} - a\frac{\partial}{\partial a} = 1.$$

Further,

$$[\Pi^* \mathbf{a}^*, i\sqrt{u}(\Phi^*\Phi - \bar{a}^2)\tau^+\mathbf{a}]_+$$

$$= [\Pi^* \mathbf{a}^*, i\sqrt{u}(\Phi^*\Phi - \bar{a}^2)]_-\tau^+\mathbf{a} - i\sqrt{u}(\Phi^*\Phi - \bar{a}^2)[\tau^+\mathbf{a}, \Pi^* \mathbf{a}^*]_+$$

$$= [\Pi^*, i\sqrt{u}(\Phi^*\Phi - \bar{a}^2)]_-\mathbf{a}^*\tau^+\mathbf{a} - i\sqrt{u}(\Phi^*\Phi - \bar{a}^2)\tau^-\Pi^*[\mathbf{a}, \mathbf{a}^*]_+$$

$$= i\sqrt{u}\Phi \mathbf{a}^*\tau^+\mathbf{a}$$

and

$$[i\sqrt{u}(\Phi^*\Phi - \bar{a}^2)\mathbf{a}^*\tau^-, \Pi\mathbf{a}]_+$$

$$= [\Pi\mathbf{a}, i\sqrt{u}(\Phi^*\Phi - \bar{a}^2)]_-\mathbf{a}^*\tau^- - i\sqrt{u}(\Phi^*\Phi - \bar{a}^2)[\mathbf{a}^*\tau^-, \Pi\mathbf{a}]_+$$

$$= [\Pi, i\sqrt{u}(\Phi^*\Phi - \bar{a}^2)]_-\mathbf{a}\mathbf{a}^*\tau^- - i\sqrt{u}(\Phi^*\Phi - \bar{a}^2)\tau^-\Pi[\mathbf{a}^*, \mathbf{a}]_+$$

$$= i\sqrt{u}\Phi^* \mathbf{a}\mathbf{a}^*\tau^-$$

$$= -i\sqrt{u}\Phi^* \mathbf{a}^*\tau^-\mathbf{a}.$$

The final term must be symmetrized:

$$[\mathbf{a}^*\tau^+, \tau^-\mathbf{a}]_+ \rightarrow \frac{1}{2}\{[\mathbf{a}^*\tau^+, \tau^-\mathbf{a}]_+ + [\mathbf{a}^*\tau^-, \tau^+\mathbf{a}]_+\} \tag{1.269}$$

$$= \frac{1}{2}\{[(0, a^*), (0, a)]_+ + [(b, 0), (b^*, 0)]_+\} = \frac{1}{2}\begin{pmatrix} [b, b^*]_+ & 0 \\ 0 & [a^*, a]_+ \end{pmatrix} \tag{1.270}$$

$$= \frac{1}{2}\mathbf{1}. \tag{1.271}$$

In the second line, $b$ and $b^*$ which are used in the literature are defined.

We thus obtain the fundamental relation in the supersymmetric quantum mechanics,

$$\mathcal{H} = [Q, Q^*]_+, \tag{1.272}$$

if we endow, to the constant $\Lambda^2$ in (1.264), a specified value,

$$\Lambda^2 = 2u, \tag{1.273}$$

which is reasonable [39]. However, as was mentioned before, this had to be obtained analytically from the term with $n = 4$ in Eq. (1.252). If it is assumed that the present system is the case of supersymmetry, we can avoid this tedious calculation.

The nilpotency of the charge operators,

$$Q^2 = 0, \qquad (Q^*)^2 = 0, \qquad (1.274)$$

is clearly preserved from $\mathbf{a}^2 = 0$ and $(\mathbf{a}^*)^2 = 0$. We have thus completed a supersymmetric analysis.

### 1.9.10. *Toward the Ginzburg–Landau equation*

Now we can think about the Ginzburg–Landau equation. Since the GL equation concerns only with the static part of the condensed boson field, we keep the first and third terms in Eq. (1.265). The first term is the kinetic part, while the last term is the potential part,

$$\mathcal{H}_{GL} = \Pi^* \Pi - \eta \Phi^* \Phi + \Lambda^2 (\Phi^* \Phi)^2, \qquad (1.275)$$

with

$$\eta = u(\bar{\phi}^* \bar{\phi}) \frac{(E_{k+\alpha} - E_\gamma)^2}{E_{k+\alpha} E_k}.$$

As usual, the coefficients of $\Phi^* \Phi$ and $(\Phi^* \Phi)^2$ in the Ginzburg–Landau equation are $a$ and $b$, respectively. It is crucial that $a$ is negative, which is given by $-\eta$ in the present consideration. The temperature dependence of $\eta$ is found through $u(\bar{\phi}^* \bar{\phi})$ as in Eq. (1.265). This is exactly the same as that obtained by Gorkov [31].

### 1.9.11. *Discussion*

Somebody seemed to be puzzled by a curious looking of the Ginzburg–Landau equation when it was published: the superconductivity being a macroscopic, thermodynamically stable phase, why it is predicted by a Schrödinger-like wave function, to which the microscopic phenomena are subjected. Several years later on, the microscopic theory was established by BCS, and soon Abrikosov has successfully correlated these two treatments. He has made clear that the GL function is deeply concerned with the gap function characteristic of the superconductivity and is a wave function describing the condensed Cooper pairs.

The superconducting state is certainly a stable thermodynamical state, and a phase transition from the normal state to the superconducting state is interpreted as a long-distance correlation between the Cooper pairs. Yang [40] developed a unified treatment of the phase transition in terms of the density matrix. According to his treatment, the onset of the superconductivity is understood in such a way that the off-diagonal long-range order (ODLRO) of the second-order density matrix has non-vanishing value. This concept is clearly related to the London's rigid function, the quasiboson condensation being widely seen as a powerful model of superconductivity and the variational wave function tired by BCS [3]. Recently, Dunne *et al.* [41] applied the concept of ODLRO to argue the high-temperature superconductivity in copper oxide, where the attractive interaction between electrons is assumed to be originated from Friedel oscillations in the screened potential.

However, the previous presentation of Abrikosov looks as a detour, a complicated tedious procedure; the condensed pair is characterized by an anomalous temperature Green's function which is difficult to manipulate for beginners. Therefore, the direct way to reach the GL theory from the BCS Hamiltonian should be preferable. What we have done in the present investigation is the following: the auxiliary boson field driven by the Hubbard–Stratonovitch transformation to eliminate the quartic term of electron operators is just the GL function.

Conclusively, we should like to state, from the so far treatments, the conditions which is necessary for occurrence of superconductivity.

(1) The wave function which means the ground-state average of operators of a Cooper pair must be complex or two-dimensional. If one of these two degrees of freedom gets a new stable structure, the other degree of freedom, whose direction is perpendicular to the former, offers the infinite degeneracy. In the Nambu theory, the first refers to the $\sigma^1$ direction, and the second to $\sigma^2$ in the fictitious spin space.

(2) The electron–electron interaction should be attractive, otherwise the negative coefficient of $\Phi^*\Phi$ in Eq. (1.275) cannot be obtained. In the present consideration, we have observed that this condition is established in the effective electron–electron interaction of the system involving the multiband structure.

However, we will discuss a little more about the normal state under the usual condition. In the normal species, the coupling constant between electrons is intrinsically positive. However, if we are interested in the exchange interaction, the so-called Fock term with negative coupling, this is met by condition (2). This is a short range, while the direct coupling is so strong as to overwhelm the exchange interaction.

Let us turn to the behavior of the quartic electron operators. It is unexpected that these are grouped into the pair operators of particle–particle and hole–hole, which do not conserve the particle number. Thus, it is natural to group them into a couple of particle–hole pairs. This choice makes the auxiliary boson function to be real.

The Hamiltonian which satisfies the above two conditions, looking like the BCS Hamiltonian, is the one with the dipole–dipole interaction. The simplest case is that of the intermolecular interaction due to the induced dipole–dipole interaction with a coupling constant $-d$:

$$H_d = -d\, a_r^* b_r^* b_s a_s, \tag{1.276}$$

where, for example, $a_r^*$ and $b_r^*$ are the creation operators for a particle and a hole, respectively. If the electron–electron interaction is screened enough, this may be another possibility for the superconductivity of a molecular complex.

# CHAPTER 2

# Physics of High-$T_c$ Superconductors

## 2.1. Introduction

In 1986, K.A. Müller and J.G. Bednorz discovered the phenomenon of superconductivity in Cu-oxide compounds of lanthanum and barium at the temperature $T_c = 35\,\mathrm{K}$ [2], which was marked by the Nobel's prize in 1987. This discovery gave the additional push to the intensification of the scientific activity in the field of superconductivity. For the last 35 years, the superconducting transition temperature has been increased to 140 K. Moreover, one may expect the discovery of new superconductors with higher critical temperatures. Recently, the high critical temperatures of new Mg-containing superconductors ($MgBr_2$) ($T_c = 39\,\mathrm{K}$) and superconducting oxypnictides $LnO_{(1-x)}F_xFeAs$ ($T_c = 50\,\mathrm{K}$) were registered.

The main purpose of solid-state physics is the development and the fabrication of substances which possess the superconductivity at room temperature. At present, there proceeds the great-scale search for such high-temperature superconductors. Especially promising is the method of production of superconducting materials with the help of the laser spraying of layers.

After the discovery of HTSC, the following problems become urgent:

(1) Conceptual-theoretic one: the problem of the clarification of mechanisms of high-temperature superconductivity.
(2) Engineering and technical ones: the the problem of the practical applications of HTSC.
(3) Research one: the problem of the search for materials with higher $T_c$.

The great interest in the phenomenon of superconductivity is caused by the basic possibility to use it in the future for the electric power transfer without losses and for the construction of quantum high-power generators. This phenomenon would be applied in superconductive electronics and computer technique (superconducting elements of memory). At the present time, a new trend in technology, namely the construction of quantum computers on the basis of high-temperature superconductors with $d$-pairing, is developed.

High-temperature superconductors have unique physical properties both in the normal state and the superconducting one.

To comprehend the physics of these complex compounds is one of the main tasks of the theory of superconductivity, whose solution will allow one to explain the mechanism ensuring the high-temperature superconductivity.

The disappearance of resistance on the cooling of superconductors down to a certain critical temperature is one of the most characteristic effects in superconductors. But, in order to understand the reason for the rising of superconductivity, it is necessary to study the other effects which accompany this phenomenon.

Consider briefly the main properties of high-temperature superconductors (HTSC):

(1) New HTSC have a great anisotropy across the axis **c** and possess a multilayer structure. The main block defining the metallic and superconductive properties is planes with $CuO_2$ which form the square lattices of Cu ions.

(2) HTSC are superconductors of the second kind ($l/\varkappa \ll 1$, where $l$ is the coherence length, and $\varkappa$ is the penetration depth of a magnetic field).

(3) High-temperature superconductors have high critical temperatures $T_c$.

(4) They have antiferromagnetic ordering of spins of Cu in the $CuO_2$ planes and powerful spin fluctuations with a wide spectrum of excitations.

At the present time, there exists no theoretical approach which would explain the totality of thermodynamical, magnetic, and superconductive properties of high-temperature superconductors from the single viewpoint.

The electron–phonon mechanism of pairing, being principal in standard superconductors, gives a considerable contribution to the establishment of the superconducting state in high-temperature superconductors. But, in order to attain the proper description, it is necessary to consider the other mechanisms inherent to high-temperature superconductors.

One of such mechanisms is the spin-fluctuation one proposed by D. Pines (USA). The reason for the pairing of electrons can be the scattering of electrons on spin fluctuations. This model of pairing is one of the most actual models of high-temperature superconductivity.

The thermodynamics of superconductors at low temperatures is determined by the excitation of two quasiparticles. In the traditional superconductors with pairing of the BCS type, the energy gap is isotropic ($s$-pairing), and the temperature dependence of the heat capacity has exponential form $\sim \exp^{-\Delta/k_B T}$, where $\Delta$ is the superconductor's gap. In superconductors with anisotropic pairing, the temperature dependence of the heat capacity has power character, namely, $T^n$. The appearance of such temperature dependences is related to that the superconductor's gap has zeros on the Fermi surface. The development of the thermodynamics of this model is the actual problem of high-temperature superconductors.

In Sec. 2.2, we consider the history of the development of studies of the phenomenon of superconductivity. The structure of high-temperature superconductors and their physical properties are analyzed. We make survey of the mechanisms of superconductivity and discuss the problem on the symmetry of the order parameter.

We draw conclusion that, on the whole, the most probable is the "synergetic" mechanism of pairing which includes, as components, the electron–phonon and spin-fluctuation interactions in cuprate planes.

Only if the interaction of all the degrees of freedom (lattice, electronic, and spin ones) is taken into account, a number of contradictory properties observed in the superconducting and normal phases can be explained. Moreover, one needs also to consider a complicated structure of high-temperature superconductors. It will be emphasized that new experiments should be executed in order to explain the available experimental data.

In this section, we also consider the thermodynamics of the spin-fluctuation mechanism of pairing, present the method of functional integration for the calculation of thermodynamical properties on the basis of the Pines spin-fluctuation Hamiltonian, and deduce the Schwinger–Dyson equations for Green's functions and equations for the thermodynamic potential. Based on the Schwinger–Dyson equation, we obtain equations for the superconductor gap, which are used in numerical calculations of the thermodynamics of high-temperature superconductors. We present analytic formulas for the thermodynamic potential and its jump. The numerical calculations of the temperature dependence of the electron heat capacity indicate that the temperature dependence of the heat capacity is proportional to the square of the temperature. We emphasize that such temperature dependence is related to the $d$-pairing. It is shown that the measurement of the temperature dependence of the heat capacity can be a supplementing test for the establishment of a type of the symmetry of pairing in high-temperature superconductors. The jump of the heat capacity of cuprate superconductors near the critical temperature is evaluated as well.

## 2.2.  History of the Development of Superconductivity

The phenomenon of superconductivity was discovered by the physicist H. Kamerlingh-Onnes at the Leiden Laboratory (the Netherlands) in 1911, in the same year when Rutherford discovered an atom. H. Kamerlingh-Onnes registered the disappearance of resistance of Hg at the temperature $T = 4.5\,\mathrm{K}$. This state was called superconducting. Being cooled down to a temperature less than the above-mentioned critical temperature, many conductors can be transferred in the superconducting state, in which the electric resistance is absent. The disappearance of resistance is the most dramatic effect in superconductors. But, in order to comprehend the reason for the origin of superconductivity, we need to study the other effects accompanying this phenomenon. The dissipationless current states in superconductors were a puzzle for a long period.

The phenomenon of superconductivity is a bright example of the manifestation of quantum effects on the macroscopic scale. At present, the superconductivity occupies the place of the most enigmatic phenomenon in condensed-state physics, namely in the physics of metals.

The main purpose of solid-state physics consists in the creation of super-conductors which have the superconductive property at room temperature. At the present time, the researchers continue the wide-scale search for such high-temperature superconductors by testing the variety of various substances.

A number of phenomenological models were proposed to clarify the phenomenon of superconductivity, namely the models advanced by London and Ginzburg–Landau [9, 42]. The essence of these theories which have played the important role in the development of ideas of superconductivity was presented in Chapter 1.

The success of the Ginzburg–Landau theory was related to the circumstance that it is placed in the mainstream of the general theory of phase transitions.

Among the first microscopic theories devoted to the consideration of the electron–phonon interaction, the work by H. Fröhlich [43] is of the greatest importance.

The following stage in the development of superconductivity started from the universal theory of BCS published in 1957 [148] which strongly promoted the subsequent study of superconductivity [44]. The authors of this theory were awarded by the Nobel's prize.

The BCS theory gave the possibility to elucidate a lot of experiments on superconductivity of metals and alloys. However, while developing the theory, some grounded assumptions and approximations were accepted.

Therefore, the BCS theory was needed in a substantiation with the help of more strict arguments. This was made by Academician N.N. Bogoliubov in [44]. He developed a microscopic theory of the phenomena of superconductivity and superfluidity. By using the Fröhlich effective Hamiltonian, N.N. Bogoliubov calculated the spectrum of excitations of a superconductor within the method of canonical transformations proposed by him in 1947.

In 1962, B.D. Josephson advanced the theory of tunnel effects in superconductors (the Josephson effect) which was marked by the Nobel's prize in 1973 [45]. This discovery strongly intensified the experimental studies of superconductivity and was applied to electronics and computer technologies.

In 1986, K.A. Müller and J.G. Bednorz discovered high-temperature superconductors [2], which initiated the huge "boom" manifested in the publication of plenty of works. The principal thought of K.A. Müller was the following: by selecting the suitable chemical composition, one can enhance the electron–phonon interaction and, thus, increase the critical temperature $T_c$. K.A. Müller and J.G. Bednorz found a new class of high-temperature superconductors, the so-called cuprate superconductors.

High-temperature superconductors are studied already 35 years by making significant efforts, but the while pattern is not else clear completely. This is related to a complicated structure of cuprates, difficulties in the production of perfect single crystals, and a hard control over the degree of doping. The comprehension of HTSC will be attained if our knowledge about HTSC will approach some critical level which will be sufficient for the understanding of a great amount of experimental data from the single viewpoint.

In Fig. 2.1, we present the plot of the critical temperatures of superconductivity over years [46–48].

## 2.3. Structural Properties of High-Temperature Superconductors

The properties of new high-temperature superconductors differ essentially from the properties of traditional superconductors which are described by the BCS theory. Let us briefly consider the main properties of high-temperature superconductors. As known, the atomic structure defines the character of chemical bonds in solids and a number of relevant physical properties. Even small changes of the structure lead frequently to significant changes of their electron properties; for example, at the phase transitions metal-dielectric. Therefore, the study of a crystal structure with long-range atomic order and

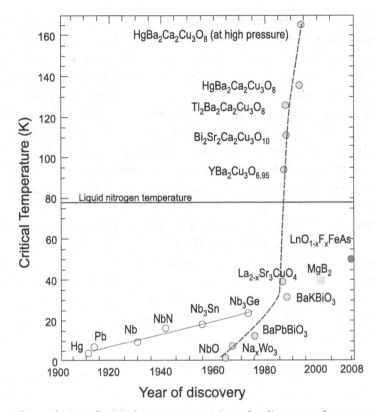

Fig. 2.1.    The evolution of critical temperatures since the discovery of superconductivity.

its dependence on the temperature, pressure, and composition is of great importance for high-temperature superconductors. These investigations are significant for the comprehension of mechanisms of high-temperature superconductivity.

At the present time, we have several families of HTSC: thermodynamically stable Cu oxides containing lanthanum, yttrium, bismuth, thallium, and mercury [47].

We note that the structure of all Cu-oxide superconductors has a block character. The main block defining metallic and superconductive properties of a compound is the plane with $CuO_2$ which form the square lattices of Cu ions coupled with one another through oxygen ions. Depending on the composition, the elementary cell of a high-temperature compound can have one, two, and more cuprate layers. In this case, the critical temperature of the superconducting transition increases with the number of cuprate layers. In Fig. 2.2, we present an elementary cell of the orthorhombic structure of yttrium ceramics $YBa_2Cu_3O_{7-\delta}$, $\delta \approx 0.1$–$0.3$. The maximum $T_c \approx 90$ K.

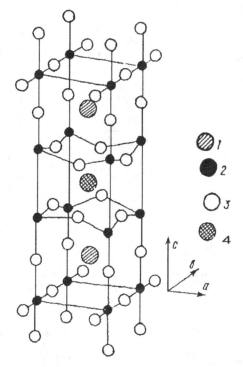

Fig. 2.2. Elementary cell of yttrium ceramics $YBa_2CuO_7$: 1-yttrium, 2-barium, 3-oxygen, 4-copper.

The size of the cell is characterized by the following parameters: $a = 3.81$ Å, $b = 3.89$ Å, $c = 11.7$ Å.

Thus, the high-temperature superconductors are characterized by both large volumes of elementary cells and a clearly manifested anisotropy of layers.

Of significant interest is the discovery of the compounds $Bi_2CaSr_2Cu_2O_8$ (Bi/2–1–2–2) and $Tl_2CaBa_2Cu_2O_8$ (Tl/2–1–2–2) with the temperature of the superconducting transition $T_c > 100$ K. These compounds can have different numbers of cuprate layers and are described by the general formula $A_2Ca_nY_2Cu_nO_{2n4}$, where A = $Bi(Tl)$, Y = $Sr(Ba)$. Their temperature $T_c$ depends on the number of cuprate layers and take values of 10, 85, and 110 K for the compounds with Bi and 85, 105, and 125 K for the compounds with Tl for $n = 1, 2, 3$, respectively. We mention also the compounds with Tl with the single layer Tl–O with the general formula $TlCa_{n-1}Ba_2Cu_nO_{2n+3}$(Tl/1...), where the number of cuprate layers reaches $n = 5$.

Below, we present the structures of the family of Tl-based superconductors in Figs. 2.3 and 2.4.

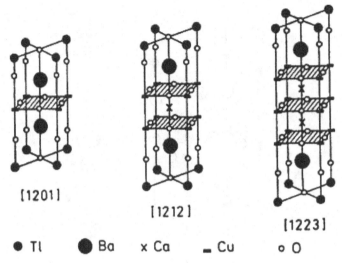

Fig. 2.3. Schematic distribution of ions in the unit cell of superconductors $Tl_1Ba_2Ca_{N-1}Cu_NO_{2N+4}$ or, in the brief notation, Tl(1: 2: N-1: N) at $N = 1, 2, \ldots, 5$.

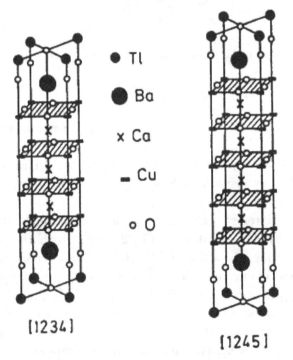

Fig. 2.4. Schematic distribution of ions in the unit cell of superconductors $Tl_1Ba_2Ca_{N-1}Cu_NO_{2N+4}$ or, in the brief notation, Tl(1: 2: $N-1$ : $N$) at $N = 1, 2, \ldots, 5$.

Table 2.1.  Tl-based systems $Tl_1Ba_2Ca_{N-1}Cu_NO_{2N+3}$.

| $N$ | 1 | 2 | 3 | 4 | 5 | 6 |
|---|---|---|---|---|---|---|
| $a$ (Å) | — | 3.8500 | 3.8493 | 3.8153 | 3.8469 | — |
| $T_c(K)$ | 13–15 | 78–91 | 116–120 | 122 | 106 | 102 |

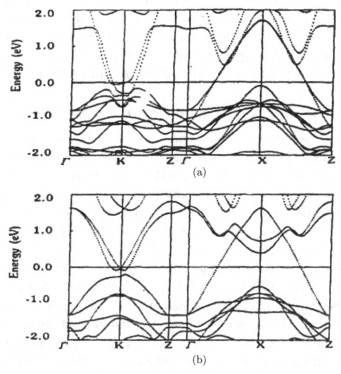

Fig. 2.5.  Structure of energy bands near the Fermi energy for, (a) high-temperature superconductor $Bi_2Sr_2CaCu_2O_8$ and, (b) low-temperature superconductor $Bi_2Sr_2CaCuO_6$.

In Table 2.1, we show the dependences of $T_c$ and the interplane distance (Å) on the number of layers in the unit cell $N$.

The explanation of the dependence of the critical temperature on the number of cuprate layers is an urgent problem [49, 50].

All high-temperature superconductors have a complicated multiband structure [139] shown in Fig. 2.5.

We note the importance of the structure of zones near the Fermi level. For example, the structures of the bands of two superconductors (high-temperature $Bi_2Sr_2CaCu_2O_8$ and low-temperature $Bi_2Sr_2CaCuO_6$) are significantly different at point K. The energy interval from the band bottom to the Fermi surface is equal to 0.7 eV at point K for the high-temperature superconductor and 0.1 eV for the low-temperature one. The explanation of the influence of the multiband structure on $T_c$ is given in Chapter 3 of this book.

It is worth noting that all copper-oxide superconductors have block character of their structure. The main block defining the metallic and superconductive properties of a compound of a plane with $CuO_2$ which contains the square lattice of Cu ions coupled with one another through oxygen ions. Depending on the composition, the elementary cell of a high-temperature compound can have one, two, three, and more cuprate layers. Moreover, the critical temperature of the superconducting transition has a nonmonotonous dependence on the number of cuprate layers [49].

### 2.3.1.   *Phase diagram of cuprate superconductors*

All electron properties of high-temperature superconductors depend strongly on the doping. High-temperature superconductors without doping are dielectrics and antiferromagnetics. As the concentration $x$ increases, these materials become metals. Superconductivity arises at large $x$, behind the limits of the magnetically ordered phase. The experiments showed that the charge carriers have the hole character for all classes of high-temperature superconductors.

It becomes clear recently that the high-temperature superconductivity is related to peculiarities of the behavior of these compounds in the normal phase. As seen from the phase diagram (Fig. 2.6), the superconducting states arise near the antiferromagnetic phase. In yttrium-containing systems, the antiferromagnetic and superconducting regions adjoin one another.

The experiments on the inelastic magnetic scattering of neutrons indicate the existence of strong magnetic fluctuations in the doped region, even beyond the limits of the antiferromagnetic phase. This points out the important role of antiferromagnetic fluctuations in the compounds with high-temperature superconductivity.

In high-temperature superconductors, the gap is present in the absence of the phase coherence, i.e., in nonsuperconducting specimens. This gap is called a pseudogap. A pseudogap is shown in Fig. 2.6. It appears at temperatures less than some characteristic temperature $T^*$ which depends on the doping. Its nature is not completely explained else.

The study of a pseudogap in the electron spectrum of high-temperature superconductors was carried out in many works [51].

Metals become superconductors, if their free electrons are bound in Cooper's pairs. Moreover, the pairs are formed in such a way that their wave functions have the same phase. The phase coherence is responsible for

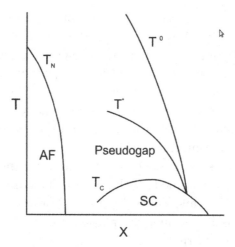

Fig. 2.6. Phase diagram of cuprate superconductors in the temperature-doping variables, $T_N$ — the Néel temperature, $T_c$ — the critical temperature of the superconducting transition, $T^*$ — the characteristic temperature of a pseudogap, $T^0$ — upper crossover temperature.

the change of the resistance on the cooling below the critical temperature $T_c$. The presence of coupled pairs in a superconductor causes the appearance of a gap in the spectrum of excitations. In the standard superconductors, the phase coherence of pairs appears simultaneously with the appearance of pairs. From one viewpoint, a pseudogap is related to the appearance of coupled pairs, which is not related to the phase coherence.

Another viewpoint consists in the following. The pseudogap arises in HTSC in connection with the formation of magnetic states which compete with superconducting states. The efforts of experimenters aimed at the solution of this dilemma are complicated by a strong anisotropy of the superconductor gap. Some physicists believe that the most probable situation is related to the creation of the superconducting state with paired electrons at a certain doping which coexists with antiferromagnetism. It is possible that just it is the "new state of matter" which has been widely discussed for the last years in connection with HTSC.

### 2.3.1.1. *Antiferromagnetism of HTSC*

An interesting peculiarity of copper-oxide compounds which has universal character consists in the presence of the antiferromagnetic ordering of spins of Cu in the $CuO_2$ planes. The sufficiently strong indirect exchange interaction of spins of Cu induces the 3D long-range antiferromagnetic order with the relatively high Néel temperatures $T_N = 300$–$500$ K [52]. Though the long-range order disappears in the metallic (and superconducting) phase, strong fluctuations with a wide spectrum of excitations are conserved. This allows one to advance a number of hypotheses on the possibility of electron pairing in copper-oxide compounds through the magnetic degrees of freedom.

Therefore, the study of the antiferromagnetic properties of high-temperature superconductors is of importance for the verification of the hypotheses on the magnetic mechanism of superconductivity. The interaction of spins of Cu in a plane has 2D character, and their small values, $S = 1/2$, lead to significant quantum fluctuations.

The first indications to the existence of antiferromagnetism in the copper-oxide compounds were obtained on the basis of macroscopic measurements of susceptibility. The detailed investigation of both the magnetic structure and spin correlations in the metallic phase became possible only with the help of the neutron scattering.

## 2.4.   Mechanisms of Pairing of High-Temperature Superconductors

To explain the high-temperature superconductivity, a lot of models and mechanisms of this unique phenomenon were proposed. The key question is the nature of the mechanism of pairing of carriers. There are available many different models of superconductivity, among which we mention the following ones: the magnon model, exciton model, model of resonant valence bonds, bipolaronic model, bisoliton model, anharmonic model, model of local pairs, plasmon model, etc. We give some classification of mechanisms of pairing for high-temperature superconductors which is shown in Fig. 2.7, according to [53].

This classification demonstrates the diverse physical pattern of high-temperature superconductors.

Together with the ordinary BCS mechanism based on the electron–phonon interaction, there exist many other mechanisms, as was mentioned above. All these models used the conception of pairing with the subsequent formation of the Bose-condensate at temperatures $T_c$ irrespective of the nature of the resulting attraction.

### 2.4.1.   *Specific mechanisms of pairing in superconductivity*

Consider the models most popular at the present time. Along with the ordinary BCS mechanism based on the electron–phonon interaction, we turn to the magnetic, exciton, plasmon, and bipolaronic mechanisms of pairing. All these models applied the conception of pairing with the subsequent

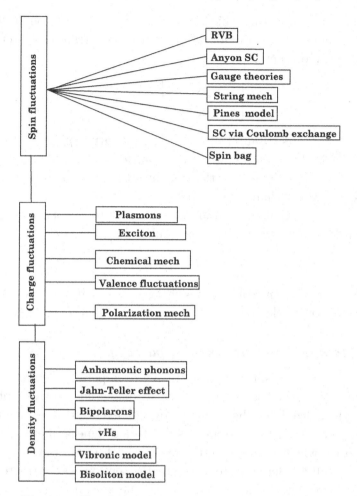

Fig. 2.7. Classification of mechanisms of pairing in HTSC.

formation of a Bose-condensate (at certain temperatures $T_c$) regardless of the reasons for the attraction.

The BCS theory presents the formula for the critical temperature $T_c$ in the case of the weak electron–phonon interaction:

$$T_c = 1.14\Theta \exp(-1/N_0 V), \tag{2.1}$$

where $\Theta = \hbar\Omega_D/k_B$, $\hbar\Omega_D$ is the Debye energy, $N_0$ is the density of states of the Fermi level, and $V$ is the attractive pairing potential acting between electrons.

The maximum value of the critical temperature given by the BCS theory is 40 K. Therefore, there arises the question about the other mechanisms of pairing. The interaction of electrons is repulsive, i.e., one needs to seek a "transfer system" in metals which is distinct from the phonon system. The general scheme of the interaction of electrons via a transfer system $X$ can be schematically presented as

$$\begin{cases} e_1 + X \to e_1^1 + X^*, \\ e_2 + X^* \to e_2^1 + X, \end{cases} \tag{2.2}$$

where $e_i$ corresponds to an electron with momentum $p_i$, $X$ is the ground state, and $X^*$ is the excited state of the transfer system.

As a result of this reaction, the system returns to the initial state, and the electrons make exchange by momenta.

It can be shown that such an interaction leads to the attraction, and the critical temperature is given by the formula

$$T_c \sim \Delta E \exp(-1/\lambda),$$

where $\Delta E$ is the difference of energies of the states $X$ and $X^*$, and $\lambda$ depends on the interaction of electrons with the system $X$ [54].

### 2.4.2. *Magnetic mechanism of pairing*

The magnetic mechanism of pairing in high-temperature superconductors was studied by many researchers. These studies used the assumption that the pairing is realized due the exchange by spin excitations — magnons, i.e., the transfer system is the system of spins in a magnetic metal. Magnon is a quasiparticle which is the quantum synonym of a spin wave of excitation in a magnetically ordered system. A great attention was attracted by the pioneer work by A.I. Akhiezer and I.Ya. Pomeranchuk (see [55, 56]). They considered the interaction between conduction electrons caused by both the exchange by acoustic phonons and an additional interaction related to the exchange by spin waves (magnons). It was shown that superconductivity and ferromagnetism can coexist in the same spatial regions. At sufficiently low concentrations of the ferromagnetic component, an increase of its concentration leads to an increase of $T_c$ in the case of the triplet pairing. Within the phonon mechanism, the pairing occurs in a singlet state, then an increase of the concentration of the ferromagnetic component induces a decrease of $T_c$. It is known that ferromagnetism competes with superconductivity. The different situation is characteristic of antiferromagnetism. High-temperature superconductors are antiferromagnetic dielectrics. As was mentioned above, high-temperature superconductors reveal strong magnetic fluctuations in the region of doping which can be responsible for the pairing. In this chapter, we will consider the spin-fluctuation model of pairing.

After the discovery of HTSC, a lot of relevant works dealt with namely the problem of the evolution of a system of Cu ions under the transition from the dielectric antiferromagnetic state to the metallic one.

It is worth also noting the model of resonance valence bonds (RVB) which was advanced by Nobel's Prize winner P. Anderson in 1987 [5] who made attempt to explain the Cooper pairing in HTSC by the participation of magnetic excitations. The Anderson model is based on the conception of magnetic ordering which was named the model of resonance valence bonds. The RVB mechanism ensures the joining of carriers in pairs with compensated spin, the so-called spinons. At the doping of HTSC compounds, there arise holes which can form the complexes with spinons—holons. Superconductivity is explained by the pairing of holons, i.e., by the creation of spinless bosons with double charge. That is, the pairing of carriers in the RVB model is realized due to the exchange by magnons. At low temperatures, the paired holons form a superconducting condensate. The RVB model has played a positive role, by attracting the attention of researchers to the study of antiferromagnetism in HTSC though spinons and holons have not been experimentally identified.

### 2.4.3. *Exciton mechanisms of pairing*

According to the general principle concerning the "transfer system" in superconductors, the electron–phonon interaction should not be obligatorily realized. Some other interaction ensuring the pairing of electrons can be suitable. In principle, the mechanism of superconductivity can be switched on by bound electrons which interact with conduction electrons. The first exciton model, in which the pairing is realized due to electron excitations, was proposed by W.A. Little [57] for organic superconductors and V.L. Ginzburg and D.A. Kirzhnitz [42] for layered systems. In the construction of this model, it was necessary to assume the existence of two groups of electrons: one of them is related to the conduction band, where the superconducting pairing occurs due to the exchange by excitons which are excitations in the second group of almost localized electrons. In view of the many-band character of the electron spectrum, layered structure, and other peculiarities of the electron subsystem in high-temperature superconductors, such a distribution of electron states is quite possible. This underlies the development of a lot of exciton models. The searches for superconductivity in organic materials were stimulated to a significant degree by the idea of W.A. Little about a possibility of high-temperature superconductivity due to the excitonic mechanism of the Cooper pairing of electrons in long conducting polymeric chains containing lateral molecular branches-polarizers. Since the mass $M$ of such excitonic excitations is small, it would be expected to observe a high value of the temperature $T_c \sim M^{-1/2}$. But this model was not practically realized, since high-energy intramolecular excitonic excitations cannot ensure the binding of electrons in pairs.

At the present time, a number of quasi-one-dimensional organic superconductors with metallic conductance have been synthesized. They pass to the superconducting state at $T = 10\,\mathrm{K}$. Crystals of similar organic superconductors consist of, as a rule, planar molecules packed in zigzag-like stacks which form chains. The good overlapping of electron wave functions of neighboring molecules in a stack ensures the metallic conductance along a chain. The overlapping of electron wave functions of neighboring chains is small, which leads to the quasi-one-dimensional character of the electron spectrum and to a strong anisotropy of electronic properties of a crystal. Up to now, no experimental proofs of a manifestation of the excitonic mechanism is such systems are available.

As the example of a laminar system, we mention a quasi-two-dimensional structure of the "sandwich" type (dielectric–metal–dielectric). In such structures, the Cooper pairing of electrons in a metal film occurs due to the attraction caused by their interaction with excitons in the dielectric plates.

### 2.4.4.    *The anharmonic model and superconductivity*

It is known that the appearance of superconductivity is often preceded by structural transformations in a crystal. They are usually explained within the anharmonic model. In the opinion of some researchers [47], such structural transformations foregoing the start of superconductivity decrease significantly the frequencies of phonons and, due to this, increase the parameter of electron–phonon interaction. The softening of the phonon spectrum is caused by the great amplitudes of displacements of ions in the two-well potential which models the structural transformations. In some works [52], the effect of a structural transformation on superconductivity in the limiting case of a weak pairing interaction and the isotropic gap was studied. The properties of high-temperature superconductivity were studied also within the model, where the superconductivity is enhanced due to the singularity of the density of electron states which appears at structural or antiferromagnetic phase transitions. The weak point of the models relating the superconductivity to structural phase transitions owes to a significant temperature interval between the known structural transitions and the temperature $T_c$. The works, where non-phonon pairing mechanisms are introduced, include the studies [52] which use the Hubbard Hamiltonian in systems, where only the interaction of repulsive type is present. It is considered that the effect of pairing is caused by the kinematic interaction at a non-complete occupation of Hubbard subbands. Unfortunately, no clear comprehension of the nature of the arising attraction is attained in this case. It is possible that the bound state of quasiparticles is virtual, i.e., it decays. The search for new mechanisms of superconductivity caused by a strong correlation of electrons in cuprate superconductors with quasi-two-dimensional electron structure is reduced to the search for new non-standard ground states. We briefly mention the polaron mechanism of pairing which is related to the Jahn–Teller effect well-known in the quantum chemistry of complex compounds. The essence of the Jahn–Teller effect consists in that a nonlinear system in the presence of the electron degeneration deforms spontaneously its structure so that this degeneration disappears or decreases. The Jahn–Teller effect leads to the rearrangement of the atomic orbitals of copper under conditions of the octahedral oxygen surrounding. A displacement of oxygen ions inside of an elementary cell which is caused by the appearance of a quasiparticle induces the displacement of equilibrium positions in the neighboring cells. In such a way, there arises the strong electron–phonon interaction of a quasiparticle with a local deformation.

### 2.4.5.  *Van Hove singularities (vHs)*

A van Hove singularity (vHs) in the density of states (DOS) $N(E)$ has been proposed as a $T_c$-enhancement mechanism for intermetallic superconductors two decades ago. All cuprate superconductors possess two-dimensional elements of their structures. In the construction of the microscopic theory of high-temperature superconductivity, it is important to clarify the specific features of the dispersion $E(k)$ and the behavior of the density of states $N(E)$. For a two-dimensional problem ($n = 2, 2D$), the density of states is independent of the energy, $N(E) = $ constant, and the band is dispersionless. The photoemission experiments indicate the existence of an almost flat band near the Fermi surface for cuprate superconductors. The presence of a flat band and an isoenergetic surface in the form of the elongated saddle leads to the existence of van Hove singularities in the density of states near the Fermi surface. In this model in the calculations of $T_c$ by the BCS formula, $N(E)$ is replaced by $N(E_F)$. The formula for $T_c$ looks like

$$T_c = 1.14\Theta \exp(-1/\lambda), \tag{2.3}$$

where $\lambda = VN(E)$.

If the function $N(E)$ has the corresponding singularity at $E = E_F$, it will be related to the van Hove singularity. In the two-dimensional case, the presence of a logarithmic singularity of the density of states on the Brillouin zone boundary is possible:

$$N(E) \sim \ln \left| \frac{D}{E - E_F} \right|, \tag{2.4}$$

where $D$ is the characteristic energy cutoff. Then $T_c$ has the form:

$$T_c \sim D \exp(-1/\sqrt{\lambda}). \tag{2.5}$$

We note that, in connection with the quasi-two-dimensionality of lattices of HTSC, the hypothesis of anyonic superconductivity is of a certain interest (Fig. 2.7). Anyons are quasiparticles with intermediate statistics (between the Bose- and Fermi-statistics) which can exist just in two-dimensional structures. The term "anyon" was introduced by F. Wilczek in the framework of the conception of supersymmetry.

## 2.4.6. *Plasmon mechanism of pairing*

Many works are devoted to attempts to explain the high-temperature superconductivity on the basis of the idea of the pairing as a result of the exchange by quanta of longitudinal plasma waves — plasmons.

Longitudinal plasma waves are formed in solids in the region of frequencies, at which the dielectric permeability of the medium becomes zero. The characteristic frequency of plasma waves in 3D crystals is defined by the formula

$$\tilde{\omega}_p = 4\pi e^2 N/m, \tag{2.6}$$

where $N$ is the concentration of electrons, and $e$ and $m$ are their charge and mass, respectively. At the electron density $N \sim (1\text{--}3) \times 10^{22}\,\text{cm}^{-3}$, the plasma frequency $\tilde{\omega}_p \sim 10^{15}\text{--}10^{16}$. We might assume that the exchange by plasmons, rather than by phonons, would induce an increase of the pre-exponential factor in the formula deduced in the BCS theory,

$$T_c = \Theta \exp \left( \frac{1}{\lambda - \mu^*} \right), \tag{2.7}$$

by two-three orders, if $\Theta = \hbar\tilde{\omega}_p/k_B$. However, such an increase does not cause a significant growth of $T_c$, because the plasmons at the frequency $\tilde{\omega}_p$ which is close to the frequency of electrons, $E_F/\hbar$, cannot cause the superconducting pairing and their role is important only for the dielectric properties of crystals [46].

## 2.4.7.  *Bipolaronic mechanism of superconductivity*

One of the attempts to explain the phenomenon of high-temperature superconductivity was named the bipolaronic theory. Bipolarons are Bose-particles like the ordinary Cooper's pairs.

In the theory of bipolarons, the superconductivity is caused by the superfluidity of the Bose-condensate of bipolarons.

The ideas of polarons and bipolarons were used by A.S. Aleksandrov and J. Ranninger [58] to clarify the high-temperature superconductivity.

The idea of a polaron is based on the assumption about the autolocalization of an electron in the ion crystal due to its interaction with longitudinal optical vibrations under the local polarization which is caused by the electron itself. The electron is confined in the local polarization-induced potential well and conserves it by the own field. The idea of the autolocalization of electrons in ion crystals was intensively developed by S.I. Pekar [59].

The efficiency of the interaction of an electron with mass $m$ and charge $e$ with long-wave longitudinal optical vibrations with frequency $\Omega$ in the medium is characterized by the dimensionless parameter

$$g = \frac{e^2}{\tilde{\varepsilon}} \sqrt{m/2\Omega\hbar^2} \tag{2.8}$$

introduced by H. Fröhlich. Here, $\tilde{\varepsilon}$ is the dielectric permeability of the inertial polarization. The interaction is assumed to be small if $g < 1$. Due to a high frequency $\Omega$, the deformation field is a faster subsystem. Therefore, it has time to follow the movement of an electron. This field accompanies the movement of the electron in the form of a weak cloud of phonons. The energy of interaction of the field and the electron is proportional to the first degree of $g$.

In the BCS theory, the pairing of conduction electrons is realized due to the interaction with acoustic phonons and is characterized by the dimensionless constant of interaction

$$\lambda = VN(E_F), \tag{2.9}$$

where $N(E_F)$ is the density of energy states of electrons on the Fermi surface, and the quantity $V$ is inversely proportional to the coefficient of elasticity of a crystal.

In the bipolaronic model [60], we have the parameter

$$\lambda* = (2\lambda^2\hbar\Omega z - V_c)/D, \tag{2.10}$$

where $\lambda$ is defined by relation (2.9), $z$ is the number of the nearest neighbors, $D$ is the width of the conduction band of free quasiparticles.

A bipolaron, like a Cooper's pair, has charge $2e$, and its effective mass is determined by the formula

$$\tilde{m} \approx m \frac{\triangle}{D} \exp(\lambda^2). \qquad (2.11)$$

The effective mass $\tilde{m}$ can be very high and can excess the mass of a free quasiparticle in the conduction band by several orders.

To calculate the superconducting transition temperature $T_c$ which corresponds to the Bose-condensation, the authors applied the formula for the ideal Bose-gas

$$k_B T_c = 3.31 \hbar^2 N^{2/3} / \tilde{m}. \qquad (2.12)$$

At $\tilde{m} = 100$ and $N = 10^{21}$ cm$^{-3}$, the last formula yields $T_c \sim 28$ K.

Analogously to the bipolaronic model, the bisoliton mechanism of high-temperature superconductivity was proposed in [47, 49].

## 2.5.   The Symmetry of Pairing in Cuprate Superconductors

Of great importance for high-temperature superconductors is the pairing symmetry or the symmetry of the order parameter. This question was considered at many conferences and seminars over the world. Several NATO-seminars and conferences on this trend which hold in Ukraine in the town of Yalta were organized by one of the authors of this book [61–64].

The development of the microscopic theory of superconductivity was followed by the interest in the question about the nontrivial superconductivity corresponding to the Cooper's pairing with nonzero orbital moment.

The system, in which the nontrivial pairing was first discovered, is He$^3$. To explain this phenomenon, it was necessary to introduce a supplementing mechanism of pairing due to spin fluctuations.

### 2.5.1.   *Superconductor's order parameter*

Most physical properties depend on the properties of the symmetry of a superconductor's order parameter which is defined by the formula

$$\Delta_{\alpha\beta}(\mathbf{k}) = \langle a_{\alpha\mathbf{k}} a_{\beta-\mathbf{k}\alpha} \rangle. \qquad (2.13)$$

The problem of pairing symmetry is the problem of the pairing of charged fermions into states with the final orbital moment.

As usual, both the standard pairing called the $s$-pairing and the nonstandard $d$-pairing are considered. They differ by the orbital moment of the pair: in the first and second cases, the moments are $L = 0$ and $L = 2$, respectively.

We note also that the continuous symmetry group in crystals is broken, and it is necessary to speak not about the orbital moment, but about the irreducible representations, by which the order parameter is classified. We will consider this question in the following subsection.

Usually, the standard pairing frequently called the $s$-pairing and the nonstandard pairing are distinguished. At the nonstandard pairing, the symmetry of the order parameter is lower than the symmetry of a crystal.

For a two-dimensional tetragonal crystal (square lattices), the possible symmetries of the superconductor's order parameter were enumerated by M. Sigrist and T.M. Rice [65] on the basis of the theory of group representations. The basis functions of relevant irreducible representations define the possible dependence of the order parameter on the wave vector.

It is worth noting that the anisotropic pairing with the orbital moment $L = 2$, i.e., $d_{x^2-y^2}$, has the following functional form in the $k$ space:

$$\Delta(\mathbf{k}) = \Delta_o[\cos(k_x a) - \cos(k_y a)], \qquad (2.14)$$

where $\Delta_o$ is the maximum value of the gap, and $a$ is the lattice constant. The gap is strongly anisotropic along direction (110) in the $k$ space. In this case, the order parameter sign is changed in the directions along $k_x$ and $k_y$.

Together with the $d$-symmetry, it is worth to consider also the $s$-symmetry, for which we can choose two collections of basis functions:

$$\Delta(\mathbf{k}) = \Delta_o. \qquad (2.15)$$

The anisotropic $s$-pairing is considered as well. This form of the pairing is analyzed in works by P.W. Anderson [5] with co-workers who have studied the mechanism of pairing on the basis of the tunneling of electrons between layers. In these states, the order parameter sign is invariable, and its amplitude is varied along direction (110):

$$[\text{anisotropic } s]\Delta(k) = \Delta_o[\cos(k_x a) - \cos(k_y a)]^4 + \Delta_1, \qquad (2.16)$$

where $\Delta_1$ corresponds to the minimum along direction (110).

It follows from the symmetry-based reasoning that the mixed states with various symmetries can be realized. We mention the states which are mostly in use. The "extended" $s$-coupled states were considered in works by D.J. Scalapino [67]. A possible functional form of these states is as follows:

$$[\text{extended } s\text{-wave}] \; \Delta(k) = \Delta_o\{(1 + \gamma^2)[\cos(k_x a) - \cos(k_y a)]^2 - \gamma^2\}. \quad (2.17)$$

They have eight parts with alternating signs and eight nodes which are split by $\pm\gamma\pi/2$ along direction (110). G. Kotliar with co-workers used mixed $s + id$ states [67]:

$$[s + id] \; \Delta(k) = \Delta_o[\varepsilon + i(1 - \varepsilon)[\cos(k_x a) - \cos(k_y a)]]. \tag{2.18}$$

R.B. Laughlin [66] analyzed the mixed states $d_{x^2+y^2} + id_{xy}$ on the basis of the anyonic mechanism of pairing:

$$[d + id]\Delta(k) = \Delta_o[(1 - \varepsilon)[\cos(k_x a) - \cos(K_y a)]$$
$$+ i\varepsilon[2\sin(k_x a)\sin(k_y a)]], \tag{2.19}$$

where $\varepsilon$ is the share of $s$ or $d_{xy}$ states mixed with the $d_{x^2+y^2}$ states, and $\varepsilon\Delta_o$ is the minimum value of the energy gap. These mixed states are of interest because they are not invariant with respect to the inversion in the time. The value and phase of the superconductor's order parameter, as functions of the direction in cuprate planes $CuO_2$ are given in Fig. 2.8 for various kinds of the pairing symmetry.

## 2.5.2.  *Classification of the superconductor's order parameter by the representations of symmetry groups*

It should be noted that there is no classification of states by the orbital moment for crystals.

The general theory of nonstandard pairing has been developed on the basis of the analysis of point symmetry groups.

J. Annett was one of the first who classified superconducting states by the irreducible representations of groups for high-temperature superconductors [68]. As for superconductors with heavy fermions, the group analysis was carried out in [69].

Of importance is the question whether the symmetry of the order parameter in HTSC is lower than the symmetry of the crystal lattice.

The order parameter can be represented as a linear combination

$$\Delta_k = \sum \eta_{\Gamma_i} f_{\Gamma_i}(k), \tag{2.20}$$

where $\eta_{\Gamma_i}$ is the irreducible representation of groups, by which the order parameter is transformed, $f_{\Gamma_i}(k)$ is basis functions of the irreducible representation, $A_{1g}$ corresponds to the anisotropic $s$-symmetry, and $B_{1g}$ corresponds to the $d$-pairing.

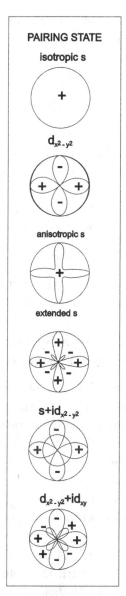

Fig. 2.8. Kinds of pairing symmetries which are considered for high-temperature super-conductors.

An analogous decomposition of the superconducting order parameter was performed for a superconductor with heavy fermions

$$UPt_3.$$

The symmetric states of the system can be presented by the indication of all possible subgroups of the full group, relative to which the order parameter is invariant [70]. The full symmetry group of a crystal includes a point symmetry group $G$, operations of inversion of the time $R$, and the group of calibration transformations $U(1)$. That is, we have the following partition into subgroups:

$$G \times R \times U(1). \tag{2.21}$$

The La- and Y-based superconductors which have the tetragonal symmetry of crystal lattices, i.e., $G = D_{4h} = D_4 \times I$ [62], are most studied. In the Y-based compounds, one observes small orthorhombic distortions of a crystal lattice.

Group $D_{4h}$ includes the operations of rotations $C_n$, around the $z$-axis by angles of $\pi n/2$ and rotations $U_n$ by angles of $\pi$:

$$x \cos(\pi n/2) + y \cos(\pi n/2), \tag{2.22}$$

where $n = 0, 1, 2, 3$ have five reducible representations: four one-dimensional $(A_{1g}, A_{2g}, B_{1g}, B_{2g})$ and one two-dimensional $(E)$.

## 2.6. Experimental Studies of a Symmetry of the Superconducting Order Parameter

At the present time, the results of three groups of experiments, in which the symmetry of the order parameter is revealed, are available.

The first group joins different low-temperature characteristics of super-conductors such as the Knight shift, the rate of relaxation in NMR, the temperature dependence of the heat capacity, the penetration depth, etc.

If the superconductor's order parameter has zeros on different areas of the Fermi surface (as in the case of the $d_{x^2-y^2}$-symmetry), the mentioned quantities will have the power temperature dependence, rather than the exponential one.

The second group of experiments is based on the direct measurement of the phase of the order parameter with the help of interference phenomena on Josephson junctions in a magnetic field.

The third group deals with the direct measurements of a value of the gap by means of spectroscopic experiments. Here, the most interesting results are presented by photoemission spectroscopy with angle resolution, and Raman and neutron spectroscopies.

### 2.6.1.  *Measurements of the Josephson tunnel current*

The most definite information about a symmetry of the superconducting order parameter can be obtained from the studies of the phase of the order parameter, in which the critical current in Josephson junctions positioned in a magnetic field is measured. The critical current for a rectangular Josephson junction oscillates with the field by the Fraunhofer diffraction law:

$$I_c(\Phi) = J_0 A \frac{\sin(\pi\Phi/\Phi_0)}{\pi\Phi/\Phi_0}, \tag{2.23}$$

where $\Phi$ is the magnetic flux through the junction, $\Phi_0$ is the quantum flux, $J_0$ is the density of the critical current in the zero field, and $A$ is the junction area. A diffraction pattern is shown in Fig. 2.9(a). Let us consider the superconductor $YBCO$ which possesses the tetragonal symmetry. Let its axis be oriented normally to the plane of the figure, and let its edges be perpendicular to the axes $A$ and $B$ of the base plane. In the tunnel junction with corner geometry, another superconductor is applied to both edges which are perpendicular to $A$ and $B$ and, moreover, are joined with each other (Figs. 2.9(b) and 2.9(c)).

This experiment can be compared with that involving a two-junction SQUID, in which there occurs a superposition of tunnel currents produced by electrons with wave vectors $k_x$ and $k_y$, so that the resulting diffraction pattern depends on the symmetry of the order parameter of a superconductor under study. At the $s$-symmetry, the order parameter on both edges of the corner junction is the same, and the resulting diffraction pattern will be such as that in the case of the standard junction. But, in the case of the $d$-symmetry, the order parameter on the corner junction edges has different signs, and this fact changes basically the diffraction pattern.

The total current is illustrated in Fig. 2.10(a). In the zero field, the critical current turns out to be zero due to the mutual compensation of its two components.

In the symmetric contact, the dependence on the field is determined by the formula [71]

$$I_c(\Phi) = J_0 A \frac{\sin^2(\pi\Phi/2\Phi_0)}{\pi\Phi/2\Phi_0}, \tag{2.24}$$

corresponding to the pattern in Fig. 2.10(b).

Thus, by the difference of a diffraction pattern from both the standard one and that corresponding to the corner Josephson junction, we can judge about a symmetry of the order parameter. In work [71], a similar experiment was carried out on the tunnel junction YBCO–Au–Pb. The results presented in Fig. 2.10 testify to the $d$-symmetry of the order parameter in superconducting YBCO.

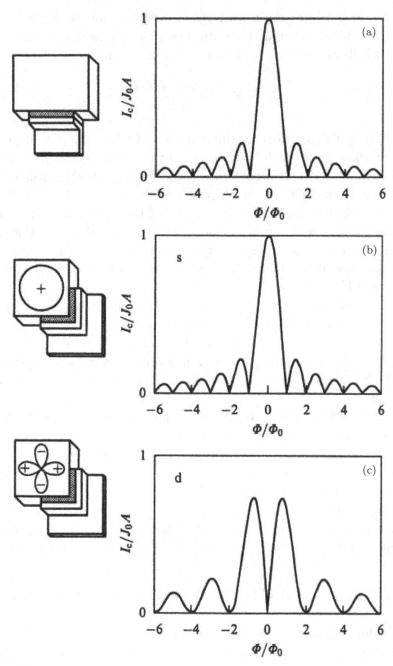

Fig. 2.9.  Critical current in a Josephson junction versus the applied magnetic field:
(a) standard tunnel contact, (b) corner tunnel junction for a superconductor with the
$s$-symmetry of the order parameter, (c) corner tunnel junction for a superconductor with
the $d$-symmetry of the order parameter [71].

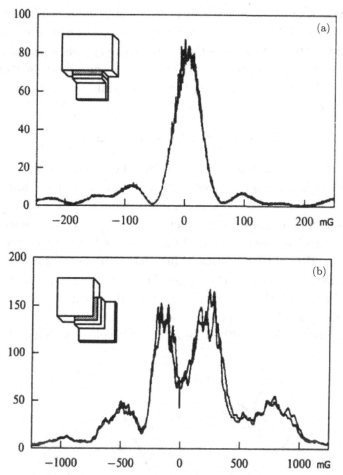

Fig. 2.10. Critical current as a function of the magnetic field in a Josephson junction YBCO–Au–Pb in two0 geometries: (a) standard and (b) corner ones [71].

All the above-mentioned experiments with Josephson junctions were performed with a single crystal $YBa_2Cu_3O_6$.

## 2.6.2. *Measurements of the quantization of a flow by the technique of three-crystal unit*

Another type of experiments on the determination of a symmetry of the order parameter is based on the measurement of a flow quantum in a superconducting ring fabricated from three superconducting single crystals of yttrium with different orientations.

The idea of such an experiment is based on the theoretical result obtained by M. Sigrist and T.M. Rice [65]: for superconductors with the $d$-symmetry, the tunnel current between two superconducting crystals separated by a thin boundary depends on the orientation of the order parameter with respect to the interface. The current between the superconductors with numbers $ij$ is given by the formula

$$I_s^{ij} = (A^{ij} \cos 2\theta_i \cos 2\theta_j) \sin \Delta\Phi_{ij}. \tag{2.25}$$

Here, $A^{ij}$ is the constant which characterizes the junction of $ij$, $\theta_i$, $\theta_j$ are the angles of the crystallographic axes with the boundary plane, and $\Phi_{ij}$ is the difference of phases of the order parameters on both sides of the boundary. It was shown that a spontaneous magnetization, which corresponds to the flow equal to a half of $\Phi_0$, appears in a superconducting ring with a single Josephson junction with the phase difference $\pi$. If the ring has odd number of $\pi$-junctions, the result is the same. The direct measurement of a half-quantum of the flow through such a ring would testify to the $d$-symmetry of the order parameter. The direct measurement of a half-quantum of the flow was realized in work [72]. The scheme of the experiment is shown in Fig. 2.11.

Fig. 2.11.   Scheme of the experiment on the measurement of a half-integer number of quanta of the flow captured by a superconducting ring with three Josephson junctions with the $d$-symmetry of the order parameter [34].

## 2.7. Thermodynamics of the *d*-Pairing in Cuprate Superconductors

### 2.7.1. *Introduction*

The basic question of the theory of superconductivity concerns the mechanism ensuring the pairing of electrons. In the BCS theory, it is the electron–phonon interaction. Some recent theoretical models postulate the mechanism of antiferromagnetic spin fluctuations [73–79], so that the electron scattering on them can be the reason for the pairing of electrons. Spin fluctuations play an important role in superconductors with heavy fermions [76, 77]. The authors of works [68, 69] performed the calculations of a value of the superconductor gap, the critical temperature, the temperature dependence of the resistance, and many other quantities, but the thermodynamics was not considered. For this reason, it is of interest to calculate the thermodynamics of antiferromagnetic spin fluctuations, namely the temperature dependence of the heat capacity, its jump near $T_c$, and the parameter $R = \Delta C/(\gamma T_c)$ equal to 1.42 by the BCS theory.

The purpose of this section is the calculation of the thermodynamics of antiferromagnetic spin fluctuations, namely the electron heat capacity and the jump of the heat capacity.

The thermodynamics of superconductors at low temperatures is determined by the excitation of two quasiparticles. In the traditional superconductors with pairing of the BCS-type, the energy gap is isotropic (*s*-pairing), and the temperature dependence of the heat capacity has the exponential form $\sim \exp^{-\Delta/k_B T}$, where $\Delta$ is the superconductor's gap. In the superconductors with the anisotropic pairing, the temperature dependence of the heat capacity has a power character, namely, $T^n$. The appearance of such temperature dependences is related to the fact that the superconductor's gap has zeros on the Fermi surface.

As was noted above, the anisotropic pairing with the orbital moment $L = 2$, i.e., with the $d_{x^2-y^2}$-symmetry, has the following functional form in the $k$ space:

$$\Delta(\mathbf{k}) = \Delta_o[\cos(k_x a) - \cos(k_y a)], \qquad (2.26)$$

where $\Delta_o$ is the maximum value of the gap, and $a$ is the lattice constant.

The gap is strongly anisotropic in direction (110) in the $k$ space, and the sign of the order parameter is changed along the directions $k_x$ and $k_y$.

The main results of this section are published in works [82–85].

## 2.7.2. *Antiferromagnetic spin fluctuations in high-temperature superconductors*

For the first time, the idea of the possibility for the electron pairing through spin fluctuations was advanced by A.I. Akhiezer and I.Ya. Pomeranchuk [85]. They showed that the indirect interaction of electrons through spin waves in a ferromagnetic metal has the character of attraction in the triplet state and, hence, can lead to the triplet pairing. Consider some experiments and facts on antiferromagnetic spin fluctuations.

The basis for the hypothesis on the spin-fluctuation mechanism of pairing consists in the fact that the stoichiometric compounds $La_2CuO_4$ and $YBa_2Cu_3O_6$ are antiferromagnetic dielectrics. The doping of superconductors leads to the appearance of the metallic state and superconductivity. The closeness of high-temperature superconductors to the antiferromagnetic transition with the wave vector $Q = (\pi/a, \pi/a)$ defines the important role of spin fluctuations, the interaction with which forms the quasiparticle spectrum of electrons and can simultaneously cause the Cooper's pairing.

High-temperature superconductors are referred to the class of strongly correlated systems which are theoretically studied in the frame of the Hubbard model. This model describes the hops of electrons in the lattice with the matrix element $t$ for the nearest neighbors with regard for the Coulomb repulsion $U$, when the electrons are positioned at the same site. The model is set by the Hamiltonian

$$H = -t \sum_{i,j,\sigma} C_{i\sigma}^{\dagger} C_{j\sigma} + U \sum_{i,} n_{i\uparrow} n_{i\downarrow}, \qquad (2.27)$$

where $C_{i\sigma}^{\dagger}(C_{j\sigma})$ is the operator of creation (annihilation) of an electron at the site $i$ with spin $\sigma$, and $n_{i\uparrow} = C_{i\sigma}^{\dagger} C_{j\sigma}$ is the number of electrons at the site. In the given region of the parameters $t$, $U$, and $n$ (the electron concentration), the appearance of magnetically ordered phases is possible. Near the boundary of the existence of such a phase from the side of the paramagnetic region, strong fluctuations of the magnetic order parameter, paramagnons, must be manifested.

In the two-dimensional system of CuO layers in cuprate superconductors, the electron spectrum is presented by the formula

$$\varepsilon(t) = -2t \cos k_x a + \cos k_y a), \qquad (2.28)$$

and the chemical potential $\mu$ is determined by the given electron concentration $n$. On the half-filling ($n = 1$), this spectrum has the nesting at the wave vector $q = Q$, which induces a sharp peak in the spin susceptibility near this point. This means an instability of the system relative to the formation of the antiferromagnetic state with the wave vector $Q$ and the intensification of spin fluctuations near the point of the magnetic phase transition.

Near the half-filling, when the system is really antiferromagnetically unstable, the numerical calculations indicate that the superconducting order parameter has $d$-symmetry, i.e., the gap depends on the wave vector by relation (2.26).

Gap (2.26) is an alternating function of the wave vector (Fig. 2.12) and has zero values on the diagonals.

Figure 2.12 shows that the wave function of a Cooper's pair is equal to zero just on the diagonals of the square. Therefore, the repulsive interaction on these diagonals does not act on the pair, and a Cooper's pair with the $d$-symmetry survives even at large values of $U$. The superconductors with the $d$-pairing should have a number of particular properties which can be observed in experiments. Many of these peculiarities are related to the zeros of the order parameter. The quasiparticle spectrum at low temperatures must give the power contribution to the thermodynamic properties, such as the heat capacity, parameters of NMR, and the penetration depth of a magnetic field, rather than the exponential one as in ordinary isotropic superconductors.

The observation of such power contributions will indicate the presence of a nontrivial order parameter with zeros on the Fermi surface. The totality of experimental data for various high-temperature superconductors indicates the certain realization of the anisotropic order parameter in them and, with a high probability, with the $d$-symmetry. The last circumstances present the important argument in favor of the spin-fluctuation mechanisms of high-temperature superconductors. One of them is intensely developed by Pines and his co-workers [73,75] who used a phenomenological form of the magnetic

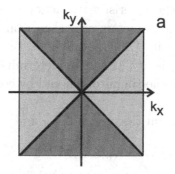

Fig. 2.12.   Distribution of signs of the gap function in the limits of the Brillouin first band.

susceptibility with the parameters determined from the experiments on cuprates.

Let us consider this model in more details.

We introduce a Hamiltonian which involves antiferromagnetic spin fluctuations, as it was made in works by Pines [73, 74]:

$$H = H_o + H_{\text{int}}, \qquad (2.29)$$

where $H_0$ is Hamiltonian of free electrons. The interaction is described by the Hamiltonian

$$H_{\text{int}} = \frac{1}{\Omega} \sum_q g(q) s(q) S(-q), \qquad (2.30)$$

where $\Omega$ is the cell volume, $g(q)$ is the interaction constant;

$$s(q) = \frac{1}{2} \sum_{\alpha,\beta,k} \Psi^+_{k+q} \sigma_{\alpha\beta} \Psi_{k\beta} \qquad (2.31)$$

is operator of spin density; $\sigma_{\alpha\beta}$ is Pauli matrix; $\Psi^+_{k+q,\alpha}$ is operator of creation of an electron with the momentum $k + q$ and the spin projection $\alpha$; $\Psi_{k\beta}$ is operator of creation of a hole with the momentum $k$ and the spin projection $\beta$; $S(-q)$ is operator of spin fluctuations, whose properties are set by the correlator $\chi(q,\omega)$ [73, 74], $\chi_{ij}(\mathbf{n}, \mathbf{m})$ is the spin susceptibility which is modeled by

$$\chi(q,\omega) = \frac{\chi_Q}{1 + \xi^2 (q - Q)^2 - i\omega/\omega_{\text{SF}}},$$

$$q_x > 0, \quad q_y > 0, \qquad (2.32)$$

where $\chi_Q$ is the static spin susceptibility with the wave vector $Q = (\pi/a, \pi/a)$, $\xi$ is the temperature-dependent antiferromagnetic correlation length, and $\omega_{\text{SF}}$ is the characteristic frequency of spin fluctuations of the paramagnon energy. All parameters are taken from experiments, including the data from NMR studies [73]. In this case, the interaction constant is a free parameter of the theory. It can be determined, by calculating some quantity with the help of the Hamiltonian and by comparing the result with experiments.

We now define the quantities $\chi_Q$ and $\omega_{\text{SF}}$ as

$$\chi_Q = \chi_0 (\xi/a)^2 \beta^{1/2}, \qquad (2.33)$$

$$\omega_{\text{SF}} = \Gamma/(\pi(\xi/a)^2 \beta^{1/2}), \qquad (2.34)$$

where $\chi_0$ is the experimentally measured long-wave limit of the spin susceptibility, $\beta = \pi^2$, and $\Gamma$ is the energy constant. The NMR data for the compounds yield $\xi(T_c) = 2.3a$, $\omega_{SF} = 8\,\text{meV}$, $\Gamma = 0.4\,\text{meV}$.

It is foreseen that the phenomenological Hamiltonian will give a self-consistent description of the spin dynamics of the system in the sense that the spin susceptibility calculated with its help (through the characteristics of a quasiparticle spectrum which themselves depend on the susceptibility) will agree with values determined by formula (2.32).

### 2.7.3. *Continual model of antiferromagnetic spin fluctuations*

The authors of works [82–85] proposed a continual model for the spin-fluctuation mechanism of pairing which allows one to efficiently solve the problem of thermodynamics for this mechanism.

Let us consider this model in more details.

We now calculate the thermodynamics which is set by the Pines spin-fluctuation Hamiltonian. We write the Hamiltonian in the lattice representation

$$H = H_0 + H_{\text{int}}, \tag{2.35}$$

$$H = -t\sum_{\mathbf{n},\mathbf{p}} \psi_\alpha^+(\mathbf{n})\psi_\alpha(\mathbf{n}+\mathbf{p}) + \frac{1}{2}\sum_{\mathbf{n},\mathbf{m}} S_i(\mathbf{n})\chi_{ij}^{-1}(\mathbf{n},\mathbf{m})S_j(\mathbf{m})$$

$$+ g\sum_{\mathbf{n}} \psi_\alpha^+(\mathbf{n})\left(\frac{\sigma_i}{2}\right)_{\alpha\beta}\psi_\beta(\mathbf{n})S_i(\mathbf{n}), \tag{2.36}$$

$$H_0 = -t\sum_{\mathbf{n},\mathbf{p}} \psi_\alpha^+(\mathbf{n})\psi_\alpha(\mathbf{n}+\mathbf{p}) + \frac{1}{2}\sum_{\mathbf{n},\mathbf{m}} S_i(\mathbf{n})\chi_{ij}^{-1}(\mathbf{n},\mathbf{m})S_j(\mathbf{m}), \tag{2.37}$$

$$H_{\text{int}} = g\sum_{\mathbf{n}} \psi_\alpha^+(\mathbf{n})\left(\frac{\sigma_i}{2}\right)_{\alpha\beta}\psi_\beta(\mathbf{n})S_i(\mathbf{n}), \tag{2.38}$$

where the sum is taken over all the sites of the infinite lattice (the lattice constant is equal to $a$), $\boldsymbol{\rho}$ is the unit vector joining the neighboring sites, $S_i$ is the spin operator, $N = \sum_{\mathbf{n}} \psi_\alpha^+(\mathbf{n})\psi_\alpha(\mathbf{n})$ is the operator of the number of particles, $t$ is the half-width of the conduction band [73,74], and $\chi_{ij}(\mathbf{n},\mathbf{m})$ — the spin correlation function.

It is necessary to calculate the grand partition function

$$\exp\{-\beta\Omega(\mu,\beta,g)\} \equiv \text{Tr}\exp\{-\beta(H - \mu N)\},$$
$$\beta = 1/kT, \tag{2.39}$$

where $\mu$ is the chemical potential, $g$ is the coupling constant, and $\Omega(\mu, \beta, q)$ is the thermodynamic potential.

It is convenient to use the formalism of continual integration for a system of Fermi-particles. The method of functional integration, i.e., the integration in the space of functions, was proposed by N. Wiener in 1925, but the physicist-theorists paid no attention to this method. Continual integrals were introduced in physics by R. Feynman [4] in the 1940s and were used for the reformulation of quantum mechanics. The continual integration is one of the most powerful methods of the contemporary theoretical physics which allows one to simplify, accelerate, and clarify the process of analytic calculations. The application of the method of continual integration to a system with infinite number of degrees of freedom allows one to develop, in such a way, the diagram theory of perturbations.

The grand partition function can be written in the form of a continual integral [87]:

$$\exp - \beta \Omega = N \int \prod_n dS_i(n) d\psi_\alpha^+(n, \tau) d\psi_\alpha(n, \tau) \exp\left\{ - \int_0^\beta d\tau L(\tau) \right\},$$

(2.40)

where

$$L(\tau) = \sum_n \psi_\alpha^+(\mathbf{n}, \tau) \left( \frac{\partial}{\partial \tau} - \mu \right) \psi_\alpha(\mathbf{n}, \tau) - t \sum_{n,p} \psi_\alpha^+(\mathbf{n}, \tau) \psi_\alpha(\mathbf{n} + \mathbf{p}, \tau)$$

$$+ g \sum_n \psi_\alpha^+(\mathbf{n}, \tau) \left( \frac{\sigma_i}{2} \right)_{\alpha, \beta} \psi_\beta(\mathbf{n}, \tau) S_i(\mathbf{n}, \tau)$$

$$+ \frac{1}{2} \sum_{n,m} S_i(\mathbf{n}, \tau) \chi_{ij}^{-1}(\mathbf{n}, \mathbf{m}, \tau) S_j(m, \tau),$$

(2.41)

where $L(\tau)$ is the Lagrangian of the system, and $N$ is the normalizing factor,

$$N^{-1} = \int \prod_n dS_i(n) d\psi_\alpha^+(n, \tau) d\psi_\alpha(n, \tau) \exp\left\{ - \int_0^\beta d\tau L(\tau, \mu = g = 0) \right\}.$$

(2.42)

We will use the matrix formalism to construct the theory of perturbations for the Green functions. It is convenient to introduce a four-component bispinor (Majorana)

$$\Psi = \begin{pmatrix} \psi \\ -\sigma_2 \psi^* \end{pmatrix},$$

$$\psi = \begin{pmatrix} \psi_1 \\ \psi_2 \end{pmatrix}, \tag{2.43}$$

where $\sigma_1$ is Pauli spin matrices. The Majorana spinor is a Weyl spinor written in the four-component form.

Now, we can write $L$ in the form

$$L = \frac{1}{2} \sum_{\mathbf{n,p}} \bar{\Psi}(\mathbf{n}, \tau) \left[ \left( \Gamma^0 \frac{\partial}{\partial \tau} - \Gamma^0 \Gamma_5 \mu \right) \hat{\delta} - t(\tau - \hat{\delta}) \Gamma^0 \Gamma_5 \right]$$

$$\times \psi(\mathbf{n} + \mathbf{p}, \tau) + \frac{g}{4} \sum_{\mathbf{n}} \bar{\Psi}(\mathbf{n}, \tau) \Gamma^i \Gamma_5 \Psi(\mathbf{n}, \tau) S^i(\mathbf{n}, \tau) \tag{2.44}$$

$$+ \frac{1}{2} \sum_{\mathbf{n,m}} S_i(\mathbf{n}, \tau) \chi_{ij}^{-1}(\mathbf{n}, \mathbf{m}, \tau) S_j(\mathbf{m}, \tau), \tag{2.45}$$

where

$$\Gamma^0 = \begin{pmatrix} 0 & I \\ I & 0 \end{pmatrix},$$

$$\Gamma^i = \begin{pmatrix} 0 & -\sigma_i \\ \sigma_i & 0 \end{pmatrix}, \tag{2.46}$$

$$I = \begin{pmatrix} 1 & 0 \\ 0 & 1 \end{pmatrix},$$

where $\Gamma^i$ are the Dirac gamma matrices, $\bar{\Psi} \equiv \Psi^+ \Gamma^0$, $\hat{\delta} = \delta_{\mathbf{n,n+p}}$, and $\Gamma^i$ are given in the chiral representation.

There exists the connection between $\Psi$ and $\bar{\Psi}$: $\Psi = C\bar{\Psi}^T$, where $C$ is the matrix of charge conjugation, $C = \Gamma^0 \Gamma^2$, $C^T = -C$. In terms of $\Psi$, the partition function takes the form

$$e^{-\beta \Omega} = N \int \prod n dS_i(n, \tau) d\Psi(n, \tau) e^{-\int_0^\beta d\tau L(\tau)}. \tag{2.47}$$

We now use the method of bilocal operators [88] to calculate the grand partition fuction. In this way, we obtain the Schwinger–Dyson equation and the equation of free energy. The details of calculations can be found in [82–85].

In order to calculate $\Omega$, we introduce a source of the bilocal operator [88]

$$e^{-\beta\Omega(J)} = N \int dS d\Psi e - \int_0^\beta d\tau$$

$$\times \left[ L(\tau) + \frac{1}{2}\mathbf{n}, \mathbf{m} \sum \int_0^\beta \bar{\Psi}(\mathbf{n}, \tau) J(\mathbf{m}, \mathbf{n}, \tau, \tau') \Psi(\mathbf{m}, \tau') \right]. \quad (2.48)$$

The full Green fermion function is determined by the formula

$$G_{nm}(J) = e^{\beta\Omega(J)} \int dS \, d\Psi \Psi(n, \tau) \bar{\Psi}(m, \tau') \exp\{\dots\}$$

$$= \langle 0 | T\Psi(n, \tau) \bar{\Psi}(m, \tau') | 0 \rangle. \quad (2.49)$$

Let us write the Schwinger–Dyson equation for the Green function $G$, by using the method of bilocal operator [85]:

$$\frac{\delta F}{\delta G} = 0, \quad (2.50)$$

$$F(G) = \beta\Omega(J = 0). \quad (2.51)$$

First, we consider the case of a free system. By integrating over the fermion fields in the functional integral, we obtain

$$F_0 = \beta\Omega(J) = -\frac{1}{2}\mathrm{Tr}\,\mathrm{Ln}(G_0^{-1} - \Gamma^0\Gamma_5\mu + J) + \frac{1}{2}\mathrm{Tr}\,\mathrm{Ln}\,G_0^{-1}, \quad (2.52)$$

$$G_0 = \left[ \Gamma^0 \frac{\partial}{\partial\tau} - t(1 - \hat{\delta})\Gamma^0\Gamma_5 \right]^{-1}. \quad (2.53)$$

For the interacting system, the free energy takes the form

$$F = F_0 + F_{\text{int}}. \quad (2.54)$$

We represent $F_{\text{int}}$ as a series in $g$: $F_{\text{int}} = \sum_{n=1}(g^2)F_n$. In the lowest order in $g$, we get the relation

$$F_1 = -\frac{g^2}{32}\mathrm{Tr}\left\{ \Gamma^i G \, Tr\Gamma^j G - 2G\Gamma^i G\Gamma^j \right\} \chi_{ij}, \quad (2.55)$$

where the Green function for fermions and spins is as follows:

$$\langle \Psi(x, \tau)\bar{\Psi}(y, \tau') \rangle = G(x, \tau; y, \tau'),$$
$$\langle S_i(x, \tau)S_j(y, \tau') \rangle = \chi_{ij}(x, \tau; y, \tau'). \quad (2.56)$$

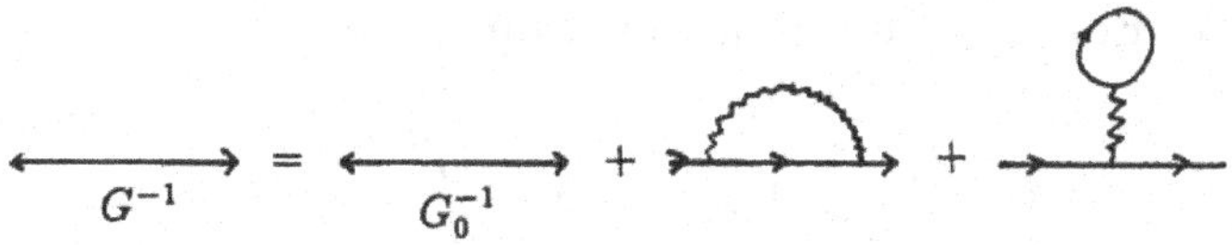

Fig. 2.13. Graphical form of Eq. (2.57).

$$\bigcirc\!\!\!\sim\!\!\!\bigcirc \quad -2 \quad \bigcirc\!\!\!\sim\!\!\!\bigcirc$$

Fig. 2.14. Graphical form of the right-hand side of Eq. (2.58).

Taking the condition $\delta F/\delta G = 0$ into account, we obtain the Schwinger–Dyson equation

$$G^{-1} = G_0^{-1} - \Gamma^0\Gamma_5\mu + \frac{g^2}{12}\{\Gamma^i Tr\Gamma^j G\chi_{ij} - 2\Gamma^i G\Gamma^j\chi_{ij}\}, \qquad (2.57)$$

where $\Gamma_0$ and $\Gamma^i$ are Dirac matrices (Fig. 2.13).

Here, the continuous line corresponds to the Green function for fermions $G$, and the wavy line does to the Green function for spins $\chi_{i,j}(= \delta_{ij}\chi)$. The equation for free energy

$$F_1 = -\frac{g^2}{32}\mathrm{Tr}\left\{\Gamma^i G\,\mathrm{Tr}\Gamma^j G - 2G\Gamma^i G\Gamma^j\right\}\chi_{ij}, \qquad (2.58)$$

where $G$ and $\chi_{ij}$ are, respectively, the Green fermion and spin functions, corresponds to the contribution of two vacuum diagrams (Fig. 2.14).

### 2.7.4. *Equation for superconducting gap*

From Eq. (2.57), it is easy to deduce the equation for a gap which can be found in [74]. By executing the Fourier transformation

$$G(x,\tau) = \sum_{n=-\infty}^{\infty} \int_{-\pi/a}^{\pi/a} \frac{d^2k}{(2\pi)^2} G(\mathbf{k},\omega_n)e^{i\omega_n\tau - i\mathbf{kx}}$$

$$\times \begin{cases} \omega_n = \dfrac{(2n+1)\,\pi}{\beta} & \text{for fermions,} \\[2mm] \omega_n = \dfrac{2n\pi}{\beta} & \text{for bosons,} \end{cases}$$

we rewrite Eq. Eq. (2.57) in the momentum space as

$$G^{-1}(\mathbf{k}, i\omega_n) = \widetilde{G_0^{-1}}(\mathbf{k}, i\omega_n) + \frac{g^2}{4\beta} \sum_{m=-\infty}^{\infty} \int_{-\pi/a}^{\pi/a} \frac{d^2p}{(2\pi)^2}$$

$$\times \left[ \Gamma^i \Gamma_5 Tr \Gamma^i \Gamma_5 G\left(\mathbf{p}, i\omega_m\right) - 2\Gamma^i \Gamma_5 G\left(\mathbf{p}, i\omega_m\right) \Gamma^i \Gamma_5 \right]$$

$$\times \chi\left(\mathbf{k} - \mathbf{p}, i\omega_n - i\omega_m\right), \tag{2.59}$$

where we set $\chi_{ij} = \delta_{ij}\chi$,

$$G_0^{-1}(\mathbf{k}, i\omega_n) = \begin{pmatrix} 0 & \{i\omega_n - (\varepsilon(\mathbf{k}) - \mu)\}I \\ \{i\omega_n + \varepsilon(\mathbf{k}) - \mu\}I & 0 \end{pmatrix}$$

$$\equiv \Gamma^0 i\omega_n - \Gamma^0 \Gamma_5(\varepsilon(\mathbf{k}) - \mu), \tag{2.60}$$

$$\varepsilon(\mathbf{k}) = -2t[\cos k_x a + \cos k_y a]. \tag{2.61}$$

According to the standard relation of the diagram technique, we denote the free energy by $\Sigma(\mathbf{k}, i\omega_n)$. Then we have

$$G^{-1}(\mathbf{k}, i\omega_n) = \widetilde{G_0^{-1}}(\mathbf{k}, i\omega_n) - \Sigma(\mathbf{k}, i\omega_n), \tag{2.62}$$

and Eq. (2.57) is the equation for $\Sigma$. We write the solution for $\Sigma$ in the form

$$\Sigma = \Gamma^0 A + \Gamma^0 \Gamma_5 B + \Delta + \Gamma^0 \Gamma_5 \Gamma^i \Delta_i \tag{2.63}$$

$(A, B, \Delta,$ and $\Delta_i$ are functions of $\mathbf{k}$ and $i\omega_n$).

By determining the matrix which is inverse to (2.62), we get

$$G(\mathbf{k}, i\omega_n) = \frac{\Gamma^0(\omega_n - A) - \Gamma^0 \Gamma_5 \left(\varepsilon(\mathbf{k}) - \mu + B\right) + \Delta + \Gamma^0 \Gamma_5}{(\omega_n - A)^2 - (\varepsilon(\mathbf{k}) - \mu + B)^2 - \Delta^2 - \Delta_i^2}. \tag{2.64}$$

Substituting (2.64) and (2.63) in (2.63), we obtain the system of equations for the functions $A, B, \Delta,$ and $\Delta_i$:

$$A(\mathbf{k}, i\omega_n) = \frac{3g^2}{4\beta} \sum_m \int \frac{d^2p}{(2\pi)^2} \frac{i\omega_n - A(\mathbf{p}, i\omega_m)}{D(i\omega_m, \mathbf{p})}$$

$$\times \chi(\mathbf{k} - \mathbf{p}, i\omega_n - i\omega_m), \tag{2.65}$$

$$B(\mathbf{k}, i\omega_n) = \frac{3g^2}{4\beta} \sum_m \int \frac{d^2p}{(2\pi)^2} \frac{\varepsilon(\mathbf{p}) - \mu + B(\mathbf{p}, i\omega_m)}{D(i\omega_m, \mathbf{p})}$$

$$\times \chi(\mathbf{k} - \mathbf{p}, i\omega_n - i\omega_m), \tag{2.66}$$

$$\Delta(\mathbf{k}, i\omega_n) = \frac{3g^2}{4\beta} \sum_m \int \frac{d^2p}{(2\pi)^2} \frac{\Delta(\mathbf{p}, i\omega_m)}{D(i\omega_m, \mathbf{p})}$$

$$\times \chi(\mathbf{k} - \mathbf{p}, i\omega_n - i\omega_m), \qquad (2.67)$$

$$\Delta_i(\mathbf{k}, i\omega_n) = -\frac{g^2}{4\beta} \sum_m \int \frac{d^2p}{(2\pi)^2} \frac{\Delta_i(\mathbf{p}, i\omega_m)}{D(i\omega_m, \mathbf{p})}$$

$$\times \chi(\mathbf{k} - \mathbf{p}, i\omega_n - i\omega_m), \qquad (2.68)$$

where

$$D(i\omega_m, \mathbf{p}) = (\omega_n - A)^2 - (\varepsilon(\mathbf{k}) - \mu + B)^2. \qquad (2.69)$$

Equations (2.67) and (2.68) correspond, respectively, to the singlet and triplet pairings. Moreover, the singlet channel is characterized by the repulsion, whereas the triplet one by the attraction. Therefore, it is necessary to take the trivial solution, $\Delta = 0$, of Eq. (2.67). In the following calculations, we will neglect the contributions from the functions $A$ and $B$, which leads only to the renormalization of the wave function and the chemical potential, and will study only Eq. (2.67). By neglecting $\Delta^2$ in the denominator, we get the linearized equations for $\Delta_0$ which determine the critical temperature $T_c$.

For the correlation function $\chi(q, i\omega_n)$, we will use the dispersion relations

$$\chi(q, i\omega_n) = -\frac{1}{\pi} \int_{-\infty}^{\infty} \frac{d\omega' \operatorname{Im} \chi(q, \omega')}{i\omega_n - \omega'}$$

$$= -\frac{1}{\pi} \left[ \int_0^{\infty} \frac{d\omega' \operatorname{Im} \chi(q, \omega')}{i\omega_n - \omega'} + \int_0^{\infty} \frac{d\omega' \operatorname{Im} \chi(q, -\omega')}{i\omega_n + \omega'} \right]. \qquad (2.70)$$

In order to sum over $m$ in (2.70), we consider the contour $C = C_1 + C_2$ (Fig. 2.15).

The following formula for $\omega_n = \frac{(2n+1)\pi}{\beta}$ is valid:

$$\sum_{n=-\infty}^{\infty} F(i\omega_n) = -\frac{\beta}{2\pi i} \int_C \frac{F(\omega)d\omega}{\exp \beta\omega + 1} = -\frac{\beta}{2\pi i} \int_C \frac{F(\omega)d\omega}{\exp -\beta\omega + 1}, \qquad (2.71)$$

or

$$\sum_{n=-\infty}^{\infty} F(i\omega_n) = -\frac{\beta}{2\pi i} \frac{1}{2} \int_C F(\omega) \operatorname{th} \frac{\beta\omega}{2} d\omega. \qquad (2.72)$$

Taking into account that $\operatorname{Im} \chi(q, -\omega) = -\operatorname{Im} \chi(q, \omega)$, Eq. (2.70) can be reduced to

$$\chi(q, i\omega_n) = -\frac{1}{\pi} \int_0^{\infty} d\omega' \operatorname{Im} \chi(q, \omega') \left[ \frac{1}{i\omega_n - \omega'} - \frac{1}{i\omega_n + \omega'} \right]. \qquad (2.73)$$

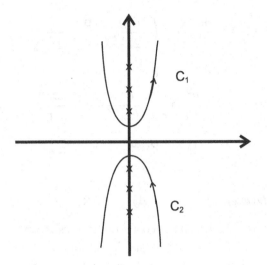

Fig. 2.15.    Integration contour $C = C_1 + C_2$.

With the help of formula (2.72), we can rewrite (2.67) in the form

$$\Delta(\mathbf{k}, i\omega_n) = \frac{g^2}{8\pi\,(2\pi i)} \int \frac{d^2p}{(2\pi)^2} \int_C d\omega F(\mathbf{p}, \omega) \int_0^\infty d\omega' \mathrm{Im}\, \chi(\mathbf{k} - \mathbf{p}, \omega')$$

$$\times \left[ \frac{1}{i\omega_n - \omega - \omega'} - \frac{1}{i\omega_n - \omega + \omega'} \right] th\frac{\beta\omega}{2}, \qquad (2.74)$$

where

$$F(\mathbf{p}, \omega) = \frac{\Delta(\mathbf{k}, i\omega_n)}{\omega^2 - (\varepsilon(\mathbf{p}) - \mu)^2}.$$

After some calculations and transformations, we get the general formula for the superconducting parameter [83–85]:

$$\mathrm{Re}(\mathbf{k}, \omega) = \frac{g^2}{8} \int_{-\pi/a}^{\pi/a} \frac{d^2p}{(2\pi)^2} \left\{ \frac{th\frac{\beta(\varepsilon(\mathbf{p}) - \mu)}{2}}{2(\varepsilon(\mathbf{p}) - \mu)} \right.$$

$$\times \left[ \mathrm{Re}\,\chi(\mathbf{k} - \mathbf{p}, \omega - (\varepsilon(\mathbf{p}) - \mu)) \right.$$

$$\left. + \mathrm{Re}\,\chi(\mathbf{k} - \mathbf{p}, \omega + (\varepsilon(\mathbf{p}) - \mu)) \right]$$

$$\times \mathrm{Re}\Delta(\mathbf{p}, \nu) - 2 \int_0^\infty \frac{d\nu}{\pi} cth\frac{\beta\nu}{2} \mathrm{Im}\, \chi(\mathbf{k} - \mathbf{p}, \nu)$$

$$\times \left[ \frac{(\omega - \nu)^2 - (\varepsilon(\mathbf{p}) - \mu)^2 - \delta^2}{((\omega - \nu)^2 - (\varepsilon(\mathbf{p}) - \mu)^2 - \delta^2)^2 + 4(\omega - \nu)^2 \delta^2} \right.$$

$$\times \operatorname{Re}\Delta(\mathbf{p}, \omega - \nu)$$

$$+ \frac{(\omega + \nu)^2 - (\varepsilon(\mathbf{p}) - \mu)^2 - \delta^2}{((\omega + \nu)^2 - (\varepsilon(\mathbf{p}) - \mu)^2 - \delta^2)^2 + 4(\omega - \nu)^2 \delta^2}$$

$$\left. \times \operatorname{Re}\Delta(\mathbf{p}, \omega + \nu) \right] \Bigg\}, \tag{2.75}$$

$$\operatorname{Re} F(\mathbf{p}, \omega + i\delta) = \operatorname{Re} \Delta(\mathbf{p}, \omega) \frac{\omega^2 - (\varepsilon(\mathbf{p}) - \mu)^2 - \delta^2}{[\omega^2 - (\varepsilon(\mathbf{p}) - \mu)^2 - \delta^2]^2 + \omega^2 \delta^2}. \tag{2.76}$$

If the superconducting gap depends weakly on the frequency, then $\Delta(\mathbf{k}, \omega) \simeq \Delta(\mathbf{k}, 0)$. In this case, we set $\omega = 0$ in Eq. (2.75). Considering the relation $\operatorname{Re} \chi(q, \omega) = \operatorname{Re} \chi(q, -\omega)$ and making the simple transformations, we obtain the equation similar to that in [74]:

$$\Delta(\mathbf{k}) = \frac{g^2}{8} \int_{-\pi/a}^{\pi/a} \frac{d^2 p}{(2\pi)^2} \left\{ \operatorname{Re} \chi(\mathbf{k} - \mathbf{p}, \varepsilon(\mathbf{p}) - \mu) \frac{\operatorname{th} \frac{\beta(\varepsilon(\mathbf{p}) - \mu)}{2})}{(\varepsilon(\mathbf{p}) - \mu)} \right.$$

$$+ 2 \int_0^\infty \frac{d\nu}{\pi} \operatorname{cth} \frac{\beta \nu}{2} \operatorname{Im} \chi(\mathbf{k} - \mathbf{p}, \nu)$$

$$\left. \times \frac{(\varepsilon(\mathbf{p}) - \mu)^2 - \nu^2 + \delta^2}{[(\varepsilon(\mathbf{p}) - \mu)^2 - \nu^2 + \delta^2]^2 + 4\nu^2 \delta^2} \right\} \Delta(\mathbf{p}). \tag{2.77}$$

It is important that our method of calculations yields a more general equation for the superconducting gap than that obtained by D. Pines. Our equation coincides with the Pines equations after some simplification, which is the test for our calculations.

## 2.7.5. *Thermodynamic potential of antiferromagnetic spin fluctuations*

For the free energy, we have the equation

$$F(G) = \beta \Omega = -\frac{1}{2} \operatorname{Tr}[\ln G_o G^{-1} + (G_o^{-1} + \Gamma^o \Gamma_5 \mu) G^{-1}]$$

$$- \frac{g^2}{32} \operatorname{Tr}\{\Gamma^i G T r \Gamma^j G \chi_{ij} - 2G \Gamma^i G \Gamma^j \chi_{ij}\}, \tag{2.78}$$

where $G$ satisfies Eq. (2.57). We now multiply Eq. (2.58) by $G$ and take the trace, Tr. This allow us to obtain

$$\frac{g^2}{8} \text{Tr} \left\{ \Gamma^i G Tr \Gamma^j G \chi_{ij} - 2G\Gamma^i G\Gamma^j \chi_{ij} \right\}$$
$$= -\text{Tr}\{(G_o^{-1} + \Gamma^o\Gamma_5\mu)G - 1\}. \tag{2.79}$$

Using (2.57), we can rewrite relation (2.79) for the functional of free energy calculated with the use of solutions of the Schwinger–Dyson equation in the form

$$\beta\Omega = -\frac{1}{2}\text{Tr} \left[ \ln G_o\Gamma^{-1} + \frac{1}{2}(G_o^{-1} + \Gamma^o\Gamma_5\mu)G - \frac{1}{2} \right]. \tag{2.80}$$

By implementing the Fourier transformation, we get

$$\Omega = \frac{1}{2\beta}V \sum_{n=-\infty}^{\infty} \int_{-\pi/a}^{\pi/a} \frac{d^2k}{(2\pi)^2} \text{Tr} \left[ \ln G_o(\mathbf{k}, iw_n)G^{-1}(\mathbf{k}, iw_n) \right.$$
$$\left. + \frac{1}{2}(G_o^{-1}(\mathbf{k}, iw_n) + \Gamma^o\Gamma_5\mu)G(\mathbf{k}, iw_n) - \frac{1}{2} \right], \tag{2.81}$$

where $V$ is the two-dimensional volume (the area of the cuprate plane), and Tr stands for the trace of a matrix.

For the free energy functional $\Omega(\Delta)$, we have the formula

$$\Omega = -\frac{V}{2\beta} \sum_{n=-\infty}^{\infty} \int \frac{d^2k}{(2*\pi)^2} \text{Tr} \left[ \ln G_o(\mathbf{k}, i\omega_n)^{-1}(\mathbf{k}, i\omega_n) \right.$$
$$\left. + \frac{1}{2}(G_o^{-1}(\mathbf{k}, i\omega_n) + r^o r_5\mu)G(\mathbf{k}, i\omega_n) - \frac{1}{2} \right], \tag{2.82}$$

where $G_o^{-1}(\mathbf{k}, i\omega_n) = r^o i\omega_n - r^o r_5\varepsilon(\mathbf{k})$, and $G(\mathbf{k}, i\omega_n)$ is given by formula (2.57). We note that the functional $\Omega(\Delta, \mu)$ is normalized so that $\Omega(\Delta = 0, \mu = 0) = 0$.

By calculating the trace of the $r$-matrix, we obtain the expression

$$\Omega(\Delta, \mu) = -\frac{V}{\beta} \sum_{n=-\infty}^{\infty} \int \frac{d^2k}{(2*\pi)^2} \left[ \ln \frac{w_n^2 + (\varepsilon - \mu)^2 + \Delta^2}{w_n^2 + \varepsilon^2} \right.$$
$$\left. - \frac{\Delta^2}{w_n^2 + (\varepsilon - \mu)^2 + \Delta^2} \right]. \tag{2.83}$$

By using the formula

$$\sum_{n=-\infty}^{\infty} \ln \frac{\omega_n^2 + b^2}{\omega_n^2 + a^2} = \int_0^{\infty} dx \left[ \frac{\beta}{2\sqrt{a^2 + x}} \operatorname{th} \frac{\beta\sqrt{a^2 + x}}{2} \right.$$

$$\left. - \frac{\beta}{2\sqrt{b^2 + x}} \operatorname{th} \frac{\beta\sqrt{b^2 + x}}{2} \right] = 2 \ln \frac{\operatorname{ch}\frac{\beta b}{2}}{\operatorname{ch}\frac{\beta a}{2}}, \quad (2.84)$$

we get the equation for the thermodynamic potential:

$$\Omega(\Delta) = V \int \frac{d^2 k}{(2\pi)^2} \left\{ -\frac{2}{\beta} \ln \frac{\operatorname{ch}\frac{\beta}{2}\sqrt{(\varepsilon - \mu)^2 + \delta^2}}{\operatorname{ch}\frac{\beta\varepsilon}{2}} \right.$$

$$\left. + \frac{\Delta^2}{2\sqrt{(\varepsilon - \mu)^2 + \Delta^2}} \operatorname{th} \frac{\beta\sqrt{(\varepsilon - \mu)^2 + \Delta^2}}{2} \right\}. \quad (2.85)$$

By making some transformations and calculating the traces of $\Gamma$ matrices, we arrive at the equation

$$\Omega(\Delta) - \Omega(0) = \frac{V}{2} \int_C \frac{dw}{2\pi i} \int \frac{d^2 k}{(2\pi)^2} \Delta_i^4(\mathbf{k}, w)$$

$$\times \frac{1}{e^{Bw} + 1} \frac{1}{[w^2 - (\varepsilon(\mathbf{k} - \mu)^2]^2}. \quad (2.86)$$

By expanding the contour $C$ and calculating the contribution to the integral at poles $w = \pm(\varepsilon(k) - \mu)$, we get

$$\Omega(\Delta) - \Omega(0) = \frac{V}{8} \int \frac{d^2 k}{(2\pi)^2} \frac{\Delta_i^4(\mathbf{k})}{[\varepsilon(\mathbf{k}) - \mu]^2}$$

$$\times \left\{ \frac{\beta}{2} \left( 1 - \operatorname{th}^2 \frac{\beta(\varepsilon(\mathbf{k}) - \mu)}{2} \right) \right.$$

$$\left. - \frac{1}{(\varepsilon(\mathbf{k}) - \mu)} \operatorname{th} \frac{(\varepsilon(\mathbf{k}) - \mu)}{2} \right\}, \quad (2.87)$$

where $\varepsilon(k)$ describes the spectrum of two-dimensional electrons, the free energy $F$ is connected with the thermodynamic potential $\Omega$ by the relation $F = \beta\Omega$; $\Omega(\Delta)$ and $\Omega(0)$ are the thermodynamic potentials at $T < T_c$ and $T > T_c$, respectively; and $V$ is the two-dimensional volume (the area of a cuprate layer).

It is easy to verify that, despite the presence of the factors $(\varepsilon(\mathbf{k}) - \mu)$ in the denominator in formula (2.87), no singularity on the Fermi surface $\varepsilon(\mathbf{k}) - \mu$ is present.

Equations (2.85) and (2.87) (at $T \sim T_c$) can be a basis for calculations of various thermodynamic quantities, including a jump of the heat capacity.

A similar method of calculations was developed by St. Weinberg [89].

## 2.7.6.   *Heat capacity of the d-pairing*

Equation (2.85) yields the following formula for the thermodynamic potential $\Omega(\Delta)$ [83]:

$$\Omega(\Delta) = V \int \frac{d^2k}{(2\pi)^2} \left\{ -\frac{2}{\beta}\ln\frac{\text{ch}\frac{\beta}{2}(\sqrt{(\varepsilon(\mathbf{k}) - \mu)^2 + \Delta(\mathbf{k})^2})}{\text{ch}\frac{\beta\varepsilon}{2}} \right.$$

$$\left. + \frac{\Delta(\mathbf{k})^2}{2\sqrt{(\varepsilon(\mathbf{k}) - \mu)^2 + \Delta(\mathbf{k})^2}}\text{th}\frac{\sqrt{(\varepsilon(\mathbf{k}) - \mu)^2 + \Delta(\mathbf{k})^2}}{2} \right\}. \quad (2.88)$$

Here, $V$ is the two-dimensional volume (the area of a cuprate layer), $\Delta(\mathbf{k})$ is the superconductor gap, $\mathbf{k}$ is the momentum of an electron, and $\varepsilon(\mathbf{k}) = -2t[\cos(k_x a) + \cos(k_y a)]$ gives the spectrum of two-dimensional electrons. The heat capacity is calculated by the formula

$$C = -T\frac{\partial^2\Omega}{\partial T^2}. \quad (2.89)$$

The results of calculations of the heat capacity $C$ are given in work [45]. We have carried out the computer-based calculations of the heat capacity of the compound $YBa_2Cu_3O_{6.63}$. The equation for the superconductor gap $\Delta$ was obtained in Sec. 2.7.5. The task of solving the integral equation was reduced to that of an algebraic equation which was solved by the method of iterations.

The equation for the superconductor gap depends on the spin correlation function. For it, the necessary data were taken from works [73, 74]. In calculations, we used the following values of parameters of the correlator: $\omega_{SF}(T_C) \approx 7.7\,\text{meV}$, $\chi_S(Q) = 44\,\text{eV}$, $\xi/a \approx 2.5$, $t = 2\,\text{eV}$, $\mu = 0.25$, $T_c = 95\,\text{K}$. We chose such a value of the constant $g$ which satisfies the relation $2\Delta/kT_c = 3.4$ like that in works [70, 71]. The results of numerical calculations are presented in Fig. 2.15, where we can see the temperature dependence of the electron heat capacity: 1 — the curve which is an approximation of the results of computer-based calculations (crosses); 2 — the curve which describes the exponential BCS dependence. These results are in good agreement with experiments made by the group of A. Kapitulnik [90] at low temperatures.

The calculations showed a square dependence of the heat capacity on the temperature in the temperature range from zero to the superconducting temperature. It was shown [45, 46, 83, 84] that the dependence has the form

$$C = AT^2,\qquad(2.90)$$

where $A = 0.126\,\text{J/K}^3\,\text{mole}$.

### 2.7.7.  *Heat capacity jump in superconductors*

In what follows, we present the results of calculations of a jump of the heat capacity near the critical temperature,

$$\Delta C = -T\frac{\partial^2 \Delta\Omega}{\partial T^2},\quad \beta = \frac{1}{T},\qquad(2.91)$$

and evaluate the parameter $R = \Delta C/\gamma T_c$. Omitting the awkward and quite complicated details, we will give the final results.

We obtained $R = 1.6$. This value of the parameter is larger than that in the BCS theory, where $R = 1.43$. It is worth noting that the heat capacity jump is very sensitive to the doping. Figure 2.16 shows the results of recent calculations of one of the authors [85] for $R$ as a function of the doping. It is seen that this parameter depends strongly on the doping. By using these

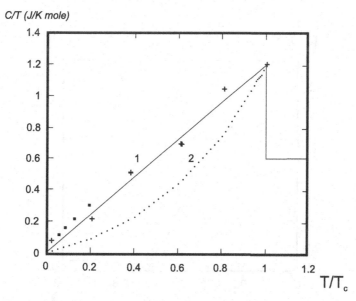

Fig. 2.16. Temperature dependence of the electron heat capacity, where 1 — the curve which is an approximation of the results of computer-based calculations (crosses); 2 — the curve which describes the exponential BCS dependence, points give the experimental data [90].

results of calculations, it is possible to evaluate the condensation energy for high-temperature superconductors which is proportional to $R^2$ (see [86]).

The calculations imply that the heat capacity depends on the temperature as $T^2$. Analogous results were obtained for the physics of heavy fermions [95]. It was shown in this work that $d$-symmetry leads to a linear dependence of $C/T$ on the temperature, whereas the $s$-symmetry is related to the exponential dependence of this quantity on the temperature. The linear dependence of $C/T$ on the temperature was observed in the experimental works [91–94] for YBaCuO superconductors, The experimental results are presented in Fig. 2.17. As was mentioned above, the problem of the determination of the symmetry of a gap in cuprate superconductors is urgent at the present time. Many experiments have confirmed the $d$-symmetry of the pairing [71, 72]. In particular, we mention new experiments [72], in which the researchers have studied the quantization of magnetic flows in a ring which includes three Josephson junctions. These works supporting the existence of $d$-pairing in high-temperature superconductors were marked by the Barkley prize. It was also shown that the sign of the order parameter in YBa$_2$Cu$_3$O$_{7-\delta}$ depends on the direction. This dependence corresponds to the $d_{x^2-y^2}$-symmetry.

The temperature dependence of the electron heat capacity obtained in our calculations [83, 84] is related to the $d$-pairing. These thermodynamic

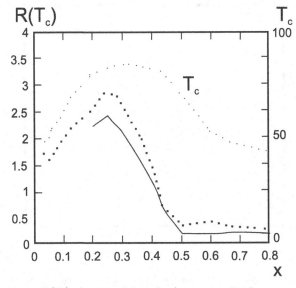

Fig. 2.17.   Parameter $R(T_c)$ characterizing the heat capacity jump as a function of the doping for Y$_{0.8}$Ca$_{0.2}$Ba$_2$Cu$_3$O$_{7-x}$. The solid curve presents the results of theoretical calculations, the points mark the experimental data [93].

calculations can be a supplementing test in the determination of the pairing symmetry in high-temperature superconductors.

It should be noted that the thermodynamic calculations clarifying the behavior of high-temperature superconductors have been also performed in [96–99] in the frame of the other mechanisms of pairing.

## 2.8. Summary

The high-temperature superconductivity is a dynamical field of solid-state physics which is intensively developed by theorists and experimenters. By summarizing the physical properties and the mechanisms of superconductivity in new high-temperature superconductors, we should like to separate the main properties and the theoretical problems arising in the studies of high-temperature superconductors.

In order to comprehend the nature of the superconducting state, it is necessary to construct a consistent microscopic theory which would be able to describe superconductive and normal properties of high-temperature superconductors. It is seen from the above-presented survey that many mechanisms of pairing in high-temperature superconductors, which pretend to the explanation of this phenomenon, have been advanced. On the whole, the most probable seems to be the "synergetic" mechanism of pairing, whose constituents are the electron–phonon interaction, spin-fluctuation, and other types of interaction in cuprate planes.

We believe that the known challenging properties, which are contradictory to a certain extent, of many chemical compounds in the superconducting and normal phases can be explained only by considering the interaction of all the degrees of freedom such as lattice-, electron-, and spin-related ones. In this case, it is also necessary to take the complicated structure of high-temperature superconductors into account. The further development of the theory will require not only the execution of bulky numerical calculations, but also the solution of a number of fundamental problems concerning the strong electron correlations.

It seems to us that one of the key problems in the field of superconductivity is the mechanism and the symmetry of pairing.

Above, we have shown that some thermodynamical problems of high-temperature superconductivity, in particular the problem of antiferromagnetic spin fluctuations, can be efficiently solved within the method of continual integrals. We would like to conclude by the following remarks:

(1) The thermodynamic theory of antiferromagnetic spin fluctuations in high-temperature superconductors is constructed. The method of

functional integration is applied to the calculations of the superconductor gap and the thermodynamic potential. The Schwinger–Dyson equation and the equation for the free energy are deduced.

(2) From the Schwinger–Dyson equation, the equations for the superconductor gap, which are used in the numerical calculations within the thermodynamics of high-temperature superconductors, are constructed. The analytic formulas for the thermodynamic potential and its jump are obtained.

(3) The numerical calculations of the temperature dependence of the electron heat capacity are carried out, and it is shown that the electron heat capacity is proportional to the square of the temperature. It is emphasized that such a temperature dependence is related to the *d*-pairing. It is shown that the measurement of the temperature dependence of the heat capacity can be a supplementing test in the determination of the type of a pairing symmetry in high-temperature superconductors. The jump of the heat capacity of cuprate superconductors near the critical temperature is evaluated as well.

# CHAPTER 3

# Iron Superconductors

## 3.1. Introduction

At the beginning of 2008, the huge interest was induced by the discovery of a new class of high-temperature superconductors, namely, layered compounds on the basis of iron. It gave hope for the progress in the synthesis of new high-temperature superconductors up to room-temperature superconductors [100]. At present, the nature of the high-temperature superconductivity (HTS) is not yet completely understood. The main problem of the solid-state physics is the production of superconductors that ensure the superconductivity at room temperatures. Now, the large-scale search for such high-temperature superconductors is continued.

This discovery broke the "monopoly" of cuprates in the physics of high-temperature superconductors and inspired new hopes for the close success in the synthesis of new superconductors and for a more profound theoretical comprehension of the mechanisms of HTS.

The "iron age" was short, and the term "HTS" is now equally related to cuprate and iron superconductors.

Three decades ago, the researchers moved by touch. But, at the time of the discovery of iron superconductors, they have accumulated a significant experience of studies of complex compounds and have had the newest experimental methods and the algorithms of calculations of the electronic structure. Moreover, more powerful computational facilities appear, and, what is the principal point, the physical ideas developed in the studies of cuprates can be applied to the new field of studies. The progress in experimental and theoretical studies in this field is quite impressive. The goal of the present chapter is to give a short introduction into the physics of new superconductors on the basis of layered compounds of iron. We will analyze the physical properties and electronic models of the new class of high-temperature superconductors in layered compounds on the basis of iron. Despite the different chemical compositions and the difference in the crystal structures, they have similar physical properties determined by electron carriers in FeAs layers. The exceptional interest in them is explained by perspectives of the practical application, though the critical temperature of iron compounds did not exceed the liquid nitrogen temperature. Now, the record among the monocrystals belongs to $SmFeAsO_{1-x}F_x$ with $T_c = 57.5\,K$. A high optimism was also induced by the discovery of the superconductivity at $T_c = 100\,K$ in one-layer films of FeSe [101, 102].

## 3.2.   Structural Properties of Iron Superconductors

At the present time, three classes of compounds containing FeAs planes are under study: LaOFeAs, $AFe_2As_2$, LiFeAs and the analogous compound FeSe, in which the superconductivity was observed at high $T_c$. The physical properties of these compounds are similar in many aspects and are determined by a one-type crystal structure including the common element such as FeAs layers. We now consider the physical properties of these compounds. Among the iron compounds possessing the superconductivity, we also separate the following two classes: pnictides and chalcogenides. Pnictides are compounds containing elements of the V group of the periodic table such as nitrogen, phosphorus, and arsenic. The basic element of those compounds is a square lattice of iron. Pnictides can be one-layer such as 1111 (LaFeAsO, LaFePO, $Sr_2VO_3FeAs$) and 111 (LiFeAs, LiFeP) and two-layer of type 122 with two layers of FeAs ($BaFe_2As_2$, $KFe_2As_2$). Chalcogenides include compounds of type 11 ($Fe_{1-d}Se$, $Fe_{1+y}Te_{1-x}Se_x$ films of FeSe) and type 122 ($KFe_2Se_2$). The structure and physical properties of compounds of iron are considered in many reviews [103–105] in details.

## 3.3.   Compounds of the Type ReFeAsO

The crystal structure of compound ReFeAsO is shown in Fig. 3.1 [106]. Such substances have a tetragonal structure at room temperature with the space group P4/nmm. The crystal structure is formed by alternating layers of FeAs manifesting the antiferromagnetism that are separated by layers of ReO. The FeAs layer is really represented by three closely placed atomic planes that form a square lattice of Fe atoms. From top and from bottom, we see the square lattices of As atoms that are located relative to the Fe-plane so that Fe atom is surrounded by a tetrahedron of As atoms. In other words, the FeAs layer is formed by $FeAs_4$-complexes. The distance between FeAs and LaO layers is equal to $1.8\,\text{Å}$.

The results of calculations of the electron spectrum can be found in [107]. It is worth to note that the specific features of the electron spectrum are

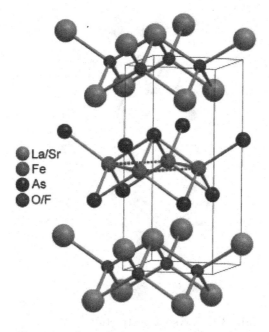

Fig. 3.1.   Crystalline structure of ReFeAsO.

connected with the quasitwo-dimensional character of the superconductivity in the FeAs planes. As an example, we mention compound $SmFeAsO_{1-x}F_x$ whose the highest temperature $T_c = 57.5\,\mathrm{K}$ is attained at $x = 0.1$.

## 3.4.   Compounds of the Type $AFe_2As_2$ (A = Ba, Sr, Ca)

In a wake of compounds of the type ReFeAsO which are input ones for the creation of superconductors with high $T_c$, compounds $BaFe_2As_2$ [108] and $SrFe_2As_2$ [109] were synthesized. After the doping, they became superconductors. The discovery of the superconductivity on compound $Ba(1-x)K_xFe_2As_2$ caused a new splash of studies of the systems on the basis of $FeAs_4$. In Fig. 3.2, we present the crystal structure of $BaFe_2As_2$. It is a tetragonal structure with the space group I4/mm and is composed from AsFe planes separated by Ba planes. Compound LaFeAsO has one Fe plane in the elementary cell, wheres $BaFe_2As_2$ has two such planes. By the data in [111], the parameters of the crystal lattice in the last compound are: $a = 3.9090\,\text{Å}$, $c = 13.2121\,\text{Å}$. Thus, the cell size in the basis plane is close to that in compound ReFeAsO, whereas the size in the direction along the $c$-axis is significantly larger.

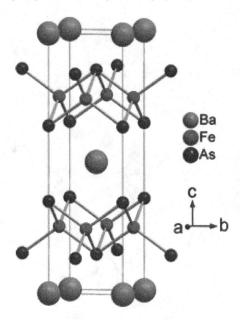

Fig. 3.2.   Crystalline structure of $AFe_2As_2$.

## 3.5.   Compounds of the Type AFeAs (A = Li)

After compounds ReFeAsO and $AFe_2As_2$, a new compound from the FeAs-group, LiFeAs, was synthesized. In it, the superconductivity with a sufficiently high $T_c = 18\,K$ was measured [110]. The detailed structural and physical studies of the compound [110] showed that LiFeAs belongs to the class of compounds containing the layers of LiFeAs such as in ReFeAsO and $AFe_2As_2$. The FeAs planes are separated by planes with Li. (Fig. 3.3). Contrary to all available undoped FeAs-compounds, LiFeAs reveals no magnetic instability, and the superconductivity in it exists in the absence of any doping. Its charge carriers are electrons. The important information about properties of LiFeAs compounds is given by the results of measurements of $T_c$ under pressure [111]. It is of interest that the superconductivity in these compounds decreases under the applied pressure. It is worth to note that the lattice parameter and the volume of an elementary cell in LiFeAs are comparatively less than in other iron superconductors.

## 3.6.   Compounds of the Type FeSe and FeTe (iron selenide and iron telluride)

A new family of superconductors on the basis of iron was found in FeSe and FeTe compounds whose $T_c$ increases at the doping and under pressure [112].

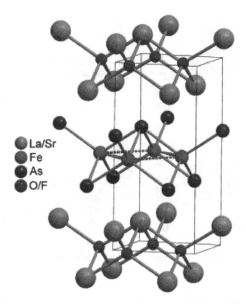

La/Sr
Fe
As
O/F

Fig. 3.3. Crystal structure of LiFeAs [110].

Among all superconductors on the basis of iron, the highest attention is attracted to FeSe due to the simplicity of the crystal structure and remarkable physical properties (Fig. 3.4). The group of scientists from the Oak Ridge National Laboratory carried out the theoretical study of properties of a new group of superconductors FeX, by using the widely spread method of solid state physics, namely, the density functional theory, and showed that those substances have the same structural, magnetic, and electron–phonon effects as iron oxypnictides that were opened earlier.

It was also shown that FeSe is a representative of the whole superconducting family FeX (X = Se, Te, S). The simplicity of the crystal lattice and the nontoxicity of compounds FeX favor the study and prediction of the superconductivity characteristics of substances containing iron.

The crystal structure of iron selenide has many common features with the earlier considered compounds with FeAs layers. Compound FeSe also consists of FeSe-planes positioned one above another. Moreover, nothing is placed between those planes. The atoms of Fe form a square lattice, and each Fe atom is located in the tetrahedral environment of Se atoms (or Te).

These compounds are not superconductors at the atmospheric pressure. The superconductivity arises at the doping or under pressure. Compound $FeSe_{1-x}$ for $x = 0.12$ reveals the superconductivity with $T_c = 8\,K$. In compounds FeSe at the pressure $P = 1.48\,GPa$, the superconductivity was observed at $T_c = 28\,K$.

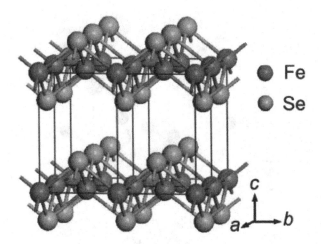

Fig. 3.4.    Crystalline structure of FeSe.

The detailed studies of the structural and superconductivity properties of FeSe at pressures up to 26 GPa were executed in work [113]. At $P = 12$ GPa, the phase transition from the tetragonal to orthorhombic phase was found. The highest value $T_c = 34$ K was obtained in the orthorhombic phase at the pressure $P = 22$ GPa.

### 3.6.1.    *FeSe films*

FeSe films deserve a separate consideration. Recently, a number of high-temperature superconductors on the basis of FeSe monolayers were discovered. First of all, we mention the systems of the following type: a monoatomic FeSe layer on a substrate of the type $SrTiO_3$ (FeSe/STO).

At the present time, the record value of the superconductivity temperature among one-layer FeSe films equals $T_c = 100$ K [101, 102].

Apparently, the question about the superconductivity in films was first considered by Gor'kov [114, 115] who advanced the idea of a possible mechanism of enhancement in the superconductivity transition temperature $T_c$ in the system FeSe/STO due to the interaction with high-energy phonons in $SrTiO_3$ [103].

If the experimental sample is made in the form of a thin film, like famous graphene, then the transition temperature becomes as high as 100 K. This is of great importance, since we can go beyond the boiling temperature of liquid nitrogen, which looks tremendously, by opening the amusing perspectives. It should be added that the just film technologies have high potentialities for the applications.

## 3.7. Superconductivity

Soon after the discovery of the superconductivity in $LaO_{1-x}F_xFeAs$, it was reported that the external pressure applied to compounds with $x = 0.11$ increases $T_c$ maximally to 43 K at a pressure of 4 GPa. It was assumed that the compression of the lattice is responsible for the observed effect. Indeed, in compound ReFeAsO, the atoms of the rare-earth element have a less radius than that of La atom, and $T_c$ is significantly higher than 55 K. In many works, an increase in $T_c$ caused by the replacement of La by other rare-earth elements is frequently called the "chemical pressure effect" [103] characteristic of high-temperature superconductors [104]. The replacement of various rare-earth ions in a number of compounds ReFeAsO, as distinct from cuprate superconductors ReBaCuO, leads to a rather large dispersion of values of $T_c$. In order to explain this phenomenon, two scenarios can be proposed. First, the effect can be caused by the disorder arising at the replacement of rare-earth elements and the realization of an anomalous pairing in FeAs (e.g., anisotropic pairing, $d$-pairing). Second, we may consider the magnetic mechanism of pairing in iron superconductors. The spins of rare-earth ions strongly interact with the Fe spins and can significantly affect the spectrum of spin fluctuations in FeAS planes, which increases $T_c$ by the magnetic mechanism. In addition to thee doping, the physical properties of iron superconductors are changed significantly under pressure. The pressure applied to FeAs compound induces the bulk structural and magnetic phase transition. In this case, we observe the tendency to the appearance of the superconductivity. In work [116], the dependence of the superconductivity transition temperature on the height of an anion for typical superconductors on the basis of iron was determined. It is shown in Fig. 3.5. The large symbols denote the initial temperature. This dependence is used in one of the key strategies aimed at the comprehension of the mechanism of superconductivity in substances on the basis of iron and the search for new superconductors on the basis of iron with higher $T_c$.

## 3.8. Electronic Structure

### 3.8.1. *Multiorbital and multizone cases*

In the explanation of the physical properties of superconductors, the significant role is played by first-principles calculations of the electronic structure. It is worth to note that one of the basic properties of high-temperature superconductors is the presence of many orbitals and many zones [117]. The superconductors on the basis of iron are represented by quasitwo-dimensional

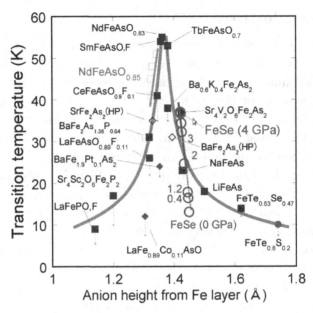

Fig. 3.5. Dependence of $T_c$ on the height of an anion for typical superconductors on the basis of Fe. Large symbols denote the initial temperature. The temperatures with zero resistance at the ambient pressure are shown by small blue circles. Filled symbols indicate the data at a certain pressure. Open symbols correspond to $SrFe_2As_2$ and $BaFe_2As_2$ at the optimum pressure. Open squares denote the data on $NdFeAsO_{0.85}$ at a high pressure. The data on FeSe at a high pressure are given by uncolored circles. The filled circle indicates the data on $FeTe_{0.8}S_{0.2}$.

substances in which the conducting plane is the square lattice of Fe atoms. At once after the discovery of iron superconductors, the detailed calculations of the zone structure of those superconductors were carried out, first of all, within the general method of local density approximation (LDA). In [107], the *ab initio* calculations of the zone structure were executed for a number of oxypnictides of the type ReFeAsO. In Fig. 3.6, we present the electron spectra of LaFeAsO and PrFeAsO along the basic directions in the Brillouin zone. It is seen that the difference in spectra related to the replacement of a rare-earth ion is quite insignificant. In a rather interval of energies near the Fermi level that is of importance from the viewpoint of the formation of the superconducting state, the spectra are practically identical. The forms of the partial densities of states (Fig. 3.7) imply that the density of states near the Fermi level is practically completely determined by the $d$-state of Fe (at a rather insignificant contribution of the $p$-state of As). E may say that all phenomena related to the superconductivity are realized on the square Fe lattice of compounds FeAs.

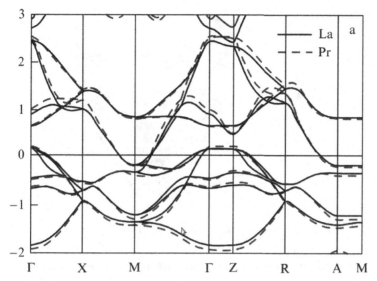

Fig. 3.6. Electron spectra of LaFeAsO and PrFeAsO along the directions with high symmetry in the Brillouin zone obtained by LDA.

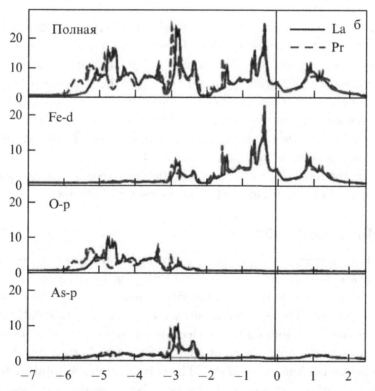

Fig. 3.7. Comparison of the total density of states of electrons and the partial densities of states of LaFeAsO and PrFeAsO.

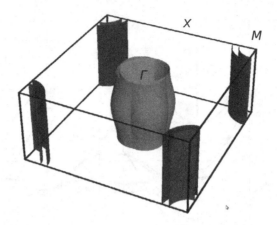

Fig. 3.8.    The Fermi surface of compound LaFeAsO.

The results of the first calculations within the density functional theory showed that $3d_6$ states of $Fe_2$ are dominant. In this case, all five orbitals $d_{x^2-y^2}$, $d_{xy}$, $d_{xy}$, $d_{xz}$, $d_{yz}$ lie on the Fermi surface or near it. Such situation leads to the existence of many orbitals and many zones in the low-energy electronic structure.

The structure of dispersion curves crossing the Fermi level determines the Fermi surface that is a multisheet one for compounds of the type LaFeAsO. The Fermi level crosses two hole zones outgoing from point Γ and two electron zones starting from point M. It is worth to mention the planar character of the curves in direction ΓZ, which indicates a weak dependence of hole quasiparticles on the momentum kF. Therefore, the Fermi surface in a vicinity of point Γ has hole character. The same is true for sheets of the Fermi surface in a vicinity of point M (Fig. 3.8). Thus, two hole cylindrical sheets with the axis along direction ΓZ and two electron sheets along axis MA are formed on the Fermi surface of compound LaFeAsO.

## 3.9.   Minimum Model

The relative simplicity of the spectrum of FeAs superconductors near the Fermi surface forces one to think about the construction of an analytic model of a spectrum that would describe the properties of electrons near the Fermi surface. The above-presented results of zone calculations imply that the highest contribution to the electron density of states near the Fermi level is given by $3d$-states of Fe atoms. Among five $d$-orbitals, we separate two degenerate orbitals $d_{xz}$ and $d_{yz}$. The overlapping of orbitals determines the values of matrix elements of the hops between them for the nearest and next-nearest neighbors. If we consider only these two orbitals, we get the two-orbital model of FeAs compounds. It is called the minimum model characterized by the Hamiltonian

$$H_0 = \sum_{\mathbf{k}\sigma} \psi_{\mathbf{k}\sigma}^+ T(\mathbf{k}) \psi_{\mathbf{k}\sigma}, \tag{3.1}$$

where $\psi_{\mathbf{k}\sigma}^+ = (c_{x\mathbf{k}\sigma}^+, c_{y\mathbf{k}\sigma}^+)$ is the two-component spinor composed of the operators of creation $c_{x\mathbf{k}\sigma}^+$ and $c_{y\mathbf{k}\sigma}^+$ of electrons, respectively, on the orbitals $d_{xz}$ and $d_{yz}$, and $T(\mathbf{k})$ is a two-row matrix composed of matrix elements of the transitions shown in the following equation:

$$T(\mathbf{k}) = \begin{pmatrix} \varepsilon_x(\mathbf{k}) - \mu & \varepsilon_{xy}(\mathbf{k}) \\ \varepsilon_{xy}(\mathbf{k}) & \varepsilon_y(\mathbf{k}) - \mu \end{pmatrix}. \tag{3.2}$$

Here,

$$\varepsilon_x(\mathbf{k}) = -2t_1 \cos k_x - 2t_2 \cos k_y - 4t_3 \cos k_x \cos k_y,$$

$$\varepsilon_y(\mathbf{k}) = -2t_2 \cos k_x - 2t_1 \cos k_y - 4t_3 \cos k_x \cos k_y, \tag{3.3}$$

$$\varepsilon_{xy}(\mathbf{k}) = -4t_4 \sin k_x \sin k_y.$$

Hamiltonian (3.1) that is a quadratic form of Fermi operators is diagonalized with the help of the canonical transformation of the operators $c_{s\mathbf{k}\sigma}$ $(s = x, y)$ to the $\gamma_{s\mathbf{k}\sigma}$:

$$c_{s\mathbf{k}\sigma} = \sum_{\nu=\pm} a_\nu^s(\mathbf{k}) \gamma_{\nu\mathbf{k}s}. \tag{3.4}$$

Here,

$$a_-^x(\mathbf{k}) = \left[ \frac{1}{2} \left( 1 + \frac{\varepsilon_-(\mathbf{k})}{\sqrt{\varepsilon_-^2(\mathbf{k}) + \varepsilon_+^2(\mathbf{k})}} \right) \right]^{1/2} = a_+^y,$$

$$a_+^x(\mathbf{k}) = \left[ \frac{1}{2} \left( 1 - \frac{\varepsilon_-(\mathbf{k})}{\sqrt{\varepsilon_-^2(\mathbf{k}) + \varepsilon_+^2(\mathbf{k})}} \right) \right]^{1/2} = -a_+^y. \tag{3.5}$$

With regard for the chemical potential, the Hamiltonian of the model takes the form

$$H_0 = \sum_{\mathbf{k}\sigma} \sum_{\nu=\pm} E_\nu(\mathbf{k}) \gamma_{\nu\mathbf{k}\sigma}^+ \gamma_{\nu\mathbf{k}\sigma}, \tag{3.6}$$

where the zone energies

$$E_\pm(\mathbf{k}) = \varepsilon_+(\mathbf{k}) \pm \sqrt{\varepsilon_-^2(\mathbf{k}) + \varepsilon_{xy}^2(\mathbf{k})} - \mu, \tag{3.7}$$

and $\varepsilon_\pm(\mathbf{k}) = \frac{1}{2} [\varepsilon_x(\mathbf{k}) \pm \varepsilon_y(\mathbf{k})]$.

We now introduce the one-electron Matsubara Green function

$$\hat{G}^{\sigma}(\mathbf{k}, \tau) = -\langle \hat{T}_{\tau} \psi_{\mathbf{k}\sigma}(\tau) \psi_{\mathbf{k}\sigma}^{+}(\tau) \rangle, \tag{3.8}$$

After some calculations, it takes the following form:

$$\hat{G}^{\sigma}(\mathbf{k}, i\omega_n) = \frac{1}{(i\omega_n - E_+(\mathbf{k}))(i\omega_n - E_-(\mathbf{k}))}$$

$$= \begin{pmatrix} i\omega_n - \varepsilon_x(\mathbf{k}) - \mu & -\varepsilon_{xy}(\mathbf{k}) \\ -\varepsilon_{xy}(\mathbf{k}) & i\omega_n - \varepsilon_y(\mathbf{k}) - \mu \end{pmatrix}. \tag{3.9}$$

The poles $E_+(\mathbf{k})$ and $E_-(\mathbf{k})$ of the Green function determine the spectrum of quasiparticles on the model. There are two branches whose the mutual position and the shape of dispersion curves depend on the values of matrix elements $t_1$, $t_2$, $t_3$, $t_4$.

They should be chosen from the condition that the formulas give the spectrum consistent with the results of numerical calculations for FeAs compounds. Spectrum (3.7) is shown in Figs. 3.9(a) and 3.9(b) in the extended and reduced Brillouin zones. The Fermi surface is approached by

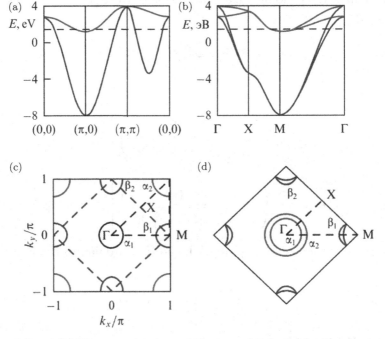

Fig. 3.9.   (a) and (b) The zone structure of the two-orbital model with parameters $t_1 = -1$, $t_2 = 1.3$, $t_3 = t_4 = -0.85$ and the chemical potential $\mu = 1.45$. (c) and (d) The Fermi surface in the two-orbital model for the extended (a) and reduced (b) Brillouin zones. (Here $\alpha_{1,2}$ are hole Fermi surfaces determined by the condition $(E - (k_F) = 0$ and $\beta_{1,2}$ are electronic Fermi surfaces determined by $E + (k_F) = 0$.

two branches of hole quasiparticles in a vicinity of point $\Gamma$ and two branches of electron ones in a vicinity of point M. The Fermi surface is shown in Figs. 3.9(c) and 3.9(d) in the extended and reduced Brillouin zones. The model under study gives two hole pockets in a vicinity of point $\Gamma$ and two electron ones in a vicinity of point M. Therefore, it can be called the minimum model of FeAs compounds. We dealt only with the kinetic part of the energy of the system. To describe various orderings of the model phases, it is necessary supplement the Hamiltonian by terms describing the interaction, first of all, the Coulomb repulsion of electrons on one node or the exchange interaction on different nodes. The first-principles calculation of the electron structure plays a crucial role in the explanation of physical properties of superconductors. The presence of many zones is one of the principal properties of high-temperature superconductors [117]. It was shown in [104] that the information about the electron spectrum and the density of states near the Fermi surface is of a high importance for the formation of a superconducting state.

## 3.10. Experimental Study of the Fermi Surface

There are available two methods of experimental determination of the Fermi surface in metals: one method is based on the de Haas–van Alphen (dHvA) effect, and the second one is referred to photoelectron spectroscopy with angular resolution (ARPES). In the first case by the measured osciillations of the magnetization in a magnetic field, we can get the information about the cross-section areas of the Fermi surface. This method is of high accuracy, but does not attach the measured area to the $k$-space of the Brillouin zone. The ARPES method has a less accuracy, but allows one to directly restore the Fermi surface in the $k$-space. Both methods have own advantages and shortcomings, supplement each other, and are significant for the construction of the pattern of superconducting pairing. LDA calculations of compounds ReOFeAs show the structure of the Fermi surface: it consists of two cylindrical hole pockets centered at the nodes of the Brillouin zone. In addition, the three-dimensional hole pocket centered at point M is observed near point $\Gamma$.

The experimental verification of these conclusions was performed with the help of ARPES [118–120]. For example, the ARPES spectra were measured in [121] on a monocrystal $NdO_{1-x}F_xFeAs$. They show the presence of pockets of the Fermi surface near points $\Gamma$ and M of the Brillouin zone in agreement with numerical first-principles calculations (Fig. 3.6). In parallel, the Fermi surface was studied with the help of the measurement of the dHvA effect in which the quantum oscillations of the magnetization in a magnetic field are determined. It was shown that the iron superconductors are characterized by cylindrical hole surfaces centered at point Γ and two electron ones centered at point M. The detailed comparison of the results of LDA calculations with ARPES spectra shows a good agreement of theory and experiment, which is confirmed by the studies of the dHvA effect.

## 3.11.  Antiferromagnetism

The possibility of the antiferromagnetic ordering in the studied systems was noticed before the implementation of neutronographic studies. Compounds ReOFeAs are antiferromagnetics. The first indications of the possibility of the magnetic ordering in LaOFeAs were given by the measurements of the temperature dependences of the electric resistance and the magnetic susceptibility that demonstrated some anomalies at the temperature $T = 150$ K. At this temperature, the structural transition from the tetragonal into orthorhombic phase was found. The complete neutronographic studies refined the situation [121]. It turned out that the magnetic phase transition arises somewhat lower at $T_N = 137$ K. The main result of studies of LaOFeAs is the magnetic structure that is realized on Fe atoms. This magnetic structure is the antiferromagnetic alternation of ferromagnetic chains in the basis plane and in the perpendicular direction. This structure is in agreement with theoretical predictions following from the calculation of the zone structure of a compound [122, 123]. In this case, the compound itself is a metal. The magnetic ordering in FeAs compounds is different from that in cuprates. The cuprate superconductors demonstrate the ordering in the basic layers, where the magnetic moments of Cu are antiparallel to moments on the nearest neighbors of the Cu sublattice. The cuprate superconductors are Mott dielectrics. The doping of cuprates causes the fracture of the long-range magnetic order and the appearance of the superconducting state. An analogous situation holds in FeAs compounds. The neutronographic studies of the doped superconductor LaOFeAs showed the absence of long-range magnetic order. Thus, the superconductivity arises, like in cuprates, near the magnetic phase transition, which indicates the significant role of antiferromagnetic fluctuations at the electron pairing.

At low temperatures, the spin density wave (SDW) arises in iron superconductors. SDW is some ordering that affects the electron spectrum, density of states, and dispersion curves. A significant task is the calculation of the electron spectrum at a doping. The electron doping should increase the electron pockets of the Fermi surface, whereas the hole doping causes an increase in hole pockets. In this case, we need to calculate the compounds with a given level of doping.

## 3.12. Phase Diagrams

Analogously to superconducting cuprates, the properties of superconductors on the basis of iron are sharply changed at the doping. The parent compounds FeAs are, as a rule, metals (as distinct from cuprates that are dielectrics), but, like cuprates, are ordered antiferromagnetically. Such order is called frequently the spin density waves (SDWs). The superconductivity arises at any electron doping, which causes the appearance of SDWs. Generally, the phase diagram is analogous to that of cuprates.

At the doping of compounds in which the magnetic structure SDW is realized, the temperature $T_N$ of the magnetic ordering gradually decreases, and the superconducting state appears in the system. It is of interest to establish the boundaries of all three phases — magnetic, structural, and superconducting ones. The local methods such as Mössbauer spectroscopy give information about the long-range magnetic order, but are not sensitive to structural distortions. The X-ray diffraction analysis enables one to study the crystal structure, but is not sensitive to the magnetic order. Only the neutron spectroscopy gives possibility to study both aspects simultaneously.

The five-orbital model from work [105] well reproduces the results of calculations of the Fermi surface consisting of four pockets: two hole pockets near the point $(0, 0)$ and two electron ones at the points $(0, p)$ and $(p, 0)$ (Fig. 3.11). Such geometry in the $k$-space leads to the possibility of the appearance of a spin density wave due to the nesting between the hole and electron Fermi surfaces on the wave vector $\mathbf{Q} = (p; 0)$ or $\mathbf{Q} = (0; p)$.

In Fig. 3.10, we show schematically the phase diagram and Fermi stufaces for various levels of doping. Depending on the topology and the mutual

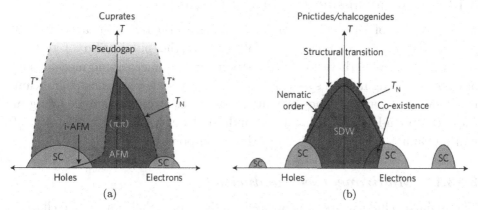

Fig. 3.10. Schematic phase diagrams of cuprates and pnictides at the hole or electron doping.

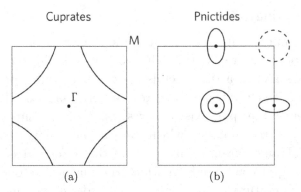

Fig. 3.11.   Schemes of two-dimensional cross-sections of the Fermi surfaces for cuprates and pnictides: (a) for weakly doped cuprates, the Fermi surface has one sheet, and the filled states (that are closer to point $\Gamma$) occupy about a half of the Brillouin zone; (b) for pnictides, the Fermi surface consists of several sheets: two pockets centered at $\Gamma$, electron pockets centered at $(0, \pi)$ and $(\pi, 0)$, and a hole pocket centered at $M = (\pi, \pi)$.

exchange of hole and electron pockets, we can observe the appearance of the competition between gaps of the $s$- and $d$-types.

The phase diagram of iron superconductors differs rather noticeably by the concentration of a doping element from the phase diagram of cuprates (Fig. 3.10). In Fig. 3.11, we present the schemes of two-dimensional cross-sections of the Fermi for cuprates and pnictides [124]. We recall that the HTS cuprates are antiferromagnetic dielectrics (Mott strongly correlated ones), and the new Fe superconductors are antiferromagnetic metals.

The superconductivity in systems on the basis of iron can be induced not only by the doping, but also under pressure.

## 3.13.   Mechanisms and Types of Pairing

The discovery of the high-temperature superconductivity in layered compounds on the basis of iron was followed by the publication of tens of theoretical works with various propositions on the microscopic mechanisms of Cooper pairing in such systems. We dwell only on several most significant mechanisms of pairing. The calculations of the electron–phonon coupling in those compounds showed that the standard electron–phonon mechanism of pairing cannot ensure so high $T_c$ in those compounds.

### 3.13.1.   *Multizone superconductivity*

The main specific feature of new superconductors is their multizone character. The electron structure in a sufficiently narrow vicinity of the Fermi level

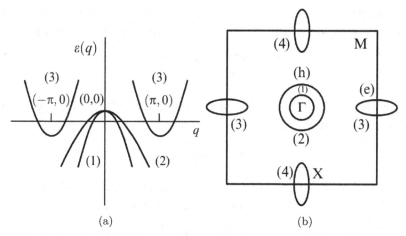

Fig. 3.12.   Schematic view of the electron spectrum: (a) Fermi surfaces; (b) iron superconductor in the scheme of extended zones. Around point $\Gamma$, there are two hole surfaces, and the electron surfaces are at points X.

is formed practically only by $d$-states of Fe. The Fermi surface consists of several hole and electron cylinders. On each cylinder, the energy gap can be formed. Such situation is not new and will be considered in Chapter 4 "Multizone model of superconductivity". The schematic views of the electron spectrum and Fermi surface are shown in Fig. 3.12 [111]. The electron doping decreases the sizes of hole channels, and the hole doping decreases the sizes of electron ones.

In work [125], the symmetry analysis of possible types of the super-conductivity parameters of order is presented, and the BCS solutions are constructed. The electron spectrum of the system is given in Fig. 3.12.

The BCS pairing interaction in the problem under study can be represented by the matrix

$$V = \begin{pmatrix} u & u & t & t \\ u & u & t & t \\ t & t & \lambda & \mu \\ t & t & \mu & \lambda \end{pmatrix}, \tag{3.10}$$

where $\lambda = V^{eX,eX} = V^{eY,eY}$ sets the interaction on one electron pocket at the point X, $\mu = V^{eX,eY}$ connects the electrons from different pockets at the points $(\pi, 0)$ and $(0, \pi)$, $u = V^{h1,h1} = V^{h2,h2} = V^{h1,h2}$ characterizes the BCS interaction in two hole pockets surrounding point $\Gamma$, and $t = V^{h,eX} = V^{h,eY}$ connects the electrons at points X and $\Gamma$.

The critical temperature of the superconductivity transition is determined by the solution of the system of linearized equations for a gap

$$\Delta_i = \sum_j \bar{V}^{i,j} \Delta_j \ln \frac{2\gamma\bar{\omega}}{\pi T_c}, \tag{3.11}$$

where $\bar{\omega}$ is the ordinary frequency of a cutting of the logarithmic divergence in the Cooper channel,

$$\bar{V}^{i,j} \equiv -\frac{1}{2} V^{i,j} \nu_j, \tag{3.12}$$

and $\nu_j$ is the density of states on the $j$-th pocket of the Fermi surface.

Introducing the effective coupling constant $g$ and writing $T_c$ as

$$T_c = \frac{2\gamma\bar{\omega}}{\pi} \exp\left(-\frac{2}{g}\right), \tag{3.13}$$

we find the solutions of three types:

(1) a solution corresponding to the $d_{x^2-y^2}$-symmetry, when the gaps on different pockets at points X differ by sign, and the gaps on hole pockets are equal to zero:

$$\Delta_1 = \Delta_2 = 0, \quad \Delta_3 = -\Delta_4 = \Delta, \tag{3.14}$$

$$g = (\mu - \lambda)\nu_3; \tag{3.15}$$

the possibility of a solution of such type follows from the total symmetry analysis [125];

(2) two solutions corresponding to the so-called pairing of the $s^{\pm}$-type for which the gaps at points X have the same sign, and the gaps on the Fermi surfaces surrounding point $\Gamma$ can have another sign. In this case,

$$2g_{+,-} = -u(\nu_1 + \nu_2) - (\lambda + \mu)\nu_3$$
$$\pm \sqrt{(u(\nu_1 + \nu_2) - (\lambda + \mu)\nu_3)^2 + 8t^2\nu_3(\nu_1 + \nu_2)}, \tag{3.16}$$

and

$$\Delta_1 = \Delta_2 = \kappa\Delta, \quad \Delta_3 = \Delta_4 = \Delta, \tag{3.17}$$

where $\kappa^{-1} = -(g_{+,-} + u(\nu_1 + \nu_2))/(t\nu_3)$.

For the first time, the possibility of a pairing of the $s^{\pm}$-type in compounds on the basis of FeAs was indicated in [126]. This solution is in the qualitative correspondence with the data of ARPES. At a difference of the BCS constants on hole cylinders, it is easy to get different values of the gaps and to attain the agreement of the theoretical results and experimental data.

This case will be considered in the section on the symmetry of a pairing in this chapter in more details.

### 3.13.2. *Electron–phonon mechanism*

As the mechanism of pairing, we consider the electron–phonon one. In connection with the discovery of the direct and inverse isotopic effects, the superconductivity transition temperature was calculated in work [127] in the model with $s^{\pm}$-symmetry of the parameter of order by means of the solution of the Éliashberg equations for a superconductor with hole and electron pockets on the Fermi surface. The coarse solution of the equations in the spirit of the BCS approximation led to the following formulas for $T_c$ and the isotopic parameter $\alpha$:

$$T_c = 1.14\omega_{\text{ph}} \exp\left(\frac{1 + \lambda_{\text{AF}}^+ + \lambda_{\text{ph}}^+}{\Lambda - \lambda_{\text{ph}}^-}\right), \tag{3.18}$$

$$\alpha = 1/2 \left[1 - \frac{1 + \lambda_{\text{AF}}^+ + \lambda_{\text{ph}}^+}{1 + \lambda_{\text{AF}}^+}\left(\frac{\Lambda}{\Lambda - \lambda_{\text{ph}}^-}\right)^2\right], \tag{3.19}$$

where

$$\Lambda = \frac{\lambda_{\text{AF}}^-}{1 - \frac{\lambda_{\text{AF}}^-}{1 + \lambda_{\text{AF}}^+} \ln\left(\frac{\omega_{\text{AF}}}{\omega_{\text{ph}}}\right)}.$$

Here, $\omega_{\text{AF}}$ and $\omega_{\text{ph}}$ are the limiting frequencies of spin fluctuations and phonons, and the quantities $\lambda_{\text{AF}}^{\pm}$ and $\lambda_{\text{ph}}^{\pm}$ represent linear combinations of the corresponding quantities related to the pairing inside one pocket and between different pockets:

$$\lambda_{\text{AF}}^{\pm} = \lambda_{\text{AF}}^{\text{inter}} \pm \lambda_{\text{AF}}^{\text{intra}}, \quad \lambda_{\text{ph}}^{\pm} = \lambda_{\text{ph}}^{\text{inter}} \pm \lambda_{\text{ph}}^{\text{intra}}. \tag{3.20}$$

Soon after the discovery of the superconductivity in pnictides, some estimates of the pairing owing to the electron–phonon interaction were executed. The coupling constant turned out to be lower than that for aluminium, which allows one to conclude that the electron–phonon interaction is not dominant.

### 3.13.3. *Spin-fluctuation mechanism of pairing*

Such situation led to the search for new mechanisms of pairing. Let us focus on the more promising mechanism of pairing, namely, the spin-fluctuation one. We mention the following causes to separate it as a perspective one:

(1) this theory is based on the model of strongly delocalized electrons, which serves a good starting point in the description of iron;

(2) to describe the various observable properties of pnictides and chalcogenides, we should not introduce some additional parameters into the theory but need to consider the specific features of the interaction of a zone structure and the interaction in various classes of Fe compounds.

At the present time, popular are the models of pairing based on the defining role of spin fluctuations. At many aspects, they are analogous to those in the physics of HTS-cuprates which were considered in Chapter 2. The detailed analysis of possible electron mechanisms of pairing in the framework of a Hubbard generalized Hamiltonian was carried out in [128, 129].

It is worth to note that most works devoted to the mechanism of pairing and based on the exchange of spin fluctuations contain yet no direct calculations of $T_c$ and comparison with experiments.

The spin fluctuations are connected with the jittering of spins in the electron subsystem. Such excitations distort the lattice, which forces the spins to jitter more consistently and to form chains (stripes) along some of the directions in a crystal. This induces, in turn, the nematicity that should be understand as a self-organized electron state violating the rotational symmetry of the lattice. Such general jittering of spins, the lattice, and distribution of electrons leads to the creation of Cooper pairs in the space. Such pairs can move through the crystal lattice, as if not noticing its presence.

The spin fluctuations tend to the self-organization in order to form strings (in the solid-state physics, they are called stripes). In the plane with Fe atoms, the chains of spins are formed and are self-organized. These chains cause, in turn, the so-called the nematic transition. The nematic state is referred to the physics of liquid crystals. But, in this case, it is related to the loss of the rotational symmetry of the electron fluid, which is caused by the appearance of the chains of spins in iron superconductors.

For the further lowering in the temperature, it was shown that a strong interaction is formed in those stripes and increases at the transition to the superconducting state. So, we may conclude that the spin fluctuations have played the key role in the formation of the superconductivity, in particular, in iron selenides.

Thousands of physicists deal with the superconductivity. They can be partitioned into different camps. Some physicists assert that the key mechanism is a phonon one, whereas others say about the exchange by virtual magnons, i.e., by spin oscillations. As was mentioned in Chapter 2, the synergetic mechanism of pairing accounting for all degrees of freedom on the same footing is of importance as well. We emphasize that the smallest changes in the chemical composition can cause a change in the ratios of different mechanisms.

### 3.13.4.  *Symmetry of pairing*

We note that there exists a consensus in the comprehension of the crystal, magnetic, and electron structures of FeAs systems. But the problems of the symmetry of the superconductivity parameters of order and the mechanism of electron pairing are far from their solution. The different experimental methods lead to contradictory results. This is true for the question about the number of superconductivity gaps for the given compound. There are available two groups of experimental methods. In the first one, the value of the superconductivity parameters of order is determined, whereas the second group gives its phase that is related to the presence or absence of zeros in the superconductivity parameter of order. The first group includes the thermodynamic methods based on measurements of the electron heat capacity, London depth of penetration of the field, and NMR. Close to such thermodynamic methods are the tunnel measurements, including the Andreev reflection.

The second type of methods is based on measurements of the phase of the superconductivity parameters of order in Josephson's contacts in a magnetic field. Such experiments allowed one to uniquely determine the symmetry of a pairing in cuprates as $d_{x^2-y^2}$. As for the FeAs systems, few experiments of such type are known. Most experiments on various classes of FeAs compounds were performed by thermodynamic and tunnel methods. Unfortunately, the obtained data are frequently contradictory, and their interpretation is ambiguous. The experiments showed that the pairing is a siglet one. Three versions of the symmetry of the superconductivity gap are considered: the ordinary pairing (mainly with several gaps), $s^{\pm}$-type, and $d$-type.

The totals of the studied executed till now are such that it is impossible to make an unambiguous conclusion about the symmetry of the superconductivity parameters of order in FeAs compounds. The various electron models of FeAs compounds lead to the $s^{\pm}$ superconducting state as the most energetically optimum under conditions where the hole and electron pockets exist.

### 3.13.5.  *Multiorbital model*

Consider the results of calculations in a two-zone model of FeAs systems. The mean-field approximation in the spirit of the BCS theory leads to the

(a)

(b)

Fig. 3.13.   Symmetry of the superconductivity parameters of order of the $s^{\pm}$-type (a) and $d_{x^2-y^2}$ (b).

possibility to realize the superconducting states in the system with different symmetries of parameters of order. One of the possibilities is the realization of the so-called $s^{\pm}$-symmetry of parameters of order. It is characterized by the appearance of the isotropic superconductivity gap (delta) on the hole and electron sheets of the Fermi surfaces with opposite signs on sheets. Considering such parameter of order as a function of the momentum in the Brillouin zone, it is seen that it changes its sign at the transition from one sheet onto the second one (Fig. 3.13). Since the sizes of the hole sheet and electron sheet are small, the line of zeros of the parameter of order with $s^{\pm}$-symmetry passes outside the Fermi surface, and those zeros do not lead to specific dependences of some properties of a superconductor as, for example, in the case of $d_{x^2-y^2}$-symmetry of parameters of order.

For the first time, the assumption about the superconductivity with the $s^{\pm}$-symmetry of parameters of order for FeAs systems was advanced in [130, 131]. The idea of the $s^{\pm}$-symmetry was introduced prior to the discovery of iron superconductors in theoretical descriptions of the superconductivity in other compounds described by the multiorbital models [132]. It was shown that, in such superconductors, the temperature dependence of the spin-lattice relaxation can have a quasipower character (not the exponential one, as in the BCS theory). In works [130, 131], the authors started from a more realistic idea of the electron structure of iron superconductors, rather from the minimum two-orbital model.

## 3.14.   Properties of a Superconductor with $s^{\pm}$-Symmetry of Parameters of Order

In the BCS model, let us consider the appearance of the the superconducting state with $s^{\pm}$-symmetry in the case where the Fermi surface has hole and electron pockets.

Consider the Hamiltonian

$$H = \sum_{ka\sigma} \varepsilon_k^a c_{ka\sigma}^+ c_{ka\sigma}$$

$$+ \sum_{kk'q} \sum_{aa'} V_{a,a'} c_{k+\frac{q}{2}\,a\uparrow}^+ c_{-k+\frac{q}{2}\,a\downarrow}^+ c_{-k'+\frac{q}{2}\,a'\downarrow} c_{k'+\frac{q}{2}\,a'\uparrow}, \tag{3.21}$$

where the index $a = (h, e)$ enumerates the hole and electron pockets, $V_{a,a'}$ is the effective interaction electrons by means of spin fluctuations (its specific form is completely insignificant; therefore, we omit a dependence on the wave vector).

In the mean-field approximation, Hamiltonian (3.21) takes the form

$$H_{MF} = \sum_{ka\sigma} \varepsilon_k^a c_{ka\sigma}^+ c_{ka\sigma} + \sum_{ka} [\Delta_a c_{ka\uparrow}^+ c_{-ka\downarrow}^+ + h.c.]. \tag{3.22}$$

Here, the parameters $\Delta_h$ and $\Delta_e$ are given by the following formulas:

$$\Delta_h = V_{hh} \sum_k \langle c_{-kh\downarrow} c_{kh\uparrow} \rangle + V_{he} \sum_k \langle c_{-ke\downarrow} c_{ke\uparrow} \rangle, \tag{3.23}$$

$$\Delta_e = V_{ee} \sum_k \langle c_{-ke\downarrow} c_{ke\uparrow} \rangle + V_{eh} \sum_k \langle c_{-kh\downarrow} c_{kh\uparrow} \rangle. \tag{3.24}$$

Choosing the Hamiltonian in the form (3.22), we account for the pairing of electrons located in the limits of one pocket. Consider the simple situation where the interaction of pairs inside of one pocket can be neglected, i.e., we set $V_{hh} = V_{he} = 0$. Then Eqs. (3.23) and (3.24) are reduced to two equations of the form:

$$\Delta_h = V_{he} \sum_k \frac{\Delta_e}{\sqrt{\varepsilon_k^2 + |\Delta_e|^2}} \, \text{th} \frac{1}{2T} \sqrt{\varepsilon_k^2 + |\Delta_e|^2}, \tag{3.25}$$

$$\Delta_e = V_{he} \sum_k \frac{\Delta_h}{\sqrt{\varepsilon_k^2 + |\Delta_h|^2}} \, \text{th} \frac{1}{2T} \sqrt{\varepsilon_k^2 + |\Delta_h|^2}. \tag{3.26}$$

It is known that the effective interaction of electrons by means of spin fluctuations in singlet Cooper pairs has the character of repulsion, i.e., $V_{eh} = V_{he} > 0$. In this case, Eqs. (3.25) and (3.26) have a solution, if $\Delta_h$ and $\Delta_e$ are opposite by sign. Let us write the linearized equations for the parameters of order $\Delta_h$ and $\Delta_e$ in the form:

$$|\Delta_h| = V_{he} N_e(0) \, |\Delta_e| \ln\left(\frac{2\gamma\omega_0}{\pi T_c}\right),$$

$$|\Delta_e| = V_{he} N_h(0) \, |\Delta_h| \ln\left(\frac{2\gamma\omega_0}{\pi T_c}\right). \tag{3.27}$$

They yield the formula for the superconductivity transition temperature:

$$T_c = \frac{2\gamma\omega_0}{\pi} \exp\left(-1/V_h\sqrt{N_h(0)N_e(0)}\right). \tag{3.28}$$

Here, $\omega_0$ is the cutting energy for the pairing interaction, and $\gamma = 1.78$.

Formula (3.28) is a generalization of the BCS formula to the case of a two-component parameter of order [163]. It was derived in work [163] under the condition that the pairing interaction $V_{he} < 0$, i.e., it has character of attraction. In this case, both parameters of order have the same sign. The above-presented simple calculation shows that, in the case of repulsion ($V_{he} < 0$), the signs of the parameters of order on different sheets of the Fermi surface are opposite. In other words, we have the superconducting state with the $s^{\pm}$-symmetry of the parameters of order.

The anisotropy of the superconductivity gap for the doped superconductor FeTe$_{0.6}$Se$_{0.4}$ ($T_c = 14.5$ K) was studied by the ARPES method [133]. The single gap on the hole sheet of the Fermi surface was found near point Γ.

The studies executed by the ARPES method for superconductors FeAs showed some anisotropy (Fig. 3.14). The measured anisotropy of the

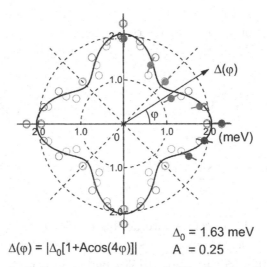

$$\Delta(\varphi) = |\Delta_0[1 + A\cos(4\varphi)]| \qquad \Delta_0 = 1.63 \text{ meV}$$
$$A = 0.25$$

Fig. 3.14.   Superconductivity gap on the hole sheet of the Fermi surface centered at point Γ by data of the ARPES method.

superconductivity gap indicates definitely that the binary interaction is not the traditional phonon one. The results indicate the importance of electron interactions of the nearest neighbors of iron in the framework of the $s^{\pm}$-wave superconductivity.

## 3.15. $t–J_1–J_2$ Model

In order to simplify the theoretical study based on the analytic methods, it is expedient to use the minimum model of FeAs compounds. The minimum electron model of FeAs systems follows from the crystal structure of FeAs layers. Atoms of Fe form a square plane lattice. From top and from bottom of such plane, the planes containing the square lattices are positioned. Thus, each Fe atom is surrounded by a tetrahedron of As atoms (Fig. 3.15(a)).

In FeAs systems, the effective exchange interaction arises not only between the nearest neighboring atoms.

We now present the explanation to Fig. 3.15. Here, $t_{2g}$ orbitals $d_{xz}$, $d_{yz}$, and $d_{xz}$ are four-lobe structures located in the planes $xz$, $yz$, and $xz$, respectively. The four lobes at a definite node of the lattice is composed of two degenerate orbitals ($d_{xz}$ and $d_{yz}$). The signs "+" and "−" indicate the sign of the wave function for a separate lobe of the orbital. The dotted lines coming from two lobes of the orbitals of the nearest neighbors indicate the corresponding matrix element of hops. Its value depends on the signs of the wave function.

For the FeAs systems, the effective exchange interaction arises not only between the nearest-neighboring atoms of Fe, but also between the next-nearest ones due to the complex structure of hoppings (Fig. 3.7). We should

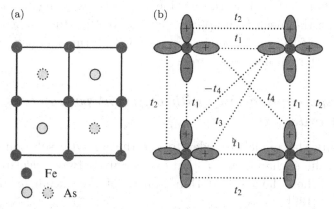

Fig. 3.15. Crystal and electron structures of a FeAs layer. (a) Location of atoms of Fe and As in the planes positioned above the Fe plane and under it. (b) Location of $d_{xz}$, $d_{yz}$ orbitals of Fe atoms and the matrix elements of electron hops between the nearest neighbors.

include two types of interaction in the Hamiltonian of the two-orbital model $H_0$ [133–135]: $H_1$ describes the exchange by electrons on different nodes, and $H_2$ corresponds to the Hund exchange in a single node. Below, we give both expressions:

$$H_1 = \sum_{iab} \sum_n J_n^{ab} [\mathbf{S}_{ai}\mathbf{S}_{bi+\delta_n} - n_{ai}n_{bi+\delta_n}], \qquad (3.29)$$

$$H_2 = \sum_{ia} J\mathbf{S}_{ai}\mathbf{S}_{\bar{a}i}. \qquad (3.30)$$

Here, $\mathbf{S}_{ai}$ is the electron spin operator on the node $i$ and orbital $a$. It can be represented in terms of the pair of Fermi-operators $c_{ia\sigma}^+$ and $c_{ia\sigma}$ of creation and annihilation of an electron in this state by means of the well-known formula:

$$\mathbf{S}_{ai} = \sum_{\sigma\sigma'} c_{ia\sigma}^+ \sigma_{\sigma\sigma'} c_{ai\sigma'}, \qquad (3.31)$$

where $\sigma$ is the vector composed of Pauli matrices. We recall that the index $a = 1, 2$ enumerates the orbitals $d_{xy}$ and $d_{yz}$ introduced into the model. The number of particles $n_{ai}$ in the state $ai$ can be presented in terms of Fermi operators:

$$n_{ai} = \sum_\sigma c_{ai\sigma}^+ c_{ai\sigma}. \qquad (3.32)$$

Eventually, the Hund exchange $H_2$ is determined by electrons on the opposite orbitals. The Hamiltonian $H = H_0 + H_1 + H_2$ describes the so-called $t$–$J_1$–$J_2$ model that includes the antiferromagnetic exchange on the nearest and next-nearest neighbors, as well as the motion of electrons along the lattice. In what follows in the mean-field approximation, we will study a possibility of the appearance of the superconducting state within this model with various symmetries of the parameters of order.

### 3.15.1.  *Superconductivity with different parameters of order*

Let us simplify the model in order to get the analytic solutions: we will consider that the exchange interaction exists only between electrons on the same orbital. Then the reduced Hamiltonian of the exchange interaction takes the form [136]

$$H_{\text{red}} = \sum_{\mathbf{kk'}} \sum_\alpha V_{\mathbf{kk'}} c_{\mathbf{k}\alpha\uparrow}^+ c_{-\mathbf{k}\alpha\downarrow}^+ c_{-\mathbf{k'}\alpha\downarrow} c_{\mathbf{k'}\alpha\uparrow}, \qquad (3.33)$$

where

$$V_{\mathbf{kk'}} = -2J_1[(\cos k_x + \cos k_y)(\cos k'_x + \cos k'_y)$$
$$+ (\cos k_x - \cos k_y)(\cos k'_x - \cos k'_y)]$$
$$- 8J_2(\cos k_x \cos k_y \cos k'_x \cos k'_y + \sin k_x \sin k_y \sin k'_x \sin k'_y).$$

$$(3.34)$$

We note also that the Hund term in $H_{\text{red}}$ is omitted, and only the interaction of Cooper pairs with opposite momenta and spins are retained in the exchange term. Thus, the reduced Hamiltonian corresponds to the BCS approximation.

In Hamiltonian (3.33), we separate the mean values for the operators of a Cooper pair:

$$\Delta_\alpha(\mathbf{k'}) = \langle c_{\alpha-\mathbf{k'}\downarrow} c_{\alpha\mathbf{k'}\uparrow} \rangle. \tag{3.35}$$

We now pass to the mean-field approximation. In the Nambu representation with four-component field operators

$$\Psi_{\mathbf{k}} = (c_{1\mathbf{k}\uparrow}, c^+_{1-\mathbf{k}\downarrow}, c_{2\mathbf{k}\uparrow}, c^+_{2-\mathbf{k}\downarrow}), \tag{3.36}$$

we can write the total Hamiltonian of the model that includes the kinetic term $H_0$ and the reduced interaction operator $H_{\text{red}}$ taken in the mean-field approximation as

$$H = \sum_{\mathbf{k}} \Psi^+(\mathbf{k}) A(\mathbf{k}) \Psi(\mathbf{k}), \tag{3.37}$$

where $A(\mathbf{k})$ is the $4 \times 4$ matrix:

$$A(\mathbf{k}) = \begin{pmatrix} \xi_x(\mathbf{k}) - \mu & \Delta_1(\mathbf{k}) & \varepsilon_{xy}(\mathbf{k}) & 0 \\ \Delta_1^*(\mathbf{k}) & -\varepsilon_x(\mathbf{k}) + \mu & 0 & -\varepsilon_{xy}(\mathbf{k}) \\ \varepsilon_{xy}(\mathbf{k}) & 0 & \varepsilon_y(\mathbf{k}) - \mu & \Delta_2(\mathbf{k}) \\ 0 & -\varepsilon_{xy}(\mathbf{k}) & \Delta_2^*(\mathbf{k}) & -\varepsilon_y(\mathbf{k}) + \mu \end{pmatrix}. \tag{3.38}$$

Here, $\Delta_a(\mathbf{k})$, $a = 1, 2$, presents the pairing amplitude composed of the binary mean quantities for electrons referred to the hole or electron sheet of the Fermi surface. The amplitude consists of five terms corresponding to different symmetries of the operators:

$$\Delta_\alpha(\mathbf{k}) = s_{0a} + s_{x^2+y^2a}(\mathbf{k}) + s_{x^2y^2a}(\mathbf{k}) + d_{x^2-y^2a}(\mathbf{k}) + d_{xya}(\mathbf{k}), \qquad (3.39)$$

where

$$\begin{aligned}
s_{x^2+y^2a}(\mathbf{k}) &= \Delta^0_{x^2+y^2a}(\cos k_x + \cos k_y), \\
s_{x^2y^2a}(\mathbf{k}) &= \Delta^0_{x^2y^2a} \cos k_x \cos k_y, \\
d_{x^2-y^2a}(\mathbf{k}) &= \Delta^0_{x^2-y^2a}(\cos k_x - \cos k_y), \\
d_{xya}(\mathbf{k}) &= \Delta^0_{xya} \sin k_x \sin k_y,
\end{aligned} \qquad (3.40)$$

and $s_{0a}$ is independent of the momentum $\mathbf{k}$.

Thus, the pairing amplitude consists of three contributions with $s$-symmetry and two contributions with $d$-symmetry. The amplitures of the corresponding parameters of order are as follows:

$$\Delta^0_{x^2+y^2a} = -\frac{2J_1}{N} \sum_{k'} (\cos k'_x \pm \cos k'_y)\Delta_a(\mathbf{k}'),$$

$$\Delta^0_{x^2y^2a} = -\frac{8J_2}{N} \sum_{k'} \cos k'_x \cos k'_y \Delta_a(\mathbf{k}'), \qquad (3.41)$$

$$\Delta^0_{xya} = -\frac{8J_2}{N} \sum_{k'} \sin k'_x \sin k'_y \Delta_a(\mathbf{k}').$$

To calculate the quantity $\Delta_a(\mathbf{k})$, we need to construct the equations of motion for the electron Green function in the superconducting state:

$$G(\mathbf{k}, \tau) = -\langle \hat{T}_\tau \Psi_{\mathbf{k}}(\tau) \Psi^+_{\mathbf{k}}(0) \rangle. \qquad (3.42)$$

Matrix (3.37) can be diagonalized with the help of the unitary transformation $U^+(\mathbf{k})A(\mathbf{k})U(\mathbf{k})$. The four eigenvalues are given by the formulas:

$$E_1(\mathbf{k}) = -E_2(\mathbf{k}), \quad E_3(\mathbf{k}) = -E_4(\mathbf{k}), \qquad (3.43)$$

$$E_{1,3}(\mathbf{k}) = \frac{1}{\sqrt{2}} \left( \xi_x^2 + \xi_y^2 + \varepsilon_{xy}^2 + \Delta_1^2 + \Delta_2^2 \right.$$

$$\left. \pm \sqrt{(\xi_x^2 - \xi_y^2 + \Delta_1^2 - \Delta_2^2)^2 + 4\varepsilon_{xy}^2 \left[ (\xi_x + \xi_y)^2 + (\Delta_1 - \Delta_2)^2 \right]} \right)^{1/2},$$

$$(3.44)$$

where $\xi_x = \varepsilon_x - \mu$, $\xi_y = \varepsilon_y - \mu$. Thus, we get the equations of self-consistency for the pairing amplitudes and for the occupation numbers. We have

$$0\Delta_1(\mathbf{k}) = \sum_{\mathbf{k}'m} V_{\mathbf{k}\mathbf{k}'} U_{2m}^*(\mathbf{k}') U_{1m}(\mathbf{k}') F(E_m(\mathbf{k}')),$$

$$\Delta_2(\mathbf{k}) = \sum_{\mathbf{k}'m} V_{\mathbf{k}\mathbf{k}'} U_{4m}^*(\mathbf{k}') U_{3m}(\mathbf{k}') F(E_m(\mathbf{k}')),$$

(3.45)

where

$$n_1 = \sum_{\mathbf{k}'\,m} U_{1m}^*(\mathbf{k}) U_{1m}(\mathbf{k}') F(E_m(\mathbf{k}')),$$

$$n_2 = \sum_{\mathbf{k}'\,m} U_{3m}^*(\mathbf{k}) U_{3m}(\mathbf{k}') F(E_m(\mathbf{k}')),$$

and $f(E)$ is the Fermi function. In Eq. (3.45), we sum the vectors over the index $m = 1, 2, 3, 4$ that enumerates components of superspinor (3.36).

The equation for the superconductivity transition temperature can be obtained by the linearization of the equations for the amplitudes $\Delta_1(\mathbf{k})$ and $\Delta_2(\mathbf{k})$. For $\Delta_2(\mathbf{k})$, we get

$$\Delta_2(\mathbf{k}) = \sum_{\mathbf{k}'} V_{\mathbf{k}\mathbf{k}'} \left[ W_3(\mathbf{k}') - W_1(\mathbf{k}') \right],$$

(3.46)

where

$$W_i = \frac{((\varepsilon_x - \mu)^2 - \tilde{E}_i^2)\Delta_2 + \varepsilon_{xy}^2 \Delta_1}{2\,|\varepsilon_x + \varepsilon_y - 2\mu|\,\tilde{E}_i\sqrt{4\varepsilon_{xy}^2 + (\varepsilon_x - \varepsilon_y)^2}} \mathrm{th}\,\frac{\tilde{E}_i}{2T},$$

(3.47)

and $\tilde{E}_i = E_i\,(\Delta_1 = \Delta_2 = 0)$.

On the phase diagram (Fig. 3.16), we present the numerical solutions of the equations in the framework of the two-orbital model.

In the upper left corner, where $J_2 > J_{2c} \simeq 1.2$, the pure phase with $s$-symmetry $s_{x^2y^2}$ is realized. In the right lower corner, where $J_1 > J_{1c} \simeq 1.05$, we observe the mixed phase with $d_{x^2-y^2}$ and $s_{x^2+y^2}$. The remaining large part $(J_1, J_2)$ of the area is occupied by another mixed phase $d_{x^2-y^2} + s_{x^2y^2}$. In this case, the signs of the parameter of order with $d_{x^2-y^2}$-symmetry are different for two orbitals. For example, if $\Delta_1 = a\cos k_x \cos k_y + b(\cos k_x - \cos k_y)$, then $\Delta_2 = a\cos k_x \cos k_y - b(\cos k_x - \cos k_y)$. No solution corresponding to the $d_{xy}$-symmetry of the parameter of order was found.

In Fig. 3.17, we show the result of the intraorbital and interorbital pairings for the case $J_1 = J_2$ with the chemical potential $\mu = 1.8$.

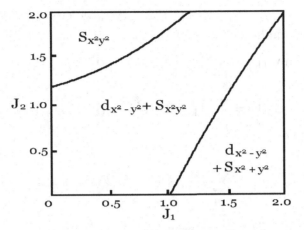

Fig. 3.16.   Phase diagram on the plane $J_1$–$J_2$ for superconductivity states with different parameters of order in the two-orbital model with the electron doping $\delta = 0.18$ [134].

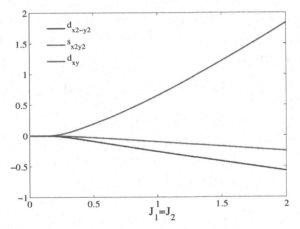

Fig. 3.17.   The intraorbital pairing order parameter with $s_{x^2y^2}$ and $d_{x^2-y^2}$ and the interorbital pairing order parameter with $d_{xy}$ as functions of $J = J_1 = J_2$, when the chemical potential is $\mu = 1.8$ [136].

## 3.16.  Conclusion

We now indicate the common features of superconductors on the basis of iron and copper oxides:

(1) Both classes of superconductors are quasitwo-dimensional (layered) systems from the viewpoint of their electron properties, which leads to the strong anisotropy.

(2) In both classes, the region of coexistence of the superconductivity on the phase diagram is adjacent to the region of antiferromagnetism.

(3) In both classes, the Cooper pairing is a singlet one.
(4) The main properties of the superconducting state are typical of super-conductors of the II kind.

What are the differences between the superconductors on the basis of iron and copper oxides?

(1) HTS cuprates are antiferromagnetic dielectrics (Mott strongly correlated ones), whereas new superconductors are antiferromagnetic metals.
(2) Cuprates in the superconducting state are one-zone metals with one Fermi surface (hole or electron), whereas new superconductors are multizone metals with several Fermi surfaces of the electron and hole types.
(3) In cuprates, the anisotropic Cooper pairing of the $d$-type is realized. In new superconductors, the isotropic pairing of the $s^{\pm}$-type is observed.
(4) The microscopic mechanisms of pairing in both classes of superconductors are different: the unambiguously electron mechanism (spin fluctuations) in cuprates, whereas the electron–phonon interaction (isotope-effect) can turn out to be quite significant in iron superconductors.

It is clear that there are more differences between cuprates and iron superconductors, than similarities. We can conclude that HTS is the unique property of cuprates, i.e., strongly correlated systems close to the dielectric state. Iron superconductors are simpler and more understandable. Their normal state is not so mysterious, as that of cuprates, though the presence of many zones introduces a certain complexity.

We may say that the phenomenon of HTS is much more spread, than it was considered two decades ago, so the community of researchers of the superconductivity can optimistically look at the future.

# CHAPTER 4

# Multiband Superconductivity

## 4.1. Introduction

In this chapter, we present the theory of superconductivity with regard for a complicated multiband structure of superconductors. Calculations of the band structure of cuprate superconductors indicate that several energy bands intersect one another on the Fermi surface in these compounds [137], and the Fermi surface passes through high-symmetry points which correspond to the Lifshitz electronic topological transition. In addition, the discovery of two-gap superconductivity in two-band superconductors $MgB_2$ allows one to consider the possibility to use a multiband theory of superconductivity. In Section 4.2, we analyze the problems of multiband superconductivity and superconductivity at room temperature. In Section 4.3, we study the physical properties of superconductors $MgB_2$ and problems of two-gap superconductivity, as well as the phase diagram for superconductors $MgB_2$ on the basis of the renormalization group approach. In the field of superconductivity, we meet the problem-maximum—it consists in the creation of room-temperature superconductors. We consider this problem in our book and give some recommendations on the search for these superconductors. In Section 4.5, we analyze the problem of high-temperature superconductivity at high pressures in hydrites. The results of this chapter were obtained by the author and published in works [137–147].

The theory of superconductivity arising from the electron–phonon interaction mechanism by Bardeen, Cooper, and Schrieffer (BCS) [148] has been well established, and it is now the standard theory for superconductivity [2,3,6,16,17,148]. In Chapter 2, we have considered non-BCS mechanisms via spin fluctuations, charge fluctuations (plasmons), and electron excitations (excitons) which have attracted a great interest, for example, in relation to the possibility of high-$T_c$ superconductivity. These mechanisms have a common characteristic that the electron–electron (e–e) interaction is an origin of the superconductivity. After the discovery of the high-$T_c$ copper oxides [2], Anderson [6] has indeed emphasized an important role of the e–e interaction. For the past decade, many non-BCS theories [150–155] have been proposed, but they do not converge as a unified and well-accepted theory yet. On the other hand, a lot of experimental studies on copper oxides have revealed the following characteristics:

(1) these species are antiferromagnets before doping, in accord with the importance of the e–e interaction,
(2) the high-$T_c$ superconductivity appears in the intermediate region of the metal-insulator transition

and disappears in the metallic or overdoped region [156–158]. Accumulated experimental results on the species and related materials suggest a guiding principle that the doping in magnetic systems, more generally charge-transfer (CT) insulators, may provide several exotic phases which are

(a) ferromagnetic metal or insulator,
(b) spin glass,
(c) paramagnetic metals,
(d) antiferromagnetic metals,
(e) ferrimagnetic metal or insulator,
(f) charge- or spin-mediated superconductor.

Relative stabilities of these phases should be dependent on several factors. This indicates, in turn, that the theoretical description of such phases and phase transitions in a systematic fashion is quite hard. Recently, the importance of multiband effects in high-$T_c$ superconductivity has been pointed out [3, 17, 144, 148]. In the framework of the two-particle Green's function techniques [139,142], it is shown that a class of new so-called coupled states arises in electron–phonon system. The model numerical calculations have shown that the superconducting (SC) gap depends on the number of bands crossing the Fermi level, and the temperature dependence of the SC gap for high-$T_c$ superconductors is more complicated than that predicted in the BCS approach. We have also investigated anomalous phases in a two-band model by using the Green's function techniques [14, 159]. The expressions for the transition temperature for several phases have been derived, and the approach has been applied to the superconductivity in molecular crystals by charge injection and field-induced superconductivity.

In this chapter, we investigate the superconductivity, by using the two-band model and the two-particle Green's function technique. In the framework of the two-band model, the coupled states in the electron system and the conditions when the coupled states can appear are investigated. We apply the model to the electron–phonon mechanism within the traditional BSC method, the electron–electron interaction mechanism for high-$T_c$ superconductivity, and the cooperative mechanism in relation to the multiband superconductivity.

## 4.2. Multiband Hamiltonian

In this section, we briefly summarize the two-band model for superconductivity, introduce a two-particle Green's function, and investigate the spectral properties of the model.

### 4.2.1. *Hamiltonian*

We start from the Hamiltonian for two-bands $i$ and $j$:

$$H = H_0 + H_{\text{int}}, \tag{4.1}$$

where

$$H_0 = \sum_{\mathbf{k},\sigma} \left[ [\varepsilon_i - \mu] \, a^+_{i\mathbf{k}\sigma} a_{i\mathbf{k}\sigma} + [\varepsilon_j - \mu] \, a^+_{j\mathbf{k}\sigma} a_{j\mathbf{k}\sigma} \right], \tag{4.2}$$

$$
\begin{aligned}
H_{\text{int}} = \frac{1}{4} \sum_{\delta(\mathbf{p}_1+\mathbf{p}_2,\mathbf{p}_3+\mathbf{p}_4)} \sum_{\alpha\beta\gamma\delta} & \left[ \Gamma^{iiii}_{\alpha\beta\gamma\delta} a^+_{i\mathbf{p}_1\alpha} a^+_{i\mathbf{p}_2\beta} a_{i\mathbf{p}_3\gamma} a_{i\mathbf{p}_4\delta} + (i \to j) \right. \\
& + \Gamma^{iijj}_{\alpha\beta\gamma\delta} a^+_{i\mathbf{p}_1\alpha} a^+_{i\mathbf{p}_2\beta} a_{j\mathbf{p}_3\gamma} a_{j\mathbf{p}_4\delta} + (i \to j) \\
& \left. + \Gamma^{ijij}_{\alpha\beta\gamma\delta} a^+_{i\mathbf{p}_1\alpha} a^+_{j\mathbf{p}_2\beta} a_{i\mathbf{p}_3\gamma} a_{j\mathbf{p}_4\delta} + (i \to j) \right],
\end{aligned}
\tag{4.3}
$$

$\Gamma$ is the bare vertex part:

$$\Gamma^{ijkl}_{\alpha\beta\gamma\delta} = \langle i\mathbf{p}_1\alpha \; j\mathbf{p}_2\beta \, | \, k\mathbf{p}_3\gamma \; l\mathbf{p}_4\delta \rangle \, \delta_{\alpha\delta}\delta_{\beta\gamma} - \langle i\mathbf{p}_1\alpha \; j\mathbf{p}_2\beta \, | \, l\mathbf{p}_4\delta \; k\mathbf{p}_3\gamma \rangle \, \delta_{\alpha\gamma}\delta_{\beta\delta}, \tag{4.4}$$

with

$$
\begin{aligned}
\langle i\mathbf{p}_1\alpha j\mathbf{p}_2\beta \, | \, k\mathbf{p}_3\gamma l\mathbf{p}_4 \rangle = \int dr_1 dr_2 \phi^*_{i\mathbf{p}_1\alpha}(\mathbf{r}_1) \phi^*_{j\mathbf{p}_2\beta}(\mathbf{r}_2) \\
\times V(\mathbf{r}_1, \mathbf{r}_2) \phi_{k\mathbf{p}_3\gamma}(\mathbf{r}_2) \phi_{l\mathbf{p}_4\delta}(\mathbf{r}_1),
\end{aligned}
\tag{4.5}
$$

and $a^+_{ip\sigma}$ ($a_{ip\sigma}$) is the creation (annihilation) operator corresponding to the excitation of electrons (or holes) in the $i$th band with spin $\sigma$ and momentum $\mathbf{p}$, $\mu$ is the chemical potential, and $\phi^*_{ip\alpha}$ is a single-particle wave function. Here, we suppose that the vertex function in Eq. (4.3) consists of the effective interactions between the carriers caused by the linear vibronic coupling in several bands and the screened Coulombic interband interaction of carriers. When we use the two-band Hamiltonian (4.1) and define the order parameters for the singlet exciton, triplet exciton, and singlet Cooper pair, the mean-field Hamiltonian is easily derived [159–163, 166, 167]. Here, we focus on four-electron scattering processes:

$$g_1 = \langle ii|ii \rangle = \langle jj|jj \rangle, \tag{4.6}$$

$$g_2 = \langle ii|jj \rangle = \langle jj|ii \rangle, \tag{4.7}$$

$$g_3 = \langle ij|ij \rangle = \langle ji|ji \rangle, \tag{4.8}$$

$$g_4 = \langle ij|ji \rangle = \langle ji|ij \rangle. \tag{4.9}$$

Here $g_1$ and $g_2$ represent the intraband two-particle normal and umklapp scatterings, respectively, $g_3$ means the interband two-particle umklapp process, and $g_4$ indicates the interband two-particle interaction on different bands (see Fig. 4.1). Note that $\Gamma$'s are given by

$$\Gamma_{\alpha\beta\gamma\delta}^{iiii} = \Gamma_{\alpha\beta\gamma\delta}^{jjjj} = g_1 \left( \delta_{\alpha\delta}\delta_{\beta\gamma} - \delta_{\alpha\gamma}\delta_{\beta\delta} \right),$$

$$\Gamma_{\alpha\beta\gamma\delta}^{iijj} = \Gamma_{\alpha\beta\gamma\delta}^{jjii} = g_2 \left( \delta_{\alpha\delta}\delta_{\beta\gamma} - \delta_{\alpha\gamma}\delta_{\beta\delta} \right), \tag{4.10}$$

$$\Gamma_{\alpha\beta\gamma\delta}^{ijij} = \Gamma_{\alpha\beta\gamma\delta}^{jiji} = g_3\delta_{\alpha\delta}\delta_{\beta\gamma} - g_4\delta_{\alpha\gamma}\delta_{\beta\delta},$$

where an antisymmetrized vertex function $\Gamma$ is considered to be a constant independent of the momenta. The spectrum is elucidated by the Green's function method. Using Green's functions, which characterize the CDW (charge-density-wave), SDW (spin-density-wave), and SSC (singlet superconducting) phases, we obtain a self-consistent equation, according to the traditional procedure [14, 137, 159, 160]. Then, we can obtain expressions for the transition temperature for some cases.

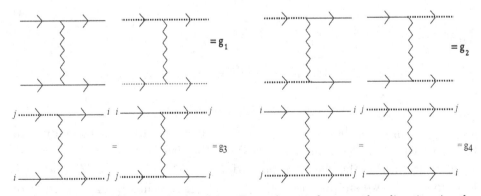

Fig. 4.1.   Electron–electron interactions. Dependence of $g$ on the direction in the momentum space is ignored in this model $g_x(\mathbf{k}) \approx g_x(x = i,j)$. We assume that $g_x$ is constant.

In the framework of the one-band model, the electronic phases are characterized by

$$-g_2 - 2g_3 + g_4 > 0, \quad \text{for CDW}$$

$$g_2 + g_4 > 0, \quad \text{for SDW}$$

$$-g_1 > 0, \quad \text{for SSC.}$$

In the framework of the two-band model, we have already derived expressions of the transition temperature for CDW, SDW, and SSC. In the previous paper [14, 159], we have investigated the dependence of $T_c$ on the hole or electron concentration for the superconductivity of copper oxides by using the two-band model and have obtained a phase diagram of $Bi_2Sr_2Ca_{1-x}Y_xCu_2O_8$ (Bi-2212) by means of the above expressions for the transition temperature. The dependence of $T_c$ on $\Delta p$ can be reproduced in agreement with the experiment [161]. Recently, we have also obtained phase diagrams of copper oxides, anthracene, oligothiophene, and $C_{60}$ crystals by using the analytic solutions [159].

### 4.2.2.  *Two-particle Green's function*

In this subsection, we introduce a two-particle Green's function [137–139] to investigate the physical properties of superconductivity in the two-band model. In statistical mechanics, Green's functions are a convenient generalization of the notion of correlation functions. Like the latter, the former are closely related to the calculations of observables and give the well-known advantages in the construction and the solution of equations. First, let us define one-particle Green's functions:

$$G_\sigma^\nu \left( \mathbf{k}, t' - t \right) = \left\langle -iT \left[ a_{\nu\mathbf{k}\sigma} (t) a_{\nu\mathbf{k}\sigma}^+ (t') \right] \right\rangle, \tag{4.11}$$

where $\sigma$ and $\nu$ mean labels for a spin and a band, respectively. The equation for a Green's function derived by using the two-band model (4.1) is written as

$$i\frac{\partial}{\partial t} \left\langle -iT \left[ a_{\nu\mathbf{k}\sigma} (t) a_{\nu\mathbf{k}\sigma}^+ (t') \right] \right\rangle$$

$$= \delta \left( t - t' \right) + \left\langle -iT \left[ i\frac{\partial a_{\nu\mathbf{k}\sigma} (t)}{\partial t} a_{\nu\mathbf{k}\sigma}^+ (t') \right] \right\rangle, \tag{4.12}$$

where

$$i\frac{\partial a_{\nu\mathbf{k}\sigma} (t)}{\partial t} = \left[ a_{\nu\mathbf{k}\sigma}, H \right]. \tag{4.13}$$

The equation for Green's function (4.12) is rewritten after inserting Eq. (4.13) as

$$i\frac{\partial}{\partial t}\left\langle-iT\left[a_{\nu\mathbf{k}\sigma}(t)\,a_{\nu\mathbf{k}\sigma}^{+}(t')\right]\right\rangle$$

$$= \delta\left(t-t'\right)+[\varepsilon_{\nu}-\mu]\left\langle-iT\left[a_{\nu\mathbf{k}\sigma}(t)\,a_{\nu\mathbf{k}\sigma}^{+}(t')\right]\right\rangle$$

$$+\frac{1}{2}\sum_{\beta\gamma\delta}\sum_{\delta(\mathbf{k}+\mathbf{p_2},\mathbf{p_3}+\mathbf{p_4})}g_1\left(\delta_{\sigma\delta}\delta_{\beta\gamma}-\delta_{\sigma\gamma}\delta_{\beta\delta}\right)$$

$$\times\,G_{2\nu\nu\nu\nu}^{\gamma\delta\beta\sigma}\left(\mathbf{p_3},\mathbf{p_4},\mathbf{p_2},\mathbf{k};t,t'\right)$$

$$+\frac{1}{2}\sum_{\nu'}\sum_{\beta\gamma\delta}\sum_{\delta(\mathbf{k}+\mathbf{p_2},\mathbf{p_3}+\mathbf{p_4})}g_2\left(\delta_{\sigma\delta}\delta_{\beta\gamma}-\delta_{\sigma\gamma}\delta_{\beta\delta}\right)$$

$$\times\,G_{2\nu'\nu'\nu\nu}^{\gamma\delta\beta\sigma}\left(\mathbf{p_3},\mathbf{p_4},\mathbf{p_2},\mathbf{k};t,t'\right)$$

$$+\frac{1}{2}\sum_{\nu'}\sum_{\beta\gamma\delta}\sum_{\delta(\mathbf{k}+\mathbf{p_2},\mathbf{p_3}+\mathbf{p_4})}\left(g_3\delta_{\sigma\delta}\delta_{\beta\gamma}-g_4\delta_{\sigma\gamma}\delta_{\beta\delta}\right)$$

$$\times\,G_{2\nu\nu'\nu'\nu}^{\gamma\delta\beta\sigma}\left(\mathbf{p_3},\mathbf{p_4},\mathbf{p_2},\mathbf{k};t,t'\right),\tag{4.14}$$

where

$$G_{2\nu\nu\nu\nu}^{\gamma\delta\beta\sigma}\left(\mathbf{p_3},\mathbf{p_4},\mathbf{p_2},\mathbf{k};t,t'\right)$$

$$= \left\langle-iT[a_{\nu\mathbf{p_3}\gamma}(t)\,a_{\nu\mathbf{p_4}\delta}(t)\,a_{\nu\mathbf{p_2}\beta}^{+}(t-0)\,a_{\nu\mathbf{k}\sigma}^{+}(t')]\right\rangle,\tag{4.15}$$

$$G_{2\nu'\nu'\nu\nu}^{\gamma\delta\beta\sigma}\left(\mathbf{p_3},\mathbf{p_4},\mathbf{p_2},\mathbf{k};t,t'\right)$$

$$= \left\langle-iT[a_{\nu'\mathbf{p_3}\gamma}(t)\,a_{\nu'\mathbf{p_4}\delta}(t)\,a_{\nu\mathbf{p_2}\beta}^{+}(t-0)\,a_{\nu\mathbf{k}\sigma}^{+}(t')]\right\rangle,\tag{4.16}$$

$$G_{2\nu\nu'\nu'\nu}^{\gamma\delta\beta\sigma}\left(\mathbf{p_3},\mathbf{p_4},\mathbf{p_2},\mathbf{k};t,t'\right)$$

$$= \left\langle-iT[a_{\nu\mathbf{p_3}\gamma}(t)\,a_{\nu'\mathbf{p_4}\delta}(t)\,a_{\nu'\mathbf{p_2}\beta}^{+}(t-0)\,a_{\nu\mathbf{k}\sigma}^{+}(t')]\right\rangle.\tag{4.17}$$

Here $\nu'$ indicates a band different from $\nu$. To calculate the density of electron states, we have to focus on the case where $t' \to t-0$. The two-particle Green's functions in Eq. (4.14) is rewritten as $G_2(\mathbf{p_3},\mathbf{p_4},\mathbf{p_2},\mathbf{k};t\text{-}t')$ ($t' \to t-0$).

In this study, we investigate only the spectral properties of two-particle Green's functions for the superconductivity. Therefore, we focus on the following two-particle Green's function:

$$G_{2\nu\nu\nu\nu}^{\gamma\delta\beta\sigma}\left(\mathbf{p_3},\mathbf{p_4},\mathbf{p_2},\mathbf{k};t,t'\right)$$

$$= \left\langle-iT[a_{\nu\mathbf{p_3}\gamma}(t)\,a_{\nu\mathbf{p_4}\delta}(t)\,a_{\nu\mathbf{p_2}\beta}^{+}(t')\,a_{\nu\mathbf{k}\sigma}^{+}(t')]\right\rangle.\tag{4.18}$$

For simplicity, we consider only three cases: (4.1) $g_1 \neq 0$ and others $= 0$, (4.2) $g_2 \neq 0$ and others $= 0$, and (4.3) $g_1 \neq 0$, $g_2 \neq 0$ and others $= 0$.

### 4.2.3. *Traditional superconductivity*

In general, in the framework of the BCS theory, the Hamiltonian is described by a single-band model. In the effective electron–electron interaction (4.1), we consider that $g_1 \neq 0$ and others $= 0$ and focus only on the single-band model. According to the approach [156] used for phonon systems which is based on the method of Bogoliubov–Tyablikov [192], we can derive the equation for a two-particle electron Green's function. The spectral features of the electron system in the mentioned region of energy are described by the Fourier transform of this function. For the simplest case of a one-electron zone crossing the Fermi energy level, it can be given as

$$
G_{2\nu\nu\nu\nu}^{\gamma\delta\beta\sigma} \left(\mathbf{p}_3, \mathbf{p}_4, \mathbf{p}_2, \mathbf{k}; t - t'\right) = \frac{f\left(\mathbf{k}, \mathbf{k}', \omega\right) \sum_{\sigma,\sigma'} \phi\left(\sigma, \sigma'\right)}{1 - g_1 \sum_{\mathbf{q}} K\left(\omega, \mathbf{k}, \mathbf{k}', \mathbf{q}\right)}, \tag{4.19}
$$

where

$$
K\left(\omega, \mathbf{k}, \mathbf{k}', \mathbf{q}\right) = \frac{2 - n_{\mathbf{k}+\mathbf{q}}^{\nu} - n_{\mathbf{k}'-\mathbf{q}}^{\nu}}{2\omega - \varepsilon_{\mathbf{k}+\mathbf{q}}^{\nu} - \varepsilon_{\mathbf{k}'-\mathbf{q}}^{\nu}}, \tag{4.20}
$$

$n_k$ indicates the filling number of electrons, and $g_1$ is the effective Fourier transform of the electron–electron interaction. If the e–e interaction constant renormalized by the electron–phonon interaction becomes negative, coupled states will appear in the electron system. In the previous papers [137–139], we have presented analysis of the spectral properties of the two-particle Green's function. According to the same procedure [17, 139], we obtain the equation for coupled states in the electron system:

$$
1 - g_1 N\left(\varepsilon_f\right) \left[-\ln\left|1 - \frac{\Delta}{a}\right|\right] = 0, \tag{4.21}
$$

where

$$
N\left(\varepsilon_f\right) = \sqrt{2}\pi m_\nu^* \sqrt{m_\nu^* \varepsilon} \left(2 - n_{\mathbf{k}+\mathbf{q}}^{\nu} - n_{\mathbf{k}'-\mathbf{q}}^{\nu}\right)|_{\varepsilon = \varepsilon_f}, \tag{4.22}
$$

$$
n_{\mathbf{k}}^{\nu} = \frac{1}{\exp\left[\left(\varepsilon_{\mathbf{k}}^{\nu} - \varepsilon_f\right)/T\right] + 1}, \tag{4.23}
$$

$a = 2(\omega - \varepsilon_f - \Delta_\nu - E)$, $E = k^2/2m$, $\varepsilon = q^2/2m$, and $m_\nu^* = m_\nu/m$. $m_\nu^*$ means the reduced effective mass of electron in the crystal energy zone.

$m$ is mass of free electron. If $g_1 < 0$, we can find solutions of Eq. (4.21) for superconductivity.

### 4.2.4.   *Copper oxides*

In copper oxides, the effective electron–electron interaction $g_2$ is efficiently important to realize the high-$T_c$ superconductivity [161, 164]. Therefore, we consider that $g_2 \neq 0$ and others $= 0$. The two-particle Green's function of Eq. (4.18) is rewritten as

$$G_{2\nu\nu\nu\nu}^{\gamma\delta\beta\sigma}\left(\mathbf{p}_3, \mathbf{p}_4, \mathbf{p}_2, \mathbf{k}; t - t'\right) = \frac{f\left(\mathbf{k}, \mathbf{k}', \omega\right) \sum_{\sigma,\sigma'} \phi\left(\sigma, \sigma'\right)}{1 - g_2^2 \sum_{\mathbf{q},\mathbf{q}'} K_2\left(\omega, \mathbf{k}, \mathbf{k}', \mathbf{q}, \mathbf{q}'\right)}, \qquad (4.24)$$

where

$$K_2\left(\omega, \mathbf{k}, \mathbf{k}', \mathbf{q}, \mathbf{q}'\right) = \frac{\left(2 - n_{\mathbf{k}+\mathbf{q}-\mathbf{q}'}^{\nu'} - n_{\mathbf{k}'-\mathbf{q}+\mathbf{q}'}^{\nu'}\right)\left(2 - n_{\mathbf{k}+\mathbf{q}}^{\nu} - n_{\mathbf{k}'-\mathbf{q}}^{\nu}\right)}{\left(2\omega - \varepsilon_{\mathbf{k}+\mathbf{q}-\mathbf{q}'}^{\nu'} - \varepsilon_{\mathbf{k}'-\mathbf{q}+\mathbf{q}'}^{\nu'}\right)\left(2\omega - \varepsilon_{\mathbf{k}+\mathbf{q}}^{\nu} - \varepsilon_{\mathbf{k}'-\mathbf{q}}^{\nu}\right)}.$$
$$(4.25)$$

According to similar procedure in previous papers [137, 139], we study the situation near the extremum (minimum or maximum) of the electron zone. Then we suppose $\mathbf{k} = \mathbf{k}' = \mathbf{k}_0 + \mathbf{k}''$, and $\varepsilon_{k0}^{\nu} = \varepsilon^{\nu}$ corresponds to the extremum of the zone. We expand the energy in the momentum $\mathbf{k} \pm \mathbf{q}$ in a series up to terms of the second order and suppose that the energy extremum is located near the Fermi-level energy. Then, the sum in the denominator of Eq. (4.24) is approximately reduced to the following expression:

$$\sum_{\mathbf{q},\mathbf{q}'} K_2\left(\omega, \mathbf{k}, \mathbf{k}', \mathbf{q}, \mathbf{q}'\right) \approx N\left(\varepsilon_f^{\nu}\right) N\left(\varepsilon_f^{\nu'}\right) \ln\left|\left(1 - \frac{2\Delta}{a_{\nu'}}\right)\left(1 - \frac{2\Delta}{a_\nu}\right)\right|.$$
$$(4.26)$$

Therefore, we obtain the equation for coupled states in the electron system:

$$1 - g_2^2 N\left(\varepsilon_f^{\nu}\right) N\left(\varepsilon_f^{\nu'}\right) \ln\left|\left(1 - \frac{2\Delta}{a_{\nu'}}\right)\left(1 - \frac{2\Delta}{a_\nu}\right)\right| = 0. \qquad (4.27)$$

From Eq. (4.27), we find the possibility of the existence of a solution for the coupled states, if $g_2 \neq 0$. Thus, the effective electron–electron interaction $g_2$ with positive value contributes to the superconductivity.

### 4.2.5. *Cooperative mechanism*

Here, we consider that $g_1 \neq 0$, $g_2 \neq 0$, and others $= 0$. In a similar way, the two-particle Green's function of Eq. (4.18) is approximately derived as

$$G_{2\nu\nu\nu\nu}^{\gamma\delta\beta\sigma}\left(\mathbf{p}_3, \mathbf{p}_4, \mathbf{p}_2, \mathbf{k}; t - t'\right)$$

$$= \frac{f\left(\mathbf{k}, \mathbf{k}', \omega\right) \sum_{\sigma,\sigma'} \phi\left(\sigma, \sigma'\right)}{[1 - (g_1 + g_2) \sum_{\mathbf{q}} K\left(\omega, \mathbf{k}, \mathbf{k}', \mathbf{q}\right)][1 - (g_1 - g_2) \sum_{\mathbf{q}} K\left(\omega, \mathbf{k}, \mathbf{k}', \mathbf{q}\right)]},$$

$$(4.28)$$

where $K(\omega, \mathbf{k}, \mathbf{k}', \mathbf{q})$ is given by Eq. (4.20). The summation in the denominator of Eq. (4.28) is performed in a similar way, and the equation for coupled states in the electron system is approximately derived as

$$\left[1 - (g_1 + g_2) N\left(\varepsilon_f\right) \ln\left|1 - \frac{\Delta}{a}\right|\right] \left[1 - (g_1 - g_2) N\left(\varepsilon_f\right) \ln\left|1 - \frac{\Delta}{a}\right|\right] = 0.$$

$$(4.29)$$

When $g_1 + g_2 < 0$ or $g_1 - g_2 < 0$, we can find solutions of this equation.

### 4.2.6. *Room-temperature superconductors*

In the previous section, we have approximately calculated two-particle Green's functions for three cases, which are the traditional superconductivity, copper oxides, and cooperative mechanism in the framework of the two-band model. From these Green's functions, we have derived the equation for coupled states for each case. In the case of a single-band model, which indicates the traditional superconductivity such as the BCS theory, it is necessary that the effective electron–electron interaction be negative ($g_1 < 0$) to realize the superconductivity. The maximal transition temperature for the superconductivity predicted by the theory is about 40 K. On the other hand, we can expect in a two-band model for negative $g_1$ that the transition temperature becomes higher than that derived within the single-band model, because of the tunneling of Cooper pairs between two bands. The tunneling of Cooper pairs causes the stabilization of the order parameter of the superconductivity [163, 164, 173]. In the framework of the two-band model, we consider that the Fermi energy level crosses with two bands. The results derived from the two-particle Green's function in the previous section suggest that the superconductivity appears for $g_2 < 0$ or $g_2 > 0$. Note that

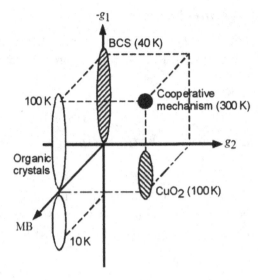

Fig. 4.2.   Schematic diagram for superconductivity. MB means multiband effects.

$g_2$ contributes to SDW. From the results derived from calculations involving $g_1$ and $g_2$ (cooperative mechanism), we expect a higher $T_c$ than that of copper oxides.

On the basis of the results obtained in this section, we present a schematic diagram for superconductivity shown in Fig. 4.2. The mechanism of high-$T_c$ superconductivity of materials such as copper oxides might be around the cooperative mechanism in Fig. 4.2. In what follows, we calculate a two-particle Green's function in the two-band model and derive an equation for coupled states. In the framework of the two-band model, the results predict that superconductivity appears even if the electron–electron interaction is positive. We can expect that the transition temperature is higher (300 K) than that for copper oxides by the cooperative mechanism (Fig. 4.3).

Let us discuss the problem of superconductors operating at room temperature (RTSC). It is obvious that the main task in the field of superconductivity is the fabrication of materials with superconducting properties at room temperature. The study of HTSC is only a stage on the way to the main purpose, namely to the development of room-temperature superconductors. At the present time, the highest known $T_c = 135$ K (at the atmospheric pressure). Now there occurs the wide-scale search for such superconductors. Since none of the known physical laws allows one to exclude the possibility for RTSC to exist, the future discovery of RTSC seems to be doubtless. New materials are created by means of physical and chemical modifications of the known compounds, including the application

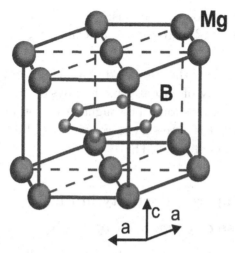

Fig. 4.3.   The structure of $MgB_2$ containing graphite-tape B layers separated by hexagonal close-packed layers of Mg.

of methods named "nanotechnological". The creation of new materials with preassigned physical properties is one of the actual problems of modern science. At present, various approaches are developed in order to avoid the labor-consuming sorting of different chemical compounds and conditions of synthesis, i.e., to optimize the solution of this problem. One of the most efficient and promising is the method of structural design. In work [165], the structural design was used in solving the problems related to the search for new HTSC on the basis of complex copper oxides.

   We now present the recommendations concerning the search for new HTSC with higher critical temperatures which were advanced by the Nobel's Prize winner K.A. Müller [166]. He divided the well-known HTSC into three classes:

(1) layered cuprates;
(2) $MgB_2$;
(3) doped fullerenes of the type $K_3C_{60}$.

K.A. Müller emphasizes that the discovery of cuprate superconductors was promoted by the conception of Jahn–Teller polarons. Two singlet polarons form a bipolaron. Bipolarons are able, in turn, to form metallic clusters (stripes). K.A. Müller indicated the following factors which Should be taken into account in the search for new HTSC:

(1) superconductivity is favored by a layered (quasi-two-dimensional) crys-
    talline structure;

(2) oxygen ions are proposed as anions;

(3) fluorine, chlorine, and nitrogen can be considered as well.

It is worth noting that the recent discovery of FeAs superconductors eliminated the monopoly of cuprates in the physics of HTSC. It is possible that new superconductors should be sought in the other rows and columns of the Periodic Table. It is necessary to concentrate efforts on the purposeful search and the creation of new HTSC.

## 4.3. Two-gap Superconductivity in $MgB_2$

### 4.3.1. *The physical properties of $MgB_2$*

Recently, the superconductivity of $MgB_2$ with $T_c = 39$ K, which is the highest temperature among two-component systems, was discovered [168]. A great interest in the study of magnesium diboride is related to the fact that $MgB_2$ has occupied the "intermediate" place between low- and high-temperature superconductors by the value of $T_c$. Therefore, the modern literature names $MgB_2$ a "medium $T_c$-superconductor" (MTSC). It is necessary to mention the cheapness of this superconductor as for the applications. We recall that wires made of cuprate superconductors include 70% of silver which is expensive. An important peculiarity of $MgB_2$ is its quasi-two-dimensional structure of the $AlB_2$ type. Interestingly, $AlB_2$ is not superconducting. Note that magnesium diboride is another example of the crucial role played by the lattice structure regarding superconductivity [167]. $MgB_2$ is an "old" material which has been known since the early 1950s, but it was only recently discovered to be superconducting. $MgB_2$ has a hexagonal structure [162], see Fig. 4.4 [24]. The results of calculations of the temperature dependence of the specific heat of $MgB_2$ given in Fig. 4.5 follow the corresponding experimental data. The band structure of $MgB_2$ has been calculated in several works since the discovery of superconductivity [172]. The electronic properties of $MgB_2$ are plotted in Fig. 4.2. The band structure of magnesium diboride is similar to that of graphite and is formed by the $\sigma$ and $\pi$ zones.

The important problem for the superconductivity in magnesium diboride is the mechanism of superconductivity. It can be the conventional electron–phonon mechanism or a more exotic mechanism of superconductivity. The presence of an isotope effect is a strong indicator of the phonon mediation of superconductivity. The large isotope effect $\alpha_B = 0.26$ [169] shows that phonons associated with B vibrations play a significant role in the magnesium diboride superconductivity, whereas the magnesium isotope effect is very

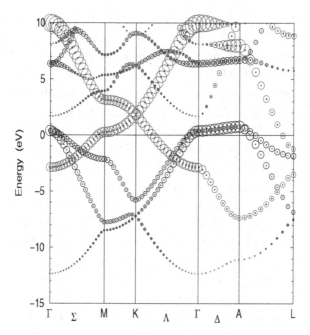

Fig. 4.4.   Band structure of $MgB_2$.

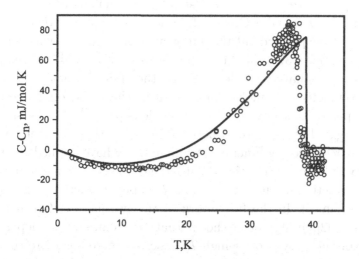

Fig. 4.5.   The $MgB_2$ specific heat vs. temperature. Solid line — theory. Points represent the experimental data.

small: $\alpha_M = 0.02$ [169]. The total isotope effect is $\alpha_B + \alpha_M = 0.3$, which supports the electron–phonon mechanism of superconductivity.

Magnesium diboride is the II-type superconductor [167]. The discovery of the superconductivity of $MgB_2$ has aroused a great interest in the multigap superconductivity [169]. Magnesium diboride has two superconducting gaps, 4 meV and 7.5 meV, due to the $\pi$ and $\sigma$ electron bands. The two-gap structure was established in a number of experiments [162]. The plot of the specific heat of $MgB_2$ vs. temperature (Fig. 4.5) demonstrates a good agreement between the theoretical and experimental data. Both gaps have $s$-symmetry and are resulted from the highly anisotropic layer structure of $MgB_2$. Although the multigap superconductivity was discussed theoretically [163, 164, 173] as early as 1958, it has been observed experimentally [170] in the 1980s. $MgB_2$ is the first material, in which the effects of the multigap superconductivity are so dominant, and its implications so thoroughly explored. The Nature has let us have a glimpse of a few her multigap mysteries and has challenged us. Recent band calculations [171,172] of $MgB_2$ with the McMillan formula for the transition temperature have supported the e–p interaction mechanism for the superconductivity. In this case, the possibility of two-band superconductivity has also been discussed in relation to two-gap functions experimentally and theoretically. Very recently, the two-band or multiband superconductivity has been theoretically investigated in relation to the superconductivity arising from Coulomb repulsive interactions. The two-band model was first introduced by Kondo [173]. We have also investigated anomalous phases in the two-band model by using the Green's function techniques [137–140, 142, 143]. Recently, we have pointed out the importance of multiband effects in high-$T_c$ superconductivity [137–139]. The expressions for the transition temperature for several phases have been derived, and the approach has been applied to the superconductivity in molecular crystals by charge injection and the field-induced superconductivity [14]. In the previous paper [137–140,142,143], we have investigated the superconductivity by using the two-band model and the two-particle Green's function techniques. We have applied the model to the electron–phonon mechanism for the traditional BCS method, electron–electron interaction mechanism for high-$T_c$ superconductivity [2], and cooperative mechanism. In the framework of the two-particle Green's function techniques [142], it has been shown that the temperature dependence of the superconductivity gap for high-$T_c$ superconductors is more complicated than that predicted in the BCS approach. In paper [172], phase diagrams for the two-band model superconductivity have been investigated, by using the renormalization group approach. Below, we will discuss the possibility of the cooperative mechanism of two-band superconductivity in relation to the high-$T_c$ superconductivity and study the effect of the increasing of $T_c$ in $MgB_2$ due to the enhanced interband pairing scattering. In this section, we will investigate our two-band model for the explanation the multigap superconductivity of $MgB_2$. We apply the model to the electron–phonon mechanism for the traditional BCS method, electron–electron interaction mechanism for high-$T_c$ superconductivity, and cooperative mechanism in relation to multiband superconductivity.

### 4.3.1.1. $Mg_{1-x}Al_xB_2$

Critical temperature and other superconducting properties of two-band superconductors depend on the doping level and on the interband and intraband scattering and can be modified by chemical substitutions. The influence of doping on the transition temperature in $Mg_{1-x}Al_xB_2$ is illustrated in Fig. 4.6. It is seen that the doping destroys the superconductivity in $MgB_2$. This can be understood as a result of the competition of two effects: the first one is a coupling effect related to the changes of the carrier concentration, and the second one depends on the introduction of new scattering centers leading to a modification of the interband and intraband scattering.

### 4.3.2. *Theoretical model*

In this section, we will use the two-band model for superconductivity. We start from the Hamiltonian for two-bands $i$ and $j$:

$$H = H_0 + H_{\text{int}}, \qquad (4.30)$$

where

$$H_0 = \sum_{\mathbf{k},\sigma} [[\epsilon_i(\mathbf{k}) - \mu]\, a^\dagger_{i\mathbf{k}\sigma} a_{i\mathbf{k}\sigma} + [\epsilon_j(\mathbf{k}) - \mu]\, a^\dagger_{j\mathbf{k}\sigma} a_{j\mathbf{k}\sigma}], \qquad (4.31)$$

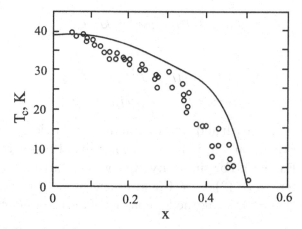

Fig. 4.6. The influence of doping on $T_c$ in $Mg_{1-x}Al_xB_2$. Solid line — theory. Points represent the experimental data.

$$H_{\text{int}} = \frac{1}{4} \sum_{\delta(\mathbf{p}_1+\mathbf{p}_2,\mathbf{p}_3+\mathbf{p}_4)} \sum_{\alpha,\beta,\gamma,\delta} [\Gamma^{iiii}_{\alpha\beta\gamma\delta} a^\dagger_{i\mathbf{p}_1\alpha} a^\dagger_{i\mathbf{p}_2\beta} a_{i\mathbf{p}_3\gamma} a_{i\mathbf{p}_4\delta} + (i \leftrightarrow j)$$

$$+ \Gamma^{iijj}_{\alpha\beta\gamma\delta} a^\dagger_{i\mathbf{p}_1\alpha} a^\dagger_{i\mathbf{p}_2\beta} a_{j\mathbf{p}_3\gamma} a_{j\mathbf{p}_4\delta} + (i \leftrightarrow j)$$

$$+ \Gamma^{ijij}_{\alpha\beta\gamma\delta} a^\dagger_{i\mathbf{p}_1\alpha} a^\dagger_{j\mathbf{p}_2\beta} a_{i\mathbf{p}_3\gamma} a_{j\mathbf{p}_4\delta} + (i \leftrightarrow j)], \qquad (4.32)$$

$\Gamma$ is the bare vertex part:

$$\Gamma^{ijkl}_{\alpha\beta\gamma\delta} = \langle i\mathbf{p}_1\alpha j\mathbf{p}_2\beta | k\mathbf{p}_3\gamma l\mathbf{p}_4\delta \rangle \delta_{\alpha\delta}\delta_{\beta\gamma} - \langle i\mathbf{p}_1\alpha j\mathbf{p}_2\beta | l\mathbf{p}_4\delta k\mathbf{p}_3\gamma \rangle \delta_{\alpha\gamma}\delta_{\beta\delta} \quad (4.33)$$

with

$$\langle i\mathbf{p}_1\alpha j\mathbf{p}_2\beta | k\mathbf{p}_3\beta l\mathbf{p}_4\alpha \rangle$$

$$= \int dr_1 dr_2 \phi^*_{i\mathbf{p}_1\alpha}(r_1) \phi^*_{j\mathbf{p}_2\beta}(r_2) V(r_1, r_2) \phi_{k\mathbf{p}_3\beta}(r_2) \phi_{l\mathbf{p}_4\alpha}(r_1), \qquad (4.34)$$

where $a_{i\mathbf{p}\sigma+}$ ($a_{i\mathbf{p}\sigma}$) is the creation (annihilation) operator corresponding to the excitation of electrons (or holes) in the $i$th band with spin $\sigma$ and momentum $\mathbf{p}$, $\mu$ is the chemical potential, and $\phi$ is a single-particle wave function. Here, we suppose that the vertex function in Eq. (4.32) involves the effective interactions between the carriers caused by the linear vibronic coupling in several bands and the screened Coulombic interband interaction of carriers.

When we use the two-band Hamiltonian (4.1) and define the order parameters for a singlet exciton, triplet exciton, and singlet Cooper pair, the mean-field Hamiltonian is easily derived [137–140, 142, 143]. Here, we focus on three electron scattering processes contributing to the singlet superconducting phase in Hamiltonian (4.1):

$$g_{i1} = \langle ii|ii \rangle, \quad g_{j1} = \langle jj|jj \rangle, \qquad (4.35)$$

$$g_2 = \langle ii|jj \rangle = \langle jj|ii \rangle, \qquad (4.36)$$

$$g_3 = \langle ij|ij \rangle = \langle ji|ji \rangle, \qquad (4.37)$$

$$g_4 = \langle ij|ji \rangle = \langle ji|ij \rangle, \qquad (4.38)$$

$g_{i1}$ and $g_{j1}$ represent the $i$th and $j$th intraband two-particle normal scattering processes, respectively, and $g_2$ indicates the intraband two-particle umklapp scattering (see Fig. 4.7). For simplicity, we consider three cases in [171]: (1) $g_1 \neq 0$ and others = 0, (2) $g_2 \neq 0$ and others = 0, and (3) $g_{i1} \neq 0$ and $g_{j1} \neq 0$, and others = 0, using the two-particle Green's function techniques (see Fig. 4.7).

It was shown that, possibly, the two-gap superconductivity arises in case (3). The superconductivity arising from the electron–phonon mechanism

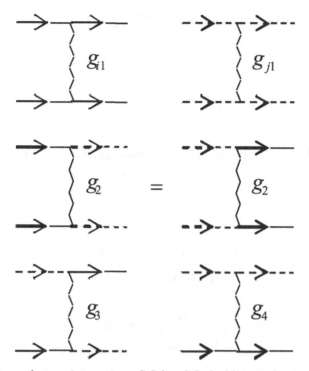

Fig. 4.7. Electron–electron interactions. Solid and dashed lines indicate $\pi$- and $\sigma$-bands, respectively; $g_{i1}$, $g_{j1}$, $g_2$, $g_3$, and $g_4$ contribute to superconductivity.

($g_1 < 0$ and $g_1 < g_2$) such as for MgB$_2$ is in the two-gap region. On the other hand, the superconductivity of copper oxides ($g_1 > g_2$) is outside the two-gap region. These results predict that we may observe two-gap functions for MgB$_2$ and only a single-gap function for copper oxides.

### 4.3.3. *Superconductivity in MgB$_2$*

We used case (3) for $g_{i1}$ and $g_{j1}$, $g_2 \neq 0$ and others $= 0$ to describe the superconductivity in MgB$_2$. We have a reduced Hamiltonian

$$H = H_0 + H_{\text{int}}, \tag{4.39}$$

where

$$H_0 = \sum_{\mathbf{k},\sigma} [[\epsilon_i - \mu]\, a_{i\mathbf{k}\sigma}^\dagger a_{i\mathbf{k}\sigma} + [\epsilon_j - \mu]\, a_{j\mathbf{k}\sigma}^\dagger a_{j\mathbf{k}\sigma}], \tag{4.40}$$

$$H_{\text{int}} = \sum g_{1i} a_{i\mathbf{k}}^\dagger a_{i-\mathbf{k}}^\dagger a_{i-\mathbf{k}} a_{i\mathbf{k}} + \sum i \to j$$
$$+ \sum g_2 a_{i\mathbf{k}}^\dagger a_{i-\mathbf{k}}^\dagger a_{j-\mathbf{k}} a_{j\mathbf{k}}. \tag{4.41}$$

We now define the order parameters which are helpful to construct the mean-field Hamiltonian:

$$\Delta_i = \sum_p \langle a_{ip\uparrow}^\dagger a_{i-p\downarrow}^\dagger \rangle, \tag{4.42}$$

$$\Delta_j = \sum_p \langle a_{jp\uparrow}^\dagger a_{j-p\downarrow}^\dagger \rangle. \tag{4.43}$$

The relation between two superconducting gaps of the system is as follows:

$$\Delta_j = \frac{1 - g_{i1}\rho_i f_i}{g_2 \rho_j f_j} \Delta_i, \tag{4.44}$$

where

$$
\begin{aligned}
f_i &= \int_\mu^{\mu - E_c} \frac{d\xi}{\left(\xi^2 + \Delta_i^2\right)^{1/2}} \tanh \frac{\left(\xi^2 + \Delta_i^2\right)^{1/2}}{2T}, \\
f_j &= \int_{\mu - E_c}^{\mu - E_j} \frac{d\xi}{\left(\xi^2 + \Delta_j^2\right)^{1/2}} \tanh \frac{\left(\xi^2 + \Delta_j^2\right)^{1/2}}{2T}
\end{aligned}
\tag{4.45}
$$

with the coupled gap equation

$$\left(1 - g_{i1}\rho_i f_i\right)\left(1 - g_{j1}\rho_j f_j\right) = g_2^2 f_i f_j. \tag{4.46}$$

We have tried to estimate the coupling constant of the pair electron scattering process between $\pi$- and $\sigma$-bands of $MgB_2$. We have calculated the parameters by using a rough numerical approximation. We focus on one $\pi$-band and $\sigma$-band of $MgB_2$ and consider electrons near Fermi surfaces. We find that the parameter $g_1 = -0.4\,eV$, by using the transfer integral between the $\pi$- and $\sigma$-bands. We estimate the coupling parameter $g_2$ of the pair electron scattering process by the expression

$$g_2 = \sum_{k1,k2} V_{k1,k2}^{1,2}, \tag{4.47}$$

$$V_{k1,k2}^{1,2} = \sum_{r,s,t,u} u_{1,r}^*(k1) u_{1,s}^*(k1) v_{rs} u_{2,t}(k2) u_{2,u}(k2), \tag{4.48}$$

where labels 1 and 2 mean the $\pi$- and $\sigma$-bands, respectively, $u_{i,r}(\xi)$ is the LCAO coefficient for the $i$th band, and $\xi$ is the moment [159, 160]. The indices $k1$ and $k2$ are summed over each Fermi surface. However, it is difficult to perform the sum exactly. In this case, we used a few points near the Fermi surface. The coupling constant of the pair electron scattering between the $\pi$-band and $\sigma$-band $g_2 = 0.025\,eV$. From numerical calculations

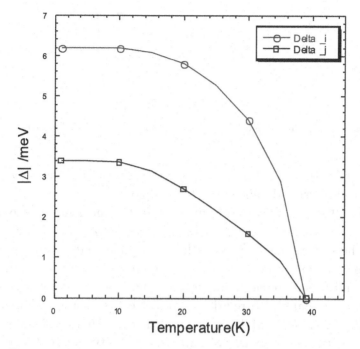

Fig. 4.8.   Temperature dependence of two superconducting gaps.

of Eqs. (4.44)–(4.46), we can also obtain the temperature dependence of the
two-gap parameters (see Fig. 4.8).

We have used the density of states of the $\pi$- and $\sigma$-bands ($\rho_i = 0.2\,\mathrm{eV}^{-1}$,
$\rho_j = 0.14\,\mathrm{eV}^{-1}$), chemical potential $\mu = -2.0$, the top energy of $\sigma$-band
$E_j = -1.0$, and the fitting parameters ($g_{i1} = -0.4\,\mathrm{eV}$, $g_{j1} = -0.6\,\mathrm{eV}$,
$g_2 = 0.02\,\mathrm{eV}$). These calculations have qualitative agreement with experiments [162, 174, 175]. The expression for the transition temperature of
superconductivity is derived in a simple approximation:

$$T_{c_+} = 1.13(\zeta - E_j) \cdot \exp(-1/g_+\rho), \qquad (4.49)$$

where

$$g_+ = \frac{1}{24}(B + \sqrt{B^2 - 4A}) \qquad (4.50)$$

and

$$A = g_{1i}g_{1j} - g_2^2, \qquad (4.51)$$

$$B = g_{1i} + g_{1j}, \qquad (4.52)$$

$$\zeta = -\mu. \qquad (4.53)$$

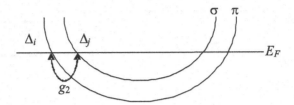

Fig. 4.9.   Schematic diagram of the mechanism of pairing for two gaps.

From the expressions for $T_{c+}$, we can see the effect of increase of $T_{c+}$ due to the enhanced interband pairing scattering ($g_2$).

Figure 4.9 shows a schematic diagram of the mechanism of pairing for two gaps. The scenario is as follows. Electrons from the $\pi$- and $\sigma$-bands make up the subsystems. For $g_2$, we have two independent subsystems with the different transition temperatures of superconductivity $T_{c\pi}$ and $T_{c\sigma}$ and two independent superconducting gaps. In our model, we have two coupled superconducting gaps with relation (16) and one transition temperature of superconductivity $T_{c+}$, which is in agreement with experiments. In this model, we have two channels of superconductivity: conventional channel (intraband $g_1$) and unconventional channel (interband $g_2$). Two gaps appear simultaneously in different bands which are like BCS gaps. The gap in the $\pi$-band is bigger as compared with that for the $\sigma$-band, because the density of state is 0.25 eV in the $\pi$-band and 0.14 eV in the $\sigma$-band. The current of Cooper pairs flows from the $\pi$-band into the $\sigma$-band, because the density of Cooper pairs in the $\pi$-band is much higher. The tunneling of Cooper pairs also stabilizes the order parameter in the $\sigma$-band. In this way, we can predict the physical properties of the multigap superconductivity, if we have the superconductors with a multiband structure, as shown in Fig. 4.9.

Thus, we have presented our two-band model with the intraband two-particle scattering and interband pairing scattering processes to describe two-gap superconductivity in $MgB_2$. We defined the parameters of our model and made numerical calculations of the temperature dependence of two gaps in a qualitative agreement with experiments. We have proposed a two-channel scenario of superconductivity : the first one is the conventional channel (intraband $g_1$) which is connected with the BCS mechanism in different bands and the unconventional channel (interband $g_2$) which describes the tunneling of Cooper pairs between two bands. The tunneling of Cooper pairs also stabilizes the order parameters of superconductivity and increases the critical temperature of superconductivity.

## 4.4. Theoretical Studies of Multiband Effects in Superconductivity by Using the Renormalization Group Approach

We present the renormalization equations by using the two-band model and construct phase diagrams for the two-band superconductivity. In the framework of the two-band model, the given results predict that superconductivity appears, even if the electron–electron interaction is positive. We discuss the possibility of a cooperative mechanism in the two-band superconductivity in relation to high-$T_c$ superconductivity. The recent discovery of superconductivity of $MgB_2$ [168] has also attracted a great interest in the elucidation of its mechanism from both experimental and theoretical viewpoints. A crucial role of the electron–phonon (e–p) interaction has been pointed out for the superconductivity of $MgB_2$. Recent band calculations of the transition temperature for $MgB_2$ [171, 172] with the McMillan formula [176] have supported the e–p interaction mechanism for the superconductivity. In this case, the possibility of the two-band superconductivity in relation to two-gap functions has also been considered experimentally and theoretically. Very recently, the two-band or multiband superconductivity has been theoretically investigated in relation to the superconductivity arising from Coulomb repulsive interactions [177, 231]. The two-band model was first introduced by Kondo [173]. Recently, we have pointed out the importance of multiband effects in high-$T_c$ superconductivity [182]. We have also investigated anomalous phases in the two-band model by using the Green's function techniques [180]. The expressions of the transition temperature for several phases have been derived, and the approach has been applied to the superconductivity in several crystals by charge injection and the field-induced superconductivity [180, 183]. In the previous section, we have investigated superconductivity by using the two-band model and the two-particle Green's function techniques [142, 192]. We have applied the model to the electron–phonon mechanism for the traditional BSC method, electron–electron interaction mechanism for high-$T_c$ superconductivity, and cooperative mechanism. In the framework of the two-particle Green's function techniques [142], it has been shown that, in the electron–phonon system, a class of new so-called coupled states arises. In this section, we investigate two- or multiband effects in superconductivity by using the two-band model within the renormalization group approach. Renormalization equations for the two-band superconductivity are derived from the response function and the vertex correction of the model. Phase diagrams numerically obtained from the renormalization equations are presented. We also discuss the superconductivity arising from the e–e repulsive interaction in relation to the two-band superconductivity.

### 4.4.1.  *Theoretical model*

In this section, we briefly summarize the two-band model for the superconductivity and introduce the renormalization group approach [186, 188].

We consider a Hamiltonian for two bands $i$ and $j$ written as

$$H = H_0 + H_{\text{int}}, \qquad (4.54)$$

where

$$H_0 = \sum_{\mathbf{k},\sigma} [[\epsilon_i(\mathbf{k}) - \mu] \, a^\dagger_{i\mathbf{k}\sigma} a_{i\mathbf{k}\sigma} + [\epsilon_j(\mathbf{k}) - \mu] \, a^\dagger_{j\mathbf{k}\sigma} a_{j\mathbf{k}\sigma}], \qquad (4.55)$$

$$
\begin{aligned}
H_{\text{int}} = \frac{1}{4} \sum_{\delta(\mathbf{p}_1+\mathbf{p}_2,\mathbf{p}_3+\mathbf{p}_4)} \sum_{\alpha,\beta,\gamma,\delta} [&\Gamma^{iiii}_{\alpha\beta\gamma\delta} a^\dagger_{i\mathbf{p}_1\alpha} a^\dagger_{i\mathbf{p}_2\beta} a_{i\mathbf{p}_3\gamma} a_{i\mathbf{p}_4\delta} + (i \leftrightarrow j) \\
&+ \Gamma^{iijj}_{\alpha\beta\gamma\delta} a^\dagger_{i\mathbf{p}_1\alpha} a^\dagger_{i\mathbf{p}_2\beta} a_{j\mathbf{p}_3\gamma} a_{j\mathbf{p}_4\delta} + (i \leftrightarrow j) \\
&+ \Gamma^{ijij}_{\alpha\beta\gamma\delta} a^\dagger_{i\mathbf{p}_1\alpha} a^\dagger_{j\mathbf{p}_2\beta} a_{i\mathbf{p}_3\gamma} a_{j\mathbf{p}_4\delta} + (i \leftrightarrow j)],
\end{aligned}
\qquad (4.56)
$$

$\Gamma$ is the bare vertex part:

$$\Gamma^{ijkl}_{\alpha\beta\gamma\delta} = \langle i\mathbf{p}_1\alpha j\mathbf{p}_2\beta | k\mathbf{p}_3\gamma l\mathbf{p}_4\delta \rangle \delta_{\alpha\delta}\delta_{\beta\gamma} - \langle i\mathbf{p}_1\alpha j\mathbf{p}_2\beta | l\mathbf{p}_4\delta k\mathbf{p}_3\gamma \rangle \delta_{\alpha\gamma}\delta_{\beta\delta}, \qquad (4.57)$$

with

$$\langle i\mathbf{p}_1\alpha j\mathbf{p}_2\beta | k\mathbf{p}_3\beta l\mathbf{p}_4\alpha \rangle$$

$$= \int dr_1 dr_2 \phi^*_{i\mathbf{p}_1\alpha}(r_1)\phi^*_{j\mathbf{p}_2\beta}(r_2) V(r_1, r_2) \phi_{k\mathbf{p}_3\beta}(r_2)\phi_{l\mathbf{p}_4\alpha}(r_1), \qquad (4.58)$$

where $a^\dagger_{i\mathbf{p}\sigma} (a_{i\mathbf{p}\sigma})$ is the creation (annihilation) operator corresponding to the excitation of electrons (or holes) in the $i$th band with spin $\sigma$ and momentum $\mathbf{p}$, $\mu$ is the chemical potential, and $\phi^*_{i\alpha\mathbf{p}_1}$ is a single-particle wave function. Here, we suppose that the vertex function in Eq. (4.56) involves the effective interactions between carriers caused by the linear vibronic coupling in several bands and the screened Coulombic interband interaction of carriers.

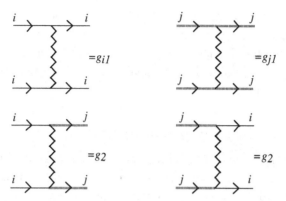

Fig. 4.10. Electron–electron interactions. Dependence of $g$ on the direction in the momentum space is ignored in this model $(g_x(\boldsymbol{k}) \approx g_x(x = i, j))$. We assume that $g_x$ is constant.

We focus on three electron scattering processes contributing to the singlet superconducting phase in Hamiltonian (4.56), as shown in Fig. 4.10,

$$g_{i1} = \langle ii|ii \rangle, \tag{4.59}$$

$$g_{j1} = \langle jj|jj \rangle, \tag{4.60}$$

$$g_2 = \langle ii|jj \rangle = \langle jj|ii \rangle, \tag{4.61}$$

$g_{i1}$ and $g_{j1}$ represent the $i$th and $j$th intraband two-particle normal scattering processes, respectively, and $g_2$ indicates the intraband two-particle umklapp scattering. Note that $\Gamma$'s are given by

$$\Gamma^{iiii}_{\alpha\beta\gamma\delta} = g_{i1} \left( \delta_{\alpha\delta}\delta_{\beta\gamma} - \delta_{\alpha\gamma}\delta_{\beta\delta} \right),$$

$$\Gamma^{jjjj}_{\alpha\beta\gamma\delta} = g_{j1} \left( \delta_{\alpha\delta}\delta_{\beta\gamma} - \delta_{\alpha\gamma}\delta_{\beta\delta} \right), \tag{4.62}$$

$$\Gamma^{iijj}_{\alpha\beta\gamma\delta} = \Gamma^{jjii}_{\alpha\beta\gamma\delta} = g_2 \left( \delta_{\alpha\delta}\delta_{\beta\gamma} - \delta_{\alpha\gamma}\delta_{\beta\delta} \right),$$

where we assume that an antisymmetrized vertex function $\Gamma$ is a constant independent of the momenta.

The spectrum is elucidated by the Green's function method. Using Green's functions which characterize the CDW, SDW, and SSC phases, we obtain a self-consistent equation according to the traditional procedure [159, 180, 182, 188–191]. Then we can obtain expressions for the transition temperature in some cases. Electronic phases of a one-dimensional system have been investigated by using a similar approximation in the framework of the one-band model [188–191]. In the framework of the mean-field approximation within the two-band model, we have already derived expressions for the transition temperature for CDW, SDW, and SSC.

In paper [180, 182], we have investigated the dependence of $T_c$ on the hole or electron concentration for the superconductivity of copper oxides by using the two-band model and have obtained a phase diagram of $Bi_2Sr_2Ca_{1-x}Y_xCu_2O_8$ (Bi-2212) by means of the above expressions for the transition temperature.

### 4.4.2.  *Renormalization group approach*

The Dyson equation is invariant under a multiple renormalization of Green's function and coupling parameters $g$. From this invariance for a scaling procedure, we obtain differential equations for the coupling parameters and the external vertex of a Cooper pair:

$$y\frac{\partial}{\partial y}\tilde{g}_i(y,u,g) = \frac{\partial}{\partial\xi}\tilde{g}_i(\xi,u/y,\tilde{g}(t,u,g))|_{\xi=1}, \qquad (4.63)$$

$$y\frac{\partial}{\partial y}\ln\Lambda(y,u,g) = \frac{\partial}{\partial\xi}\ln\Lambda(\xi,u/y,\tilde{g}(t,u,g))|_{\xi=1}, \qquad (4.64)$$

where $y$ and $u$ are parameters with the dimension of energy, $g$ means the set of original couplings, and $\Lambda$ is the external vertex.

### 4.4.3.  *Vertex correction and response function for Cooper pairs*

To solve Eqs. (4.63) and (4.64), we estimate the right-hand side of Eq. (4.63) by using the perturbation theory. We consider the lowest order correction to the vertex for a Cooper pair as shown in Fig. 4.11. From these diagrams, we obtain

$$\begin{pmatrix} \tilde{g}_{i1} \\ \tilde{g}_{j1} \end{pmatrix} = \begin{pmatrix} g_{i1} \\ g_{j1} \end{pmatrix} + \begin{pmatrix} -g_{i1}^2 & -g_2^2 \\ -g_{j1}^2 & -g_2^2 \end{pmatrix} \begin{pmatrix} L_i \\ L_j \end{pmatrix}, \qquad (4.65)$$

$$\begin{pmatrix} \tilde{g}_2 \\ \tilde{g}_2 \end{pmatrix} = \begin{pmatrix} g_2 \\ g_2 \end{pmatrix} + \begin{pmatrix} -g_{i1}g_2 & -g_2g_{j1} \\ -g_{j1}g_2 & -g_2g_{i1} \end{pmatrix} \begin{pmatrix} L_i \\ L_j \end{pmatrix}, \qquad (4.66)$$

where

$$L_i = \Pi_i(\boldsymbol{k},\omega) = \frac{T}{N}\sum_{\boldsymbol{q},\omega'} G_i(\boldsymbol{q},\omega')G_i(\boldsymbol{k}-\boldsymbol{q},\omega-\omega'),$$

$$L_j = \Pi_j(\boldsymbol{k},\omega) = \frac{T}{N}\sum_{\boldsymbol{q},\omega'} G_j(\boldsymbol{q},\omega')G_j(\boldsymbol{k}-\boldsymbol{q},\omega-\omega'),$$

$$(4.67)$$

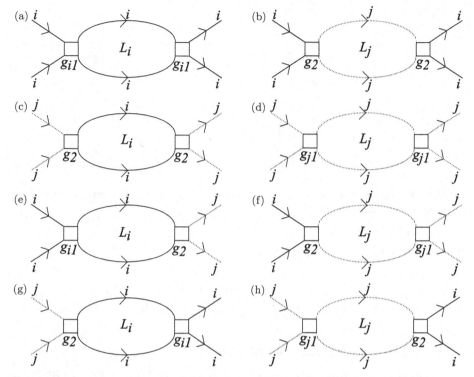

Fig. 4.11. Diagrams of the first-order vertex correction: (a) and (b) contribute to $\tilde{g}_{i1}$, (c) and (d) are diagrams for $\tilde{g}_{j1}$, (e)–(h) for $\tilde{g}_2$.

where $G$ and $T$ are the temperature Green's function and temperature, respectively. For the special case $\omega = 0$, $\boldsymbol{k} = 0$, the above functions $L_i$ and $L_j$ become

$$L_i = - \left[ \tanh \left( \frac{u_i/y}{2\xi} \right) + \tanh \left( \frac{u_i'/y}{2\xi} \right) \right] \ln \xi - 2A, \qquad (4.68)$$

$$L_j = - \left[ \tanh \left( \frac{u_j/y}{2\xi} \right) + \tanh \left( \frac{u_j'/y}{2\xi} \right) \right] \ln \xi - 2A, \qquad (4.69)$$

where

$$A = \int dx \ln x \operatorname{sech}^2 x, \qquad (4.70)$$

$u_i$ $(u_j)$ and $u_i'$ $(u_j')$ are dimensionless functions expressed by the chemical potential, cut-off energy, the top energy of $j$th band, and the density of state for the $i$th $(j$th) band.

Next, we consider a first-order response function for a singlet Cooper pair as shown in Fig. 4.12. Then the first-order vertex function $\Lambda$ for the

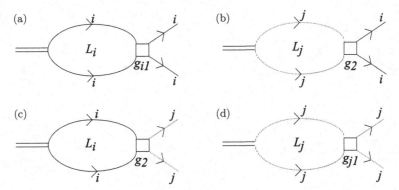

Fig. 4.12.   Diagrams of the first-order response function: (a) and (b) contribute to $\Lambda_i$, (c) and (d) show diagrams for $\Lambda_j$.

$i$th and $j$th bands can be written as

$$\begin{pmatrix} \Lambda_i + \Lambda_j \\ \Lambda_i - \Lambda_j \end{pmatrix} = \begin{pmatrix} 2 \\ 0 \end{pmatrix} + \begin{pmatrix} -g_1 & -g_2 \\ -g_1 & g_2 \end{pmatrix} \begin{pmatrix} L_i + L_j \\ L_i - L_j \end{pmatrix}. \tag{4.71}$$

### 4.4.4.  *Renormalization equation*

For simplicity, hereafter we assume $g_{i1} = g_{j1}$. From Eqs. (4.63), (4.65), and (4.66), we obtain the differential equations written as

$$\frac{\partial}{\partial x}\tilde{g}_1 = -\left(\tilde{g}_1^2 + \tilde{g}_2^2\right), \tag{4.72}$$

$$\frac{\partial}{\partial x}\tilde{g}_2 = -2\tilde{g}_1\tilde{g}_2. \tag{4.73}$$

In similar way, using Eqs. (4.64) and (4.71), we obtain the differential equations written as

$$\frac{\partial}{\partial x}\ln\Lambda_+ = -\tilde{g}_1 - \tilde{g}_2, \tag{4.74}$$

$$\frac{\partial}{\partial x}\ln\Lambda_- = -\tilde{g}_1 + \tilde{g}_2, \tag{4.75}$$

where $\Lambda_+ = \Lambda_i + \Lambda_j$ and $\Lambda_- = \Lambda_i - \Lambda_j$.

### 4.4.5.  *Phase diagrams*

In the previous section, we have derived the basic equations (4.72)–(4.75) to find the low-temperature phases. For the special case of $g_2 = 0$, we obtain an analytic solution

$$\tilde{g}_1 = \frac{1}{x + g_1^{-1}}, \qquad (4.76)$$

$$\tilde{g}_2 = 0, \qquad (4.77)$$

$$\Lambda_i = \Lambda_j = \frac{1}{g_1 x + 1}. \qquad (4.78)$$

From these solutions, we find that the superconducting phase appears only when the intraband interaction $g_1$ is negative. In the case of the traditional superconductivity described by the BCS theory, it is necessary that the effective electron–electron interaction be negative ($g_1 < 0$) to realize the superconductivity. The present result agrees with that of the traditional theory for superconductivity expressed by the one-band model. For the case of $g_2 \neq 0$, the phase diagrams numerically obtained from the above renormalization equations are shown in Fig. 4.13. Figure 4.13(a) shows the phase diagram for the sum of contributions to the superconductitivity from the $i$th and $j$th bands. Thus, this phase implies that the signs of the superconducting state for the $i$th and $j$th bands are the same. From this diagram, we find the superconductivity only for $-g_1 - g_2 > 0$ with negative $g_2$. On the other hand, the phase diagram for the difference between the superconducting states for the $i$th and $j$th bands is shown in Fig. 4.12(b). In this case, this phase means that the sign of the superconductivity for the $i$th band is different from that of the $j$th band. We can find that the superconductivity appears only for $-g_1 + g_2 > 0$ with positive $g_2$ from Fig. 4.13(b). Thus, the present results suggest that the two-band

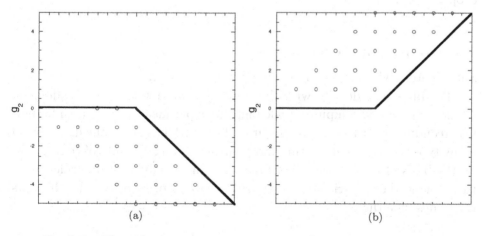

Fig. 4.13. Phase diagrams for superconductivity: (a) $\Lambda_i + \Lambda_j$, (b) $\Lambda_i - \Lambda_j$.

superconductivity appears, when the intraband umklapp repulsive scattering $g_2$ is larger than the normal repulsive scattering $g_1$. In the region of $g_2 > g_1$ with $g_2 < 0$ and $g_2 < -g_1$ with $g_2 > 0$ (two-gap region), we can expect that two-gap functions are observed. In the former region, those superconducting gaps may be expressed by $|\Lambda_+| > |\Lambda_-|$, and the latter may be $|\Lambda_+| < |\Lambda_-|$. On the other hand, we expect only a single-gap function in the other region. These results agree with the previous solutions [140] derived by using the two-particle Green's function techniques. The superconductivity arising from the electron–phonon mechanism ($g_1 < 0$ and $|g_2| < |g_1|$) such as that in $MgB_2$ is in the two-gap region. On the other hand, the superconductivity such as in copper oxides ($|g_2| > |g_1|$) is outside the two-gap region. These results predict that we may observe two-gap functions for $MgB_2$ and only a single-gap function for copper oxides. In the two-band model for negative $g_1$ with transferring or tunneling of Cooper pairs between two bands, we can expect that the transition temperature becomes higher than that derived from the single-band model. The tunneling of Cooper pairs causes the stabilization of the order parameter of superconductivity [163, 173]. We can also expect higher $T_c$ of the superconductivity than that for copper oxides in two regions ($g_1 < 0$, $g_2 < 0$ and $g_1 < 0$, $g_2 > 0$) by the cooperative mechanism. Phase diagrams for CDW, SDW, and singlet superconductivity derived from a more general Hamiltonian will be presented elsewhere.

Thus, we have derived the renormalization equations and present the phase diagrams for the two-band superconductivity. In the framework of the two-band model, the present results predict that superconductivity appears, even if the electron–electron interaction is positive. We can expect that the transition temperature becomes higher than that of copper oxides by the cooperative mechanism.

## 4.5.   Room Superconductors

Superconductivity sets up a new temperature record.

The physicists dealing with the condensed matter made huge efforts in the fields of theory, computer modeling, and experiment in the search for new superconductors at room temperature. Their work has led to the discovery of many types of high-temperature materials, including a family of compounds on the basis of copper and iron, fullerides, and magnesium diboride. In all those cases, the materials were studied in laboratories before the theorists try to describe them.

The list of physical news in 2019 [193] was rightly headed by the communications of two independently groups working at George Washington University (USA) [194] and Max-Planck Institut für Chemie (Germany) [195] on the observation of the superconductivity in lanthanium hydride $LaH_{10}$ at about 260 K. This value is well perceived already on the Celsius scale to be $-13°C$, which corresponds to the ordinary winter temperature on the northern latitudes. However, the conditions under which the superconductivity was registered (pressure of several millions of atm.) are hardly called normal. The required value (180 GPa) was attained with the use of a cell with diamond anvils.

This history is distinct from the previous breakthroughs in the superconductivity in that the discovery was not completely unpredictable: the direction of searches was set as early as in 1968 by N. Ashcroft who pointed out that, according to the BCS (Bardeen, Cooper, and Schrieffer) theory, metallic hydrogen and the compounds with its high content should possess the property of a high-temperature superconductor. The targeted seeking of the superconductivity among rare-earth hydrides started in 2017 [196] and has recently reached the success.

The scheme of the experiment and the crystalline structure of $LaH_{10}$ are shown in Figs. 4.14 and 4.15, respectively.

### 4.5.1.  *On the superconductivity of hydrogen sulfide*

The recent discovery of the superconductivity with $T_c = 203$ K in the compound $H_3S$, being an ordinary superconductor, has returned the interest

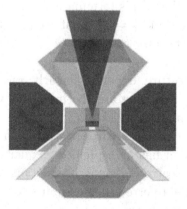

Fig. 4.14.  Scheme of the experiment, where the diamond anvils, spacer, and pistons creating a high pressure are seen.

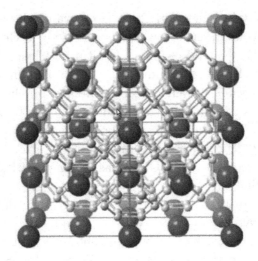

Fig. 4.15.   Crystalline structure of LaH$_{10}$.

in the BCS predictions about a possible superconductivity at room temperature. The mictoscopic BCS theory of superconductivity implies that, at a favorable combination of parameters of a material, the superconductivity at room temperature is possible. The absence of the upper theoretical limit for the critical temperature of the superconductivity transition follows from the Allen–Dynes formula

$$T_c = 0.182\omega_{\ln}\sqrt{\lambda}, \qquad (4.79)$$

where $\omega_{\ln}$ is the mean frequency of phonons, and $\lambda$ is the constant of the electron–phonon interaction. This formula is applicable only for $\lambda > 5$, but it allows one to conclude that $T_c = 300$ K, if the material has simultaneously the high Debye frequency and the large constant of the electron–phonon interaction.

The critical temperature depends linearly on the averaged phonon frequencies. Therefore, promising are the materials with light that possess high frequencies in the phonon spectrum.

It is natural to consider pure hydrogen which can be a high-temperature superconductor. Hydrogen has high phonon frequencies ($T_D = 2000$ K) and a large constant of the electron–phonon interaction ($\lambda = 1 - 2$).

V.L. Ginzburg in [199] referred the possibility to create metallic hydrogen to the most significant problems.

However, metallic hydrogen remains unattainable experimentally till now, because its metallization requires too high pressures. Nevertheless, many theoretical and experimental works are referred to this trend.

Since the high frequencies in the phonon spectrum favor the attainment of a critical temperature, it is obvious that hydrides are perspective materials with high content of hydrogen as the lightest element. N. Ashcroft [201] proposed to make experiments with such substances as $CH_4$, $SiH_4$, $GeH_4$, etc. instead of hydrogen. They can be transformed into a metal at less attainable pressures and can become high-temperature superconductors by the same reason as hydrogen due to a high Debye temperature and a strong electron–phonon interaction.

As one more promising material, we mention graphane (hydrogenated graphene) [140]. The superconductivity transition temperature is predicted for it to be 90 K and, for a multilayer system, can increase up to 160 K.

So far, the majority of predictions for hydrides were focused in two-element (binary) compounds [197]. It is worth to note that the high-temperature superconductivity can be realized by means of the use of three-element (triple) hydrides such as, for example, $Li_2MgH_{16}$ [208].

Now, we have heard about a new record. Researchers from University of Rochester and University of Nevada, Las Vegas (USA) attained the super-conductivity at room temperature (287.7 K) and a pressure of 267 GPa [200]. To stay in such room is not comfortable (a temperature of about 15°C), and the pressure is fantastic. But anyway, the progress is on hand. At once, this result aroused a storm of comments and the intense discussions in the scientific and near-scientific publications.

The authors of work [200] placed some mixture of carbon, hydrogen, and sulfur in a microscopic cavity (really, methane, hydrogen sulfide, and molecular hydrogen were mixed) compressed by diamond anvils. On the basis of the comments of experts, we may expect that the triple carbon–hydrogen–sulfur system will be very promising in the search for new superconductors.

The superconductivity was registered starting from a pressure of about 140 GPa and $T_c$ of about 150 K. Then $T_c$ increases comparatively slowly with the pressure. But, starting from approximately 190 GPa, the sharp increase in $T_c$ was observed. The presence of the superconductivity was confirmed also by measurements of the magnetic susceptibility: the system demonstrated the well-manifested Meissner effect. It is natural that $T_c$ decreased, as the magnetic field increased. The estimates of the upper critical field at the zero temperature made by the Ginzburg–Landau theory gave its value equal to about 62 T.

In the theory of hydrides as a significant success, we indicate the prediction of their crystalline structure from the first principles. As soon as the crystalline structure of a material becomes known, the phonon and electron spectra can be calculated. Then we can evaluate the critical temperature with the help of the Migdal–Éliashberg theory [201, 202].

As a result of many experimental and theoretical works, the high-temperature superconductivity with $T_c = 203\,\mathrm{K}$ in sulfur hydride was reliably established [203].

Probably, the discovery of the high-temperature superconductivity in $H_2S$ will restore the interest in superconductors described by the BCS theory, and new superconductors will be found.

In this connection, we will consider the BCS theory, as well as the Éliashberg theory describing the superconductivity in the case of strong coupling.

### 4.5.2.  *Formalism for standard superconductors*

#### 4.5.2.1.   *Microscopic theory of superconductivity: BCS theory*

The theory by Bardeen, Cooper and Schrieffer [202] was the first theory of superconductivity. It was able to explain the microscopic nature of the superconductivity state and is one of the most successful theories in the condensed matter physics. Though its predictable ability is limited relative to the modern *ab initio* methods, the BCS theory is else a significant theoretical and computational base aimed at the characterization and the understanding of superconducting materials till now.

The BCS theory was constructed with the use of a model Hamiltonian based on the assumption that, in the superconductivity state, all electrons near the Fermi surface are correlated, by forming the pairs. In each pair, the momenta and spins of two electrons are equal and opposite to each other.

It is worth to note that a more substantiated theory of superconductivity in the case of weak electron–phonon coupling was advanced by a Soviet physicist theorist N. N. Bogolyubov [44] in the framework of a generalization of the method of canonical transformations.

The BCS theory includes the following electron Hamiltonian:

$$H = H_0 + H_{\text{int}}, \tag{4.80}$$

$$H_0 = \sum_{\mathbf{k}\sigma} \varepsilon_{\mathbf{k}} c_{\mathbf{k}\sigma}^{\dagger} c_{\mathbf{k}\sigma}, \tag{4.81}$$

$$H_{\text{int}} = -\sum_{\mathbf{k}\mathbf{k}'} V_{\mathbf{k}\mathbf{k}'} (c_{-\mathbf{k}'\downarrow}^{\dagger} c_{\mathbf{k}'\uparrow}^{\dagger})(c_{\mathbf{k}\uparrow} c_{-\mathbf{k}\downarrow}), \tag{4.82}$$

where $c_{\mathbf{k}\sigma}$ and $c_{\mathbf{k}\sigma}^{\dagger}$ are the operators of annihilation and creation of a Bloch wave with the wave vector $\mathbf{k}$ and spin $\sigma$, respectively, and $V_{\mathbf{k}\mathbf{k}'}$ are matrix elements of the basic superconductivity interaction. By assuming that the electron–phonon coupling is independent of the momentum, we can write

$$V_{\mathbf{k},\mathbf{k'}} = -V\theta\left(\hbar\omega_D - |\varepsilon_{\mathbf{k}}|\right)\theta\left(\hbar\omega_D - |\varepsilon'_{\mathbf{k}}|\right), \tag{4.83}$$

where $\omega_D$ is the Debye frequency. BCS found the exact solution of the problem within this model with the help of the ansatz for the multiparticle wave function. Using the method of Green's functions developed by L.P. Gor'kov [203–205], we present the solution for Green's function:

$$G(\mathbf{k},\tau) = -\left\langle T_\tau c_{\mathbf{k}\sigma}(\tau)c^\dagger_{\mathbf{k}\sigma}(0)\right\rangle, \tag{4.84}$$

where the imaginary time dependence of the operators is defined and means the time-ordered product $X(\tau) = \exp(H\tau)X\exp(-H\tau)$ and $T_\tau$ is the time of the ordered process and the expectation of a physical quantity $A$ is calculated as

$$\langle A \rangle = \frac{\text{Tr}\left(\exp(-\beta H)A\right)}{\text{Tr}\left(\exp(-\beta H)\right)}, \tag{4.85}$$

The Green's function $G$ satisfies the equation of motion

$$\left(-\frac{d}{d\tau} - \varepsilon_{\mathbf{k}}\right)G(\mathbf{k},\tau) = \delta(\tau) - \sum_{\mathbf{k}}' V_{\mathbf{k}\mathbf{k'}}\langle T_\tau c^\dagger_{-\mathbf{k}\downarrow}(\tau)c_{\mathbf{k'}\uparrow}(\tau)c_{-\mathbf{k'}\downarrow}(\tau)c^\dagger_{\mathbf{k}\uparrow}(0)\rangle. \tag{4.86}$$

In the mean-field approximation, the solution takes the form

$$\langle T_\tau c^\dagger_{-\mathbf{k}\downarrow}(\tau)c_{\mathbf{k'}\uparrow}(\tau)c_{-\mathbf{k'}\downarrow}(\tau)c^\dagger_{\mathbf{k}\uparrow}(0)\rangle \to F(\mathbf{k},0)F^*(\mathbf{k},\tau) \tag{4.87}$$

and the anomalous Green's function is defined as

$$F(\mathbf{k},\tau) = -\left\langle T_\tau c_{\mathbf{k'}\uparrow}(\tau)c_{-\mathbf{k'}\downarrow}(0)\right\rangle.$$

This yields

$$G(\mathbf{k},i\omega_n) = \frac{-i\omega_n - \varepsilon_{\mathbf{k}}}{\omega_n^2 + E_{\mathbf{k}}^2},$$

$$F(\mathbf{k},i\omega_n) = \frac{\Delta^*_{\mathbf{k}}}{\omega_n^2 + E_{\mathbf{k}}^2},$$

where $\omega_n = (2n+1)\pi k_B T$ is the Matsubara frequency, $E_{\mathbf{k}}^2 \equiv \varepsilon_{\mathbf{k}}^2 + |\Delta_{\mathbf{k}}|^2$ and $\Delta_{\mathbf{k}}$ is defined as follows:

$$\Delta_{\mathbf{k}} \equiv \sum_{\mathbf{k'}} V_{\mathbf{k},\mathbf{k'}} F(\mathbf{k'},0). \tag{4.88}$$

This quantity determines the superconductivity gap and satisfies the BCS self-consistent equation:

$$\Delta_{\mathbf{k}} = \frac{1}{\beta} \sideset{}{'}\sum_{\mathbf{k}} \sum_{n=-\infty}^{\infty} \frac{V_{\mathbf{k}\mathbf{k}'}\Delta'_{\mathbf{k}}}{\omega_n^2 + E_{\mathbf{k}}^2} = \sideset{}{'}\sum_{\mathbf{k}} \frac{V_{\mathbf{k}\mathbf{k}'}\Delta'_{\mathbf{k}}}{2E'_{\mathbf{k}}} \tanh\left(\frac{E'_{\mathbf{k}}}{2k_B T}\right). \tag{4.89}$$

If we assume that the gap function is isotropic: $\Delta_{\mathbf{k}} = \Delta \neq 0$, then we can find the following limits:

(1) In the approximation $T \to 0$ and $VN(E_F) \ll 1$,

$$\Delta \sim 2\hbar\omega_D \exp\left(\frac{-1}{VN(E_F)}\right), \tag{4.90}$$

where $N(E_F)$ is the density of states in the Fermi level.

(2) For $T \sim T_c$, as $\Delta \to 0$, we get the critical temperature of the superconductivity transition:

$$T_c = 1.13\omega_D \exp\left(\frac{-1}{VN(E_F)}\right). \tag{4.91}$$

Comparing (4.90) and (4.91), we arrive at the universal relation of the BCS theory that connects. The energy gap width at the zero temperature and the temperature $T_c$ of the superconductivity transition is given by

$$\frac{2\Delta}{T_c} \sim \frac{4}{1.13} \sim 3.54. \tag{4.92}$$

This relation is reliably corroborated experimentally for many superconducting elements.

### 4.5.3.   *Éliashberg theory*

#### 4.5.3.1.   *Theory of superconductivity by Éliashberg–McMillan*

The Éliashberg theory of superconductivity is the most perfect approach to the microscopic description of properties of traditional superconductors with the electron–phonon mechanism of pairing. Recently, this theory has been successfully applied to the description of the record superconductivity in compounds of hydrogen at a high pressure.

The Éliashberg theory represents a multiparticle approach to the superconductivity that involves the Coulomb interaction and phonon propagators. The BCS model Hamiltonian of pairing is replaced by a Hamiltonian accounting for the interaction of electrons and phonons.

The Éliashberg equations are a system of coupled nonlinear integral equations constructed in the basis of the method of Green;s functions applied to the theory of superconductivity.

The calculations were carried out by the theory of perturbations in which the dimensionless parameter of expansion is the ratio [206]

$$\eta = \lambda \omega_D / E_F \ll 1, \qquad (4.93)$$

where $\lambda$ is the dimensionless parameter of the electron–phonon coupling, $\omega_D$ is the maximum frequency of phonons, and $E_F$ is the Fermi energy.

In calculations, the state derived in the theory of superconductivity and corresponding to the energy spectrum

$$\varepsilon_k = \sqrt{\varepsilon^2(k) + \Delta^2}$$

was taken as the zero approximation, rather than the state of noninteracting electrons. Thus, the principal effect of pairing was considered already in the zero approximation.

Then Green's function for such new zero approximation was defined and determined as the solution of the equations of perturbation theory in the parameter $(\nu)$. In the theory, the frequency dependence of the matrix element was taken into account with the introduction of both a function of the density of phonon states

$$F(\omega) = \sum_q (\omega - \omega_q),$$

where $\omega_q$ is the spectral distribution of phonon frequencies and a function $\alpha^2(\omega)$ characterizing the intensity of the electron–phonon interaction.

The spectral and tunnel measurements of properties of superconductors gave possibility to directly determine the product $\alpha^2(\omega)F(\omega)$. In 1971, F. Allen showed that, with regard for the expressions

$$\lambda = 2 \int_0^{\omega_D} \alpha^2(\omega) F(\omega) \frac{d\omega}{\omega}$$

and

$$\langle \omega^2 \rangle = \frac{2}{\lambda} \int_0^{\omega_D} \omega^2 \alpha(\omega) F(\omega) \frac{d\omega}{\omega},$$

it is possible to determine the dimensionless constant $\lambda$ of the electron–phonon interaction and the mean value $\langle \omega^2 \rangle$ of the squared frequency of the phonon spectrum.

In order to analyze the dependence of this dimensionless parameter in properties of a metal, McMillan proposed the simple formula

$$\lambda = N_0 \left( q^2/a^2 \right) / M \left\langle \omega^2 \right\rangle, \tag{4.94}$$

where $N_0$ is the energy density of states of electrons on the Fermi surface, $q/a$ is the matrix element of the electron–phonon interaction, $\left\langle \omega^2 \right\rangle$ is the mean square of the frequency of phonons, and $M$ is the ion mass.

He derived also the socalled McMillan relation for $T_c$ that includes a small number of simple parameters:

$$T_c = \frac{\omega_D}{1.45} \exp \left[ -\frac{1.04(1 + \lambda)}{\lambda - \mu_c^*(1 + 0.62\lambda)} \right]. \tag{4.95}$$

In 1972 Dynes [207] introduced the prefactor $\omega_D/1.20$ instead of $\omega_D/1.45$. Later in 1975 Allen and Dynes [210] made a careful analysis of the dependence of $T_c$ on the properties of a material $\lambda$ and $\mu_c^*$ and rewrote the relation in a somewhat different form

$$T_c = \frac{\omega_{\log}}{1.2} \exp \left[ -\frac{1.04(1 + \lambda)}{\lambda - \mu_c^*(1 + 0.62\lambda)} \right], \tag{4.96}$$

where

$$\omega_{\log} = \exp \left[ \frac{2}{\lambda} \int \log(\omega) \frac{\alpha^2 \Gamma(\omega)}{\omega} \, d\omega \right]. \tag{4.97}$$

This change corrected the contribution of phonons with low energies to the superconductivity pairing. So, the McMillan relation in the Allen–Dynes form (4.99) is widely used in the calculations of the superconductivity critical temperature from the first principles. One of the aspects making this relation so suitable, despite its formal simplicity, consists in that such parameterization of $\mu_c^*$ with its typical value of 0.11 eliminates noticeably the dependence of the results on a material. It is worth noting that this universal character of $\mu_c^*$ holds only for the McMillan formula.

The McMillan approach starts to deviate from the Éliashberg one at a very strong coupling ($\lambda > 15$), where the former underestimates the true critical temperature. This distinction was overcame by Allen and Dynes [210] by introduction of the additional parameter

$$\omega_2 = \left\{ \frac{2}{\lambda} \int \omega \alpha^2 F(\omega) \, d\omega \right\}^{1/2} \tag{4.98}$$

and the prefactor

$$f \doteq \left\{ 1 + \left[ \frac{\lambda}{2.46(1 + 3.8\mu_c^*)} \right]^{3/2} \right\}^{1/3} \left\{ 1 + \frac{(\omega_2/\omega_{\log} - 1)\lambda^2}{\lambda^2 + 3.31(1 + 6.3\mu_c^*)(\omega_2/\omega_{\log})} \right\}$$

(4.99)

on the right-hand side of relation (4.96) that becomes

$$T_c = f\frac{\omega_{\log}}{1.2} \exp\left[ -\frac{1.04(1 + \lambda)}{\lambda - \mu_c^*(1 + 0.62\lambda)} \right].$$

(4.100)

The BCS Éliashberg and NcMillan theories deal with isotropic media. This decreases significantly the region of their applicability to nonmetallic superconductors.

In subsequent studies, the theorists should find the mean allowing one to conserve the high-temperature superconductivity at lower pressures and, what is better, at the ambient one. To answer this challenge, the predictable power of computational tools must be made more reliable and exact, in particular, in the consideration of extreme conditions.

# CHAPTER 5

# Nanoscale Superconductivity

## 5.1. Introduction

Recent advances in nanoscience have demonstrated that fundamentally new physical phenomena are found, when systems are reduced in size to dimensions which become comparable to the fundamental microscopic lengths of the investigated material. Superconductivity is a macroscopic quantum phenomena, and it is therefore especially interesting to see how this quantum state is influenced when the samples are reduced to nanometer sizes. Recent developments in nanotechnologies and measurement techniques nowadays allow the experimental investigation of the magnetic and thermodynamic superconducting properties of mesoscopic samples in this regime.

In this book, we develop some theoretical models to describe such nanoscale superconducting systems and explore possible new experimental phenomena which we can predict based upon these theoretical models. In bulk samples, the standard Bardeen–Cooper–Schrieffer (BCS) theory gives a good description of the phenomenon of superconductivity [148]. However, it was noticed by Anderson in 1959 that, as the size of a superconductor becomes smaller, and the quantum energy level spacing of the electrons in the sample approaches the superconducting gap, then the BCS theory will fail [6]. The exact solution to the reduced BCS Hamiltonian for finite-size systems was developed by Richardson in the context of nuclear physics a long time ago [211]. This shows that, while the grand canonical BCS wave function gives a very accurate solution of the BCS Hamiltonian in the limit where the number of electrons is very large $N \gg 1$, for small values of $N$, one has to use exact analytical methods to obtain reliable results. The recent experimental advances in fabricating and measuring superconductivity in ultrasmall mesoscopic and nanoscale grains has renewed a theoretical interest in the Richardson solution [212–215]. Reference [10] shows the measurement for the smallest size of a superconductor in the world which is 5 A, Fig. 5.1. Such systems are interesting for nanoelectronics

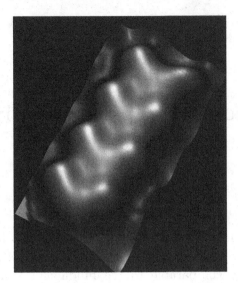

Fig. 5.1.   The smallest superconductor in the world.

and quantum computing. In this book, we propose to develop theoretical models for nanoscale superconducting systems and to apply these models to a variety of systems of current experimental interest. The Richardson solution depends on the electron energy level spacings near to the Fermi level, and so these different geometric shapes will lead to different size dependences of the thermodynamic and electronic properties. Experiments have recently demonstrated superconductivity in one-dimensional nanowires of lead [216] and carbon [217]. Our theoretical predictions will include the even-odd parity effects in tunneling spectra, which have already been observed on the nanometer scale in Al grains, Fig. 5.2, [212], but which could also be observed in nanotube superconductors. In Figs. 5.2 and 5.3, we can see the ensemble of small metallic grains which can be in the normal or superconducting state. For such systems, the important parameters are the l-granular size, electron coherence length, and the penetration depth [167]. In grains, there appears the quantum size effect (discretization of the electron energy spectrum). In Fig. 5.3, we show the size effect in grains. We note that the level spacing depends on many parameters such as the electron coherence length and the penetration depth.

Carbon nanotubes were first observed in 1991 by Iijima in Japan, Fig. 5.4.

In this chapter, we investigate properties of nanosize two-gap superconductivity by using a two-sublevel model in the framework of the mean-field approximation. A model corresponding to a nanosize two-gap superconductivity is presented, and the partition function of the nanosize system is

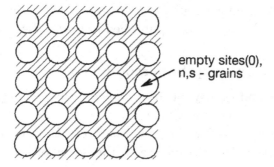

Fig. 5.2. Illustration of superconductor grains.

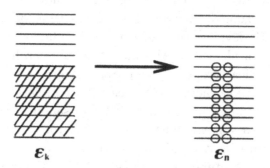

Fig. 5.3. Size effect in superconductor grains.

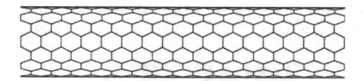

Fig. 5.4. Carbon nanotube revealing superconductivity.

analytically derived by using the path integral approach. A definition of the critical level spacing of the two-gap superconductivity is also presented, and we discuss the condensation energy and the parity gap of two-gap superconductivity in relation to the size dependence of those properties with two bulk gaps and the effective pair scattering process between two sublevels. We present the theory of interactions between two nanoscale ferromagnetic particles embedded into a superconductor and spin orientation phase transitions in such a system. We investigate the energy spectrum and wave function of electrons in hybrid superconducting nanowires. We also consider the ideas of quantum computing and quantum information in mesoscopic circuits. The theory of the Josephson effect is presented, and its applications in quantum computing are analyzed. The results of this chapter were obtained by the author and published in works [218–226].

## 5.2.   Nanosize Two-gap Superconductivity

Multiband superconductivity has been theoretically investigated in relation to $MgB_2$ superconductivity in Chapter 4.

Recent experiments [228, 232] by Black *et al.* have also generated a high interest in the size dependence of the superconductivity. Properties of ultrasmall superconducting grains have been theoretically investigated by many groups [229, 230, 233–237]. In such ultrasmall grains, the old but fundamental theoretical question was noticed by Anderson [6]. The standard BCS theory gives a good description of the phenomenon of superconductivity in large samples. However, as the size of a superconductor becomes small, the BCS theory fails. In ultrasmall Al grains, the bulk gap has been discussed in relation to physical properties in ultrasmall grains such as the parity gap [237], condensation energy [235], electron correlation [230], etc. with the size dependence of the level spacing [229] of samples.

In the recent years, great efforts have been devoted to the fabrication of $MgB_2$ nanostructures that could play a crucial role in the field of applied superconductivity, nanotubes, nanowires and nanoparticles. As for $MgB_2$, nanoparticles of approximately 40–100 nm in size are available (Fig. 5.5).

In this section, we investigate the properties of nanosize two-gap superconductivity by using a two-sublevel model in the framework of the mean-field approximation. A model corresponding to a nanosize two-gap

Fig. 5.5.   TEM micrograph of 20–100 nm nanoparticles of $MgB_2$.

superconductivity is presented, and the partition function of the nanosize system is analytically derived by using the path integral approach. A definition of the critical level spacing of the two-gap superconductivity is also presented, and we discuss the condensation energy and the parity gap of two-gap superconductivity in relation to the size dependence of those properties with two bulk gaps and the effective pair scattering process between two sublevels. In nanosize grains of a superconductor, the quantum level spacing approaches the superconducting gap. In the case of a two-gap superconductor, we can consider a model with two sublevels corresponding to two independent bands. In this section, we present a model for nanosize two-gap superconductivity and an expression for the partition function of the system.

### 5.2.1. *Hamiltonian for nanosize grains*

We consider a pairing Hamiltonian with two sublevels corresponding to two bands 1 and 2 written as

$$H = H_0 + H_{\text{int}}, \tag{5.1}$$

where

$$H_0 = \sum_{j,\sigma}[\varepsilon_{1j} - \mu]a_{j\sigma}^\dagger a_{j\sigma} + \sum_{k,\sigma}[\varepsilon_{2k} - \mu]b_{k\sigma}^\dagger b_{k\sigma}, \tag{5.2}$$

$$H_{\text{int}} = -g_1 \sum_{j,j'\in I} a_{j\uparrow}^\dagger a_{j\downarrow}^\dagger a_{j'\downarrow} a_{j'\uparrow} - g_2 \sum_{k,k'\in J} b_{k\uparrow}^\dagger b_{k\downarrow}^\dagger b_{k'\downarrow} b_{k'\uparrow}$$

$$+ g_{12} \sum_{j\in I, k\in J} a_{j\uparrow}^\dagger a_{j\downarrow}^\dagger b_{k\downarrow} b_{k\uparrow} + g_{12} \sum_{j\in I, k\in J} b_{k\uparrow}^\dagger b_{k\downarrow}^\dagger a_{j\downarrow} a_{j\uparrow}. \tag{5.3}$$

Here, $a_{j\sigma}^\dagger(a_{j\sigma})$ and $b_{j\sigma}^\dagger(b_{j,\sigma})$ are the creation (annihilation) operator in sublevels 1 and 2 with spin $\sigma$ and the energies $\varepsilon_{1j}$ and $\varepsilon_{2j}$, respectively, the operators for each sublevel satisfy the anticommutation relations, and the operators between sublevels are independent, $\mu$ is the chemical potential, the second term in Eq. (5.1) is the interaction Hamiltonian, $g_1$ and $g_2$ are the effective interaction constant for sublevels 1 and 2, and $g_{12}$ is an effective interaction constant which corresponds to the pair scattering process between two bands. The sums of $j$ and $k$ in Eq. (5.3) are over the set $I$ of $N_{1I}$ states corresponding to the half-filled band 1 with fixed width $2\omega_{1D}$ and the set $J$ of $N_{2J}$ states for band 2, respectively.

In this study, we assume that the Debye energies for two sublevels are the same: $\omega_{1D} = \omega_{2D} = \omega_D$. Within this assumption, $N_{1I}$ and $N_{2J}$ are relatively estimated by the density of state (DOS) for two bands as $N_{1I}/N_{2J} = \rho_1/\rho_2$, where $\rho_1$ and $\rho_2$ are DOS for two bands. The interaction constants $g_1$ and $g_2$ can be written as $d_1\lambda_1$ and $d_2\lambda_2$, respectively. $d_1 = 2\omega_D/N_{1I}$ and $d_2 = 2\omega_D/N_{2J}$ mean the mean energy level spacing, and $\lambda_1$ and $\lambda_2$ are the dimensionless parameters for two sublevels. We take the intersublevel interaction constant $g_{12} = \sqrt{d_1 d_2}\lambda_{12}$. In summary, we obtain a relation of $\rho_1/\rho_2 = N_{1I}/N_{2J} = d_2/d_1$.

### 5.2.2.  *Path integral approach*

It is convenient to introduce a path integral approach for the treatment of fluctuations of the order parameters. This approach gives an exact expression for the grand partition function of a superconductor

$$Z(\mu, T) = \text{Tr} \exp\left[-\frac{H - \mu N}{T}\right], \qquad (5.4)$$

where $T$ is the temperature, and $N$ is the number operator in the grain. The idea of the path integral approach is to replace the description of a system under study in terms of electronic operators by an equivalent description in terms of the superconducting order parameter.

By the path integral approach, we obtain an expression for the grand partition function for the Hamiltonian (5.1):

$$Z(\mu, T) = \int D\Delta_1 D\Delta_1^* D\Delta_2 D\Delta_2^* e^{-S[\Delta_1, \Delta_2]}. \qquad (5.5)$$

Here, the action $S[\Delta_1, \Delta_2]$ is defined as

$$S[\Delta_1, \Delta_2] = -\sum_j \left[\text{Tr} \ln G_{1j}^{-1} - \frac{\xi_{1j}}{T}\right] - \sum_k \left[\text{Tr} \ln G_{2k}^{-1} - \frac{\xi_{2k}}{T}\right]$$

$$+ \int_0^{1/T} d\tau \frac{1}{g_1 g_2 - g_{12}^2} \left[g_2 |\Delta_1(\tau)|^2 + g_1 |\Delta_2(\tau)|^2\right.$$

$$\left. + g_{12}\left(\Delta_1(\tau)\Delta_2(\tau)^* + \Delta_1(\tau)^*\Delta_2(\tau)\right)\right]. \qquad (5.6)$$

$\Delta_1$ and $\Delta_2$ are bulk gaps for sublevels 1 and 2, respectively, $\xi_{1j} = \varepsilon_{1j} - \mu$ and $\xi_{2k} = \varepsilon_{2k} - \mu$, and the inverse Green's functions

$$G_{1j}^{-1}(\tau, \tau') = \left[-\frac{d}{d\tau} - \xi_{1j}\sigma^z - \Delta_1(\tau)\sigma^+ - \Delta_1^*(\tau)\sigma^-\right]\delta(\tau - \tau'), \qquad (5.7)$$

and

$$G_{2k}^{-1}(\tau, \tau') = \left[-\frac{d}{d\tau} - \xi_{2k}\sigma^z - \Delta_2(\tau)\sigma^+ - \Delta_2^*(\tau)\sigma^-\right]\delta(\tau - \tau'), \qquad (5.8)$$

where $\sigma^{\pm} = \sigma^x \pm i\sigma^y$, and $\sigma^{x,y,z}$ are the Pauli matrices. $G_1^{-1}$ and $G_2^{-1}$ satisfy antiperiodic boundary conditions.

In the case of a stronger interaction, $\Delta_1 \gg d_1$ and $\Delta_2 \gg d_2$, we consider the mean-field approximation for the order parameters in the path integral approach. Substituting the time-independent order parameters into the action (5.6), we have

$$\Omega(\mu) = \sum_j (\xi_{1j} - \epsilon_{1j}) + \sum_k (\xi_{2k} - \epsilon_{2k})$$

$$+ \frac{1}{g_1 g_2 - g_{12}^2} \left[ g_2 \Delta_1^2 + g_1 \Delta_2^2 + g_{12} (\Delta_1^* \Delta_2 + \Delta_1 \Delta_2^*) \right], \quad (5.9)$$

where $\epsilon_{1j} = (\xi_{1j}^2 + \Delta_1^2)^{1/2}$, and $\epsilon_{2k} = (\xi_{2k}^2 + \Delta_2^2)^{1/2}$. In Eq. (5.9), the values of $\Delta_1$ and $\Delta_2$ must be chosen in such a way which minimizes $\Omega$. From the minimization of $\Omega$, we obtain a coupled gap equation at zero temperature for the two-gap system:

$$\begin{pmatrix} \Delta_1 \\ \Delta_2 \end{pmatrix} = \begin{pmatrix} g_1 \sum_j \dfrac{1}{2\epsilon_{1j}} & -g_{12} \sum_k \dfrac{1}{2\epsilon_{2k}} \\ -g_{12} \sum_j \dfrac{1}{2\epsilon_{1j}} & g_2 \sum_k \dfrac{1}{2\epsilon_{2k}} \end{pmatrix} \begin{pmatrix} \Delta_1 \\ \Delta_2 \end{pmatrix}. \quad (5.10)$$

From the coupled gap equation, Eq. (5.10), we formally obtain an expression for the bulk gap for two-gap superconductivity at zero temperature:

$$\tilde{\Delta}_1 = \omega \sinh^{-1}\left(\frac{1}{\eta_1}\right), \quad (5.11)$$

and

$$\tilde{\Delta}_2 = \omega \sinh^{-1}\left(\frac{1}{\eta_2}\right), \quad (5.12)$$

where

$$\frac{1}{\eta_1} = \frac{\lambda_2 + \alpha_{\pm}[\eta_1, \eta_2] \lambda_{12}}{\lambda_1 \lambda_2 - \lambda_{12}^2}, \quad (5.13)$$

$$\frac{1}{\eta_2} = \frac{\lambda_1 + \alpha_{\pm}^{-1}[\eta_1, \eta_2] \lambda_{12}}{\lambda_1 \lambda_2 - \lambda_{12}^2}, \quad (5.14)$$

and

$$\alpha_{\pm}[\eta_1, \eta_2] = \pm \frac{\sinh\left(\frac{1}{\eta_1}\right)}{\sinh\left(\frac{1}{\eta_2}\right)}. \quad (5.15)$$

For the two-band superconductivity, we can consider two cases for the phase of the gaps: $\text{sgn}(\tilde{\Delta}_1) = \text{sgn}(\tilde{\Delta}_2)$, and $\text{sgn}(\tilde{\Delta}_1) = -\text{sgn}(\tilde{\Delta}_2)$. For the same phase, $\alpha_+$ is used in Eqs. (5.13) and (5.14), and we use $\alpha_-$ for the opposite phase. Note that $\tilde{\Delta}_1 = -\tilde{\Delta}_2$ in the limit of strong intersublevel coupling $\lambda_{12}$, that is, the opposite phase. For $\lambda_{12} = 0$, we find the same results for two bulk gaps derived from the conventional BCS theory for two independent sublevels.

### 5.2.3. *Condensation energy*

In this section, we discuss properties such as condensation energy, critical level spacing, and parity gap of nanosize two-gap superconductivity by using the partition function derived in the previous section.

In nanosize superconductivity, the condensation energy can be defined as $E^C_{N,b}(\lambda) = E^G_{N,b}(0) - E^G_{N,b}(\lambda) - n\lambda d$, where $E^G_{N,b}$ is the ground-state energy of the $N$-electron system in the interaction band, $b$ is the number of electrons on single occupied levels, and $\lambda$ and $n$ are the dimensionless coupling parameter and the number of pair occupied level, respectively. In the case of a nanosize two-band system, the condensation energy can be written as

$$E^C_{N_1,b_1;N_2,b_2}(\lambda_1, \lambda_2, \lambda_{12}) = E^G_{N_1,b_1;N_2,b_2}(0,0,0) - E^G_{N_1,b_1;N_2,b_2}(\lambda_1, \lambda_2, \lambda_{12})$$
$$-n_1\lambda_1 d_1 - n_2\lambda_2 d_2, \tag{5.16}$$

where $E^G_{N_1,b_1;N_2,b_2}(\lambda_1, \lambda_2, \lambda_{12})$ means the ground-state energy of $(N_1 + N_2)$-electron system. From Eqs. (5.4) and (5.9), the condensation energy of the two-sublevel system can be expressed by the condensation energy of independent single-level systems:

$$E^C_{N_1,b_1;N_2,b_2}(\lambda_1, \lambda_2, \lambda_{12}) = E^C_{N_1,b_1}(\lambda_1) + E^C_{N_2,b_2}(\lambda_2) - \frac{\lambda_{12}^2}{\lambda_1\lambda_2 - \lambda_{12}^2}$$
$$\times \left( \frac{\Delta_1^2}{d_1\lambda_1} + \frac{\Delta_2^2}{d_2\lambda_2} + \frac{2(\Delta_1^*\Delta_2 + \Delta_1\Delta_2^*)}{\sqrt{d_1 d_2}\lambda_{12}} \right), \tag{5.17}$$

where $E^C_{N_1,b_1}(\lambda_1)$ and $E^C_{N_2,b_2}(\lambda_2)$ correspond to the condensation energy for the single band case. In the same phases of $\Delta_1$ and $\Delta_2$, the condensation energy (5.17) decreases, that is, there appears the instability by the coupling constant $\lambda_{12}$. On the other hand, in the opposite phases, the condensation energy becomes larger, because $\Delta_1^*\Delta_2 + \Delta_1\Delta_2^* < 0$. We can expect that the condensation energy for two-gap superconductivity leads to a higher stability, than that of two independent systems, due to the intersublevel coupling $\lambda_{12}$ and the opposite phases.

### 5.2.4. *Critical level spacing*

To discuss the critical level spacing for a two-gap system, we start from the coupled gap equation (5.10). For the case of the critical level spacing of the two-gap system, we have

$$1 = \lambda_1 \sum_j \frac{1}{2|\tilde{\xi}_{1j}|} + \lambda_2 \sum_k \frac{1}{2|\tilde{\xi}_{2k}|} - (\lambda_1\lambda_2 - \lambda_{12}^2) \sum_j \frac{1}{2|\tilde{\xi}_{1j}|} \sum_k \frac{1}{2|\tilde{\xi}_{2k}|},$$

(5.18)

where $\tilde{\xi}_i = \xi_i/d_i$ for sublevel $i = 1, 2$. For the odd or even cases, Eq. (5.18) can be approximately solved by using the digamma function: For the odd case, the critical level spacing becomes

$$d_{1c}^o = \omega_D e^\gamma \exp\left[-\frac{1}{\lambda}\right], \quad d_{2c}^o = \frac{d_2}{d_1} d_{1c}^o,$$

(5.19)

and, for the even case,

$$d_{1c}^e = 4\omega_D e^\gamma \exp\left[-\frac{1}{\lambda}\right], \quad d_{2c}^e = \frac{d_2}{d_1} d_{1c}^e.$$

(5.20)

Here, we use

$$\frac{1}{\lambda} = \frac{1}{2x}\left[\lambda_1 + \lambda_2 - ax + \sqrt{(\lambda_1 - \lambda_2 - ax)^2 + 4\lambda_{12}^2}\right]$$

(5.21)

with

$$x = \lambda_1\lambda_2 - \lambda_{12}^2,$$

(5.22)

$$a = \log\frac{d_1}{d_2}.$$

(5.23)

From these expressions, we find some relations:

$$d_{1c}^e = 4d_{1c}^o, \quad d_{2c}^e = 4d_{2c}^o$$

(5.24)

and

$$d_{1/2c}^o \approx \frac{e^\gamma}{2}\exp\left[\frac{1}{\eta_{1/2}} - \frac{1}{\lambda}\right]\tilde{\Delta}_{1/2}.$$

(5.25)

In the case of $|\lambda_1 - \lambda_2| \gg \lambda_{12}$, Eq. (5.25) can be approximately rewritten as

$$d_{1/2c}^o \approx \frac{e^\gamma}{2}\exp\left[\frac{\lambda_2 - \lambda_1 + 2\alpha\lambda_{12}}{\lambda_1\lambda_2 - \lambda_{12}^2}\right]\tilde{\Delta}_{1/2}.$$

(5.26)

On the other hand, in the limit of $|\lambda_1 - \lambda_2| \ll \lambda_{12}$, we have

$$d^o_{1/2c} \approx \frac{e^\gamma}{2} \exp\left[\frac{(1+\alpha)\,\lambda_{12}}{\lambda_1\lambda_2 - \lambda_{12}^2}\right] \tilde{\Delta}_{1/2}. \qquad (5.27)$$

For the case of $\lambda_{12} = 0$, Eq. (5.25) can be rewritten as $d^o_{1/2c} \approx \exp[\gamma]/2 \exp[1/\lambda_1 - 1/\lambda_2]\tilde{\Delta}_{1/2}$. Therefore, when the coupling constants $\lambda_1$ and $\lambda_2$ take the same value, we have a relation similar to that for a single-level system: $d^o_{1/2c} \approx 0.89\tilde{\Delta}_{1/2}$. These results suggest the critical level spacing strongly depend upon $\lambda_{12}$ and the difference between the effective interaction constants for sublevels. The relation in Eq. (5.24) is the same relation in the conventional nanosize BCS theory.

### 5.2.5.   *Parity gap*

In this subsection, we consider a parity gap in the case of two-gap superconductivity in ultrasmall grains. In the case of two sublevel spacings, the chemical potential lies halfway between the highest occupied and the lowest unoccupied levels of a smaller level spacing in the half-filled case, as shown in Fig. 5.6(a). We assume that $d_1 < d_2$ and that the numbers of occupied levels corresponding to each sublevel are $n_1$ and $n_2$, respectively. Then, the total number of electron becomes $N = 2n_1 + 2n_2$. When we consider $N = 2n_1 + 2n_2 + 1$, the chemical potential lies on the level $\varepsilon_{1n_1+1}$, as shown in Fig. 5.6(b). Figure 5.6(c) shows the position of the chemical potential in the case of $N = 2n_1 + 2n_2 + 2$. The parity gap of nanosize two-gap superconductivity is written as

$$\Delta^1_p = E^G_{2n_1+1+2n_2,1} - \frac{1}{2}(E^G_{2n_1+2n_2,0} + E^G_{2(n_1+1)+2n_2,0}). \qquad (5.28)$$

From Eq. (5.9) and for the ground-state energy $E^G_{N,b} = \Omega_{\mu_N} + \mu_N N$, we obtain

$$\Delta^1_p = \Delta_1 - \frac{d_1}{4}\left(\frac{\rho_1}{\rho_2} - 1\right). \qquad (5.29)$$

From Figs. 5.6(c)–5.6(e), we can define another parity gap:

$$\Delta^2_p = E^G_{2(n_1+1)+2n_2+1,1} - \frac{1}{2}(E^G_{2(n_1+1)+2n_2,0} + E^G_{2(n_1+1)+2(n_2+1),0}). \qquad (5.30)$$

From the latter definition of Eq. (5.30), we have

$$\Delta^2_p = \Delta_2 - \frac{d_2}{4}\left(\frac{3\rho_2}{\rho_1} - 1\right). \qquad (5.31)$$

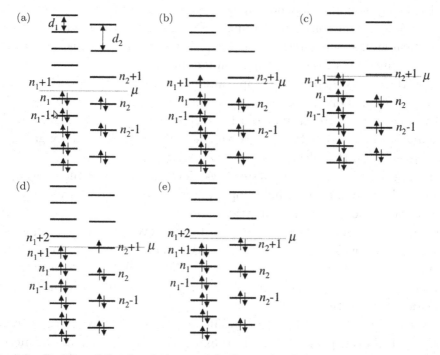

Fig. 5.6. Position of the chemical potential relative the electronic energy levels in a two-gap superconducting grain. Solid and dotted lines mean two sublevels. (a) Half-filled system with $2n_1 + 2n_2$ electrons. (b) $2n_1 + 1 + 2n_2$-electron system. (c) $2(n_1 + 1) + 2n_2$-electron system. (d) $2(n_1 + 1) + 2n_2 + 1$-electron system. (e) $2(n_1 + 1) + 2(n_2 + 1)$-electron system.

The present results suggest two kinds of the dependence of the parity gap on the level spacing. The parity gap does not depend upon the effective interaction $\lambda_{12}$. The structure around the Fermi level plays an important role of the contribution to the size dependence of the parity gap. We have investigated the properties of nanosize two-gap superconductivity by using a two-sublevel model in the framework of the mean-field approximation. From the discussion of the condensation energy in nanosize two-gap superconductivity, the phases of the gaps are very important to stabilize the superconductivity. In the same phases, the two-gap superconductivity is instable by the coupling constant $\lambda_{12}$. On the other hand, in the opposite phases, the superconductivity becomes stable. We can expect that, due to the condensation energy, the two-gap superconductivity becomes more stable than that for two independent systems due to the intersublevel coupling $\lambda_{12}$ and the opposite phases.

We have also discussed the critical level spacing for two-gap superconductivity in ultrasmall grains. These results suggest that the critical level spacing strongly depends upon $\lambda_{12}$ and the difference between the effective interaction constants for sublevels. Moreover, the relation between the critical level spacing and the bulk gaps is modified as compared with the result obtained for ultrasmall superconducting Al grains.

For the parity gap in two-gap superconductivity, the present results suggest two kinds of the dependence of the parity gap on the level spacing and that the structure around the Fermi level plays an important role by contributing to the size dependence on the parity gap. The parity gap does not depend upon the effective interaction $\lambda_{12}$.

In the case of a cluster system, we have to apply a more accurate approach beyond the mean-field approximation presented in this study by investigating the physical properties, and we have also to consider the contribution of the surface of samples to the level structure around the Fermi level. We will present these problems in the next section. On the basis of the presented results, we might expect the possibility of a new multi-gap superconductivity arising in the nanosize region with a higher critical transition temperature.

In summary, a model corresponding to a nanosize two-gap supercon-ductivity has been presented, and an expression for the partition function of the nanosize system has been analytically derived by using the path integral approach. A definition of the critical level spacing of the two-gap superconductivity has been also presented, and we discuss the condensation energy and the parity gap of the two-gap superconductivity in relation to the size dependence of those properties with two bulk gaps, as well as the effective pair scattering process between two sublevels.

## 5.3.   Exact Solution of Two-band Superconductivity in Ultrasmall Grains

Many groups have theoretically investigated the physical properties such as critical level spacing, condensation energy, parity gap, etc. in ultrasmall grains with the conventional superconductivity [229, 230, 233–237]. The question concerning such nanosize superconducting grains has been discussed by Anderson [6]. The standard BCS theory becomes false, when the level spacing approaches the superconducting gap. To investigate the properties in such nanosize systems, it is necessary to take a more accurate treatment. Braun and von Delft [233, 234, 241] have reintroduced the exact solution to the reduced BCS Hamiltonian developed by Richardson [238–240]. It is noteworthy that the Richardson's solution is applicable at distributions of single-electron energy levels. V.N. Gladilin *et al.* [235] have investigated the pairing characteristics such as the condensation energy, spectroscopic gap, parity gap, etc., by using the Richardson's exact solution for the reduced BCS Hamiltonian.

The recent discovery of superconductivity of $MgB_2$ [168] with $T_c = 39\,K$ has also been much attracted a great interest aimed at the elucidation of its mechanism from both experimental and theoretical viewpoints. Since this discovery, the possibility of two-band superconductivity has also been discussed in relation to two-gap functions experimentally and theoretically.

In this section, we investigate the two-band superconductivity in ultra-small grains. The Richardson's exact solution is extended to two-band systems, and a new coupled equation is derived according to the procedure of Richardson's works. The parity gap and the condensation energy of an ultrasmall two-band superconducting grain are numerically given by solving the coupled equation. We discuss these properties of ultrasmall grains in relation to the correlation, interband interaction, and size dependence.

### 5.3.1. *Exact solution for two-band superconductivity*

In this section, we derive an exact solution of the two-band superconductivity for a reduced BCS Hamiltonian.

### 5.3.2. *Hamiltonian*

We consider a Hamiltonian for two bands 1 and 2 written as

$$H = H_1 + H_2 + H_{\text{int}}, \tag{5.32}$$

where

$$H_1 = \sum_{j\sigma} \varepsilon_{1j} a_{j\sigma}^\dagger a_{j\sigma} - g_1 \sum_{jk} a_{j\uparrow}^\dagger a_{j\downarrow}^\dagger a_{k\downarrow} a_{k\uparrow}, \tag{5.33}$$

$$H_2 = \sum_{j\sigma} \varepsilon_{2j} b_{j\sigma}^\dagger b_{j\sigma} - g_2 \sum_{jk} b_{j\uparrow}^\dagger b_{j\downarrow}^\dagger b_{k\downarrow} b_{k\uparrow}, \tag{5.34}$$

$$H_{\text{int}} = g_{12} \sum_{jk} a_{j\uparrow}^\dagger a_{j\downarrow}^\dagger b_{k\downarrow} b_{k\uparrow} + g_{12} \sum_{jk} b_{j\uparrow}^\dagger b_{j\downarrow}^\dagger a_{k\downarrow} a_{k\uparrow}. \tag{5.35}$$

The first and second terms of Eq. (5.32) correspond to the reduced BCS Hamiltonian for bands 1 and 2, respectively. The third term means a coupling between them and corresponds to the pair scattering process between these two bands (see Fig. 5.7). $a_{j\sigma}^\dagger$ ($a_{j\sigma}$) and $b_{j\sigma}^\dagger$ ($b_{j\sigma}$) are the creation (annihilation) operator in bands 1 and 2 with spin $\sigma$ and the single-particle levels $\varepsilon_{1j}$ and $\varepsilon_{2j}$, respectively. The sums of $j$ and $k$ are taken over a set

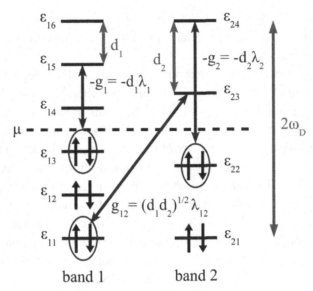

band 1          band 2

Fig. 5.7.   Two-band system. The dotted line means the chemical potential; $\varepsilon_{nj}$ is the single-particle energy for band $n$ and level $j$; $-g_1$ and $-g_2$ are the intraband pair interaction coupling constants; $g_{12}$ are the interband pair interaction coupling constant.

of $N_1$ states for band 1 with fixed width $2\hbar\omega_{1D}$ and a set of $N_2$ states for band 2 with fixed width $2\hbar\omega_{2D}$, respectively.

In this study, we assume that the Debye energies for two bands coincide with each other, i.e.,

$$\omega_{1D} = \omega_{2D} = \omega_D. \tag{5.36}$$

Within this assumption, $N_1$ and $N_2$ are relatively estimated by the density of state (DOS) for two bands as

$$\frac{N_1}{N_2} = \frac{\rho_1}{\rho_2}, \tag{5.37}$$

where $\rho_1$ and $\rho_2$ are DOS for two bands, respectively. The interaction constants $g_1$ and $g_2$ can be written as

$$g_1 = d_1\lambda_1, \quad g_2 = d_2\lambda_2, \tag{5.38}$$

where $d_1$ and $d_2$ mean the mean single-particle level spacing,

$$d_1 = \frac{2\hbar\omega_D}{N_1 - 1}, \quad d_2 = \frac{2\hbar\omega_D}{N_2 - 1}, \tag{5.39}$$

and $\lambda_1$ and $\lambda_2$ are the dimensionality interaction parameters for two bands. We define the interband interaction constant as

$$g_{12} = \sqrt{d_1 d_2}\lambda_{12}. \tag{5.40}$$

In summary, we obtain the relation

$$\frac{\rho_1}{\rho_2} \approx \frac{N_1 - 1}{N_2 - 1} = \frac{d_2}{d_1}. \tag{5.41}$$

The system we are considering consists of two half-filled bands, each of which has equally spaced $N_n$ single-particle levels and $M_n$ ($= N_n/2$) doubly occupied pair levels ($n = 1, 2$). We take the single-particle level spacing as our energy unity. Thus, the single-particle spectrum is given by

$$\varepsilon_{nj} = d_n j - \omega_D, \quad j = 1, 2, \ldots, N_n \quad (n = 1, 2). \tag{5.42}$$

Richardson has obtained his solution within the single-band model for an arbitrary set of single-particle levels. For simplicity, we assume that there are no singly occupied single-particle levels. As can be seen from Eqs. (5.33)–(5.35), these levels are decoupled from the rest of the system. They are said to be blocked and contribute with their single-particle energies to the total energy. The above simplification implies that every single-particle level $j$ is either empty (i.e. $|vac\rangle$), or occupied by a pair of electrons (i.e. $a_{j\uparrow}^\dagger a_{j\downarrow}^\dagger |vac\rangle$ and $b_{j\uparrow}^\dagger b_{j\downarrow}^\dagger |vac\rangle$). These are called as the unblocked level.

### 5.3.3. *Exact solution*

In order to extend Richardson's solution to a two-band system, we define two kinds of hard-core boson operators as

$$c_j = a_{j\downarrow} a_{j\uparrow}, \quad c_j^\dagger = a_{j\uparrow}^\dagger a_{j\downarrow}^\dagger, \tag{5.43}$$

$$d_j = b_{j\downarrow} b_{j\uparrow}, \quad d_j^\dagger = b_{j\uparrow}^\dagger b_{j\downarrow}^\dagger, \tag{5.44}$$

which satisfy the commutation relations,

$$c_j^{\dagger 2} = 0, \quad [c_j, c_k^\dagger] = \delta_{jk}(1 - 2c_j^\dagger c_j), \quad [c_j^\dagger c_j, c_k^\dagger] = \delta_{jk} c_j^\dagger, \tag{5.45}$$

$$d_j^{\dagger 2} = 0, \quad [d_j, d_k^\dagger] = \delta_{jk}(1 - 2d_j^\dagger d_j), \quad [d_j^\dagger d_j, d_k^\dagger] = \delta_{jk} d_j^\dagger, \tag{5.46}$$

which reflects the Pauli principle for the fermions they constructed from.

Hamiltonian equation (5.32) for the unblocked levels can be then written as

$$H_U = 2 \sum_{j}^{N_1} \varepsilon_{1j} c_j^\dagger c_j - g_1 \sum_{jk}^{N_1} c_j^\dagger c_k$$

$$+ 2 \sum_{j}^{N_2} \varepsilon_{2j} d_j^\dagger d_j - g_2 \sum_{jk}^{N_2} d_j^\dagger d_k$$

$$+ g_{12} \sum_{j}^{N_1} \sum_{k}^{N_2} c_j^\dagger d_k + g_{12} \sum_{j}^{N_2} \sum_{k}^{N_1} d_j^\dagger c_k. \tag{5.47}$$

We find the eigenstates $|M_1; M_2\rangle$ of this Hamiltonian with $M_1 + M_2$ pairs in the form

$$H_U |M_1; M_2\rangle_U = E(M_1; M_2)|M_1; M_2\rangle_U$$

$$= \left( \sum_{J=1}^{M_1} E_{1J} + \sum_{K=1}^{M_2} E_{2K} \right) |M_1; M_2\rangle_U, \tag{5.48}$$

where $E(M_1; M_2)$ is the eigenvalue and

$$|M_1; M_2\rangle_U = \prod_{J=1}^{M_1} C_J^\dagger \prod_{K=1}^{M_2} D_K^\dagger |vac\rangle, \tag{5.49}$$

and

$$C_J^\dagger = \sum_{j}^{N_1} \frac{c_j^\dagger}{2\varepsilon_{1j} - E_{1J}}, \quad D_J^\dagger = \sum_{j}^{N_2} \frac{d_j^\dagger}{2\varepsilon_{2j} - E_{2J}}. \tag{5.50}$$

Now, we define $C_0^\dagger$ and $D_0^\dagger$ as

$$C_0^\dagger = \sum_{j}^{N_1} c_j^\dagger, \quad D_0^\dagger = \sum_{j}^{N_2} d_j^\dagger. \tag{5.51}$$

Then we can rewrite Eq. (5.47) as

$$H_U = 2 \sum_{j}^{N_1} \varepsilon_{1j} c_j^\dagger c_j - g_1 C_0^\dagger C_0$$

$$+ 2 \sum_{j}^{N_2} \varepsilon_{2j} d_j^\dagger d_j - g_2 D_0^\dagger D_0$$

$$+ g_{12} C_0^\dagger D_0 + g_{12} D_0^\dagger C_0. \tag{5.52}$$

The commutation relations for new operators are given as

$$[c_j^\dagger c_j, C_J^\dagger] = \frac{c_j^\dagger}{2\varepsilon_{1j} - E_{1J}}, \quad [d_j^\dagger d_j, D_J^\dagger] = \frac{d_j^\dagger}{2\varepsilon_{2j} - E_{2J}}, \tag{5.53}$$

$$[C_0, C_J^\dagger] = \sum_j^{N_1} \frac{1 - c_j^\dagger c_j}{2\varepsilon_{1j} - E_{1J}}, \quad [D_0, D_J^\dagger] = \sum_j^{N_2} \frac{1 - d_j^\dagger d_j}{2\varepsilon_{2j} - E_{2J}}, \tag{5.54}$$

$$\left[H_U, C_J^\dagger\right] = E_{1J} C_J^\dagger + C_0^\dagger + g_1 C_0^\dagger \sum_j^{N_1} \frac{1 - c_j^\dagger c_j}{2\varepsilon_{1j} - E_{1J}}$$

$$+ g_{12} D_0^\dagger \sum_j^{N_1} \frac{1 - c_j^\dagger c_j}{2\varepsilon_{1j} - E_{1J}}, \tag{5.55}$$

and

$$\left[H_U, D_J^\dagger\right] = E_{2J} D_J^\dagger + D_0^\dagger + g_1 D_0^\dagger \sum_j^{N_2} \frac{1 - d_j^\dagger d_j}{2\varepsilon_{2j} - E_{2J}} + g_{12} C_0^\dagger \sum_j^{N_2} \frac{1 - d_j^\dagger d_j}{2\varepsilon_{2j} - E_{2J}}. \tag{5.56}$$

Using the above-presented commutation relations, we find

$$H_U |M_1; M_2\rangle_U$$

$$= \left( \sum_{J=1}^{M_1} E_{1J} + \sum_{K=1}^{M_2} E_{2K} \right) |M_1; M_2\rangle_U$$

$$+ C_0^\dagger \sum_{J=1}^{M_1} \left( 1 - \sum_j^{N_1} \frac{g_1}{2\varepsilon_{1j} - E_{1J}} + \sum_{J' \neq J}^{M_1} \frac{2g_1}{E_{1J'} - E_{1J}} \right) |M_1(J); M_2\rangle_U$$

$$+ D_0^\dagger \sum_{J=1}^{M_1} \left( \sum_j^{N_1} \frac{g_{12}}{2\varepsilon_{1j} - E_{1J}} - \sum_{J' \neq J}^{M_1} \frac{2g_{12}}{E_{1J'} - E_{1J}} \right) |M_1(J); M_2\rangle_U$$

$$+ C_0^\dagger \sum_{K=1}^{M_2} \left( \sum_j^{N_2} \frac{g_{12}}{2\varepsilon_{2j} - E_{2K}} - \sum_{K' \neq K}^{M_2} \frac{2g_{12}}{E_{2K'} - E_{2K}} \right) |M_1; M_2(K)\rangle_U$$

$$+ D_0^\dagger \sum_{K=1}^{M_2} \left( 1 - \sum_j^{N_2} \frac{g_2}{2\varepsilon_{2j} - E_{2K}} + \sum_{K' \neq K}^{M_2} \frac{2g_2}{E_{2K'} - E_{2K}} \right) |M_1; M_2(K)\rangle_U, \tag{5.57}$$

where

$$|M_1(L); M_2\rangle_U = \prod_{J=1}^{L-1} C_J^\dagger \prod_{J'=L+1}^{M_1} C_{J'}^\dagger \prod_{K=1}^{M_2} D_K^\dagger |vac\rangle, \qquad (5.58)$$

and

$$|M_1; M_2(L)\rangle_U = \prod_{J=1}^{M_1} C_J^\dagger \prod_{K=1}^{L-1} D_K^\dagger \prod_{K'=L+1}^{M_2} D_{K'}^\dagger |vac\rangle. \qquad (5.59)$$

Comparing Eq. (5.57) with Eq. (5.48), we obtain, for arbitrary $J$ and $K$,

$$(C_0^\dagger \ D_0^\dagger) \begin{pmatrix} 1 + g_1 A_{1J} & -g_{12} A_{2K} \\ -g_{12} A_{1J} & 1 + g_2 A_{2K} \end{pmatrix} \begin{pmatrix} |M_1(J); M_2\rangle_U \\ |M_1; M_2(K)\rangle_U \end{pmatrix} = 0, \qquad (5.60)$$

where

$$A_{nL} = -\sum_{j}^{N_n} \frac{1}{2\varepsilon_{nj} - E_{nL}} + \sum_{L' \neq L}^{M_n} \frac{2}{E_{nL'} - E_{nL}}. \qquad (5.61)$$

A nontrivial solution of Eq. (5.60) is derived from the determinantal equation

$$F_{JK} = (1 + g_1 A_{1J})(1 + g_2 A_{2K}) - g_{12}^2 A_{1J} A_{2K} = 0. \qquad (5.62)$$

This constitutes a set of $M_1 + M_2$ coupled equations for $M_1 + M_2$ parameters $E_{1J}$ and $E_{2K}$ ($J = 1, 2, \ldots, M_1; K = 1, 2, \ldots, M_2$), which may be thought of as self-consistently determined pair energies. Equation (5.62) is the exact eigenvalue equation for a two-band superconducting system and can be regarded as a generalization of the Richardson's original eigenvalue equation.

### 5.3.4.  *Preprocessing for numerical calculations*

To remove the divergences from the second term of $A_{nL}$ in Eq. (5.61), we make changes of the energy variables:

$$E_{n2\lambda} = \xi_{n\lambda} + i\eta_{n\lambda},$$

$$E_{n2\lambda-1} = \xi_{n\lambda} - i\eta_{n\lambda}, \qquad (5.63)$$

$$\lambda = 1, 2, \ldots, M_n/2,$$

where we assume that the number of pairs is even. Since the complex pair energies appear in complex conjugate pairs, the total energy is kept in real.

A further transformation is necessary in order to remove the divergences from the first term of $A_{nL}$. We define new variables $x_{n\lambda}$ and $y_{n\lambda}$ as

$$\xi_{n\lambda} = \varepsilon_{n2\lambda} + \varepsilon_{n2\lambda-1} + d_n x_{n\lambda} \quad (x_{n\lambda} \leq 0), \tag{5.64}$$

and

$$\eta_{n\lambda}^2 = -\left(\Delta\varepsilon_{n2\lambda}^2 - d_n^2 x_{n\lambda}^2\right) y_{n\lambda} \quad (y_{n\lambda} \geq 0), \tag{5.65}$$

where

$$\Delta\varepsilon_{n2\lambda} = \varepsilon_{n2\lambda} - \varepsilon_{n2\lambda-1}. \tag{5.66}$$

Considering the sign of $y_{n\lambda}$, we can express $\eta_{n\lambda}$ as

$$\eta_{n\lambda} = |\eta_{n\lambda}| e^{-i\phi_{n\lambda}},$$
$$\phi_{n\lambda} = \begin{cases} 0 & \text{for } \Delta\varepsilon_{n2\lambda}^2 - d_n^2 x_{n\lambda}^2 \leq 0, \\ \pi/2 & \text{for } \Delta\varepsilon_{n2\lambda}^2 - d_n^2 x_{n\lambda}^2 > 0. \end{cases} \tag{5.67}$$

Then we can rewrite $F_{JK}$, by using the new variables, and define the result as $F_{\alpha\beta}$. We extract the real and imaginary parts of $F_{\alpha\beta}$ as

$$\begin{aligned} F_{\alpha\beta}^+ &= \frac{1}{2}\left(F_{\alpha\beta} + F_{\alpha\beta}^*\right) \\ &= 1 + g_1 R_{1\alpha} + g_2 R_{2\beta} + \left(g_1 g_2 - g_{12}^2\right) R_{1\alpha} R_{2\beta} \\ &\quad - \left(g_1 g_2 - g_{12}^2\right) I_{1\alpha} I_{2\beta} \cos\left(\phi_{1\alpha} + \phi_{2\beta}\right) \\ &\quad - \left\{g_1 + \left(g_1 g_2 - g_{12}^2\right) R_{2\beta}\right\} I_{1\alpha} \sin\phi_{1\alpha} \\ &\quad - \left\{g_2 + \left(g_1 g_2 - g_{12}^2\right) R_{1\alpha}\right\} I_{2\beta} \sin\phi_{2\beta}, \end{aligned} \tag{5.68}$$

$$\begin{aligned} F_{\alpha\beta}^- &= \frac{1}{2i}\left(F_{\alpha\beta} - F_{\alpha\beta}^*\right) \\ &= -\left(g_1 g_2 - g_{12}^2\right) I_{1\alpha} I_{2\beta} \sin\left(\phi_{1\alpha} + \phi_{2\beta}\right) \\ &\quad + \left\{g_1 + \left(g_1 g_2 - g_{12}^2\right) R_{2\beta}\right\} I_{1\alpha} \cos\phi_{1\alpha} \\ &\quad + \left\{g_2 + \left(g_1 g_2 - g_{12}^2\right) R_{1\alpha}\right\} I_{2\beta} \cos\phi_{2\beta}, \end{aligned} \tag{5.69}$$

where

$$R_{n\lambda} = -\frac{2d_n x_{n\lambda}(1 + y_{n\lambda})}{(1 - y_{n\lambda})^2 \Delta\varepsilon_{n2\lambda}^2 - (1 + y_{n\lambda})^2 d_n^2 x_{n\lambda}^2}$$

$$+ 4\sum_{\mu \neq \lambda}^{M_n} \frac{\xi_{n\mu\lambda}(\xi_{n\mu\lambda}^2 + \eta_{n\mu}^2 + \eta_{n\lambda}^2)}{(\xi_{n\mu\lambda}^2 + \eta_{n\mu}^2 + \eta_{n\lambda}^2)^2 - 4\eta_{n\mu}^2\eta_{n\lambda}^2}$$

$$- \sum_{j \neq 2\lambda-1,2\lambda}^{N_n} \frac{2\varepsilon_{n\lambda} - \xi_{n\lambda}}{(2\varepsilon_{n\lambda} - \xi_{n\lambda})^2 + \eta_{n\lambda}^2}, \tag{5.70}$$

$$I_{n\lambda} = \left\{ \frac{1 - y_{n\lambda}^2}{(1 - y_{n\lambda})^2 \Delta\varepsilon_{n2\lambda}^2 - (1 + y_{n\lambda})^2 d_n^2 x_{n\lambda}^2} \right.$$

$$- 4y_{n\lambda}\sum_{\mu \neq \lambda}^{M_n} \frac{\xi_{n\mu\lambda}^2 - \eta_{n\mu}^2 + \eta_{n\lambda}^2}{(\xi_{n\mu\lambda}^2 + \eta_{n\mu}^2 + \eta_{n\lambda}^2)^2 - 4\eta_{n\mu}^2\eta_{n\lambda}^2}$$

$$\left. + y_{n\lambda}\sum_{n \neq 2\lambda-1,2\lambda}^{N_n} \frac{1}{(2\varepsilon_{n\lambda} - \xi_{n\lambda})^2 + \eta_{n\lambda}^2} \right\} \times \sqrt{\left| \frac{\Delta\varepsilon_{n2\lambda}^2 - d_n^2 x_{n\lambda}^2}{y_{n\lambda}} \right|}$$

$$\tag{5.71}$$

and

$$\xi_{n\mu\lambda} = \xi_{n\mu} - \xi_{n\lambda}. \tag{5.72}$$

Therefore, for an arbitrary combination of $\alpha$ and $\beta$, we must solve the following equations:

$$F_{\alpha\beta}^+ = 0,$$
$$F_{\alpha\beta}^- = 0 \quad (\alpha = 1, 2, \ldots, M_1/2; \ \beta = 1, 2, \ldots, M_2/2). \tag{5.73}$$

### 5.3.5.  *Results and discussion*

We now apply the exact solution for a two-band system to discuss the properties of the two-band superconductivity in ultrasmall grains. The single-particle level patterns of the $(2M_1 + m) + 2M_2$ electron system ($m = 0, 1$, and $2$) under consideration are represented in Figs. 5.8(a), 5.8(b), and 5.8(c), respectively. The dotted lines mean the chemical potential, and $d_1$ and $d_2$ ($d_1 < d_2$) are the mean level spacings. As seen these figures, the additional electrons first occupy band 1 and then band 2.

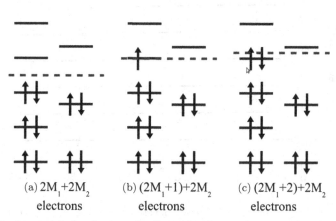

Fig. 5.8. Single-particle levels near the Fermi level in a the case of two-band superconductivity. The dotted lines mean the chemical potential. The left and right bands are band
1 and 2, respectively. $d_1$ and $d_2$ are the mean level spacings. (a) $2M_1 + 2M_2$ electron
system, where $M_n$ is a number of pair levels, (b) $(2M_1 + 1) + 2M_2$ electron system,
(c) $(2M_1 + 2) + 2M_2$ electron system.

Numerical calculations are carried out under the condition that $N_1{:}N_2 =$
3:2, $\hbar\omega_D = 50$, and $\lambda_1 = \lambda_2 = \lambda$.

## 5.3.6. *Pair energy level*

By minimizing the sum of squares of Eqs. (5.68) and (5.69)

$$F = \sum_{\alpha=1}^{M_1/2} \sum_{\beta=1}^{M_2/2} (F_{\alpha\beta}^{+2} + F_{\alpha\beta}^{-2}) \tag{5.74}$$

for various interaction parameters, we obtain a behavior of pair energy levels
$E_{nJ}$ of two bands as shown in Fig. 5.9, in which the solid and broken lines
correspond to the pair energy levels of band 1 and band 2, respectively.
Parameters used in this calculation are $N_1 = 12$, $N_2 = 8$, $M_1 = 6$, $M_2 = 4$,
$0 \leq \lambda \leq 1.0$, and $0 \leq \lambda_{12} \leq 0.2$.

As seen in the figures, band 2 condenses into degenerate levels, but the
band 1 does not. In general, we can expect that the single-particle levels in
a band, whose mean level spacing $d$ is larger than that of the other band,
degenerate faster. The behavior of the condensing band is qualitatively the
same as that in the case of calculations for the single band [234]. The coexistence of the normal band and the condensed one may be reflected in the
opposite phase of the gaps of these bands [218].

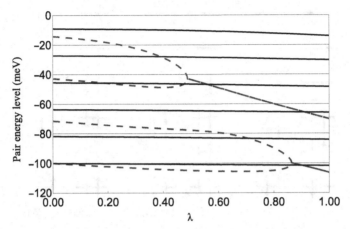

Fig. 5.9.   Typical behavior of pair energy levels of two bands for the ground state. Parameters used in calculation are $N_1 = 12$, $N_2 = 8$, $M_1 = 6$, $M_2 = 4$, $\hbar\omega_D = 50$, $0 \leq \lambda_1 = \lambda_2 \leq 1.5$, and $0 \leq \lambda_{12} \leq 0.3$. Solid and broken lines correspond to the pair energy levels of band 1 and band 2, respectively.

### 5.3.7.   *Condensation energy*

The condensation energy of band $n$ for the $(2M_1 + m) + 2M_2$ electron system can be defined as

$$E_n^C(2M_1 + m, 2M_2) = E_n(2M_1 + m, 2M_2) + \left(M_n + \frac{m}{2}\right) g_n$$
$$- E_n^0(2M_1 + m, 2M_2), \tag{5.75}$$

where $E_n(2M_1 + m, 2M_2)$ and $E_n^0(2M_1 + m, 2M_2)$ are the ground-state energies and the sum of the single-particle energies, respectively.

We calculate the condensation energies and show them in Figs. 5.10(a) and 5.10(b). The parameters used in this calculation are $\lambda = 0.5$, and $\lambda_{12} = 0.01$ for (a), $\lambda_{12} = 0.1$ for (b). Values are normalized by the bulk gap, $\Delta = \omega_D \sinh^{-1}\left(\frac{\lambda}{\lambda^2 - \lambda_{12}^2}\right)$. The solid and broken lines correspond to the condensation energy for bands 1 and 2, respectively. Lines plotted by squares, by triangles, and by circles are for the $2M_1 + 2M_2$ electron system, $(2M_1 + 1) + 2M_2$ electron system, and $(2M_1 + 2) + 2M_2$ electron system, respectively.

As seen in the figures, we can understand that band 2 condenses, but band 1 does not because of the sign of values. This difference of signs may also be reflected in the opposite phases of the gaps of these bands. The behavior of the results for the condensed band (band 2) is qualitatively the same as in the case of the single-band calculations. The condensation energy of band 2 for the $(2M_1 + 2) + 2M_2$ electron system is, however, different from

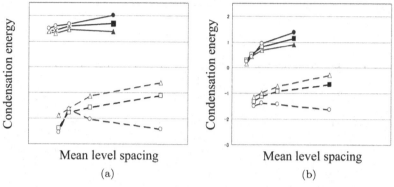

Fig. 5.10. Condensation energy. Parameters used in calculation are $\hbar\omega_D = 50$, and $\lambda_1 = \lambda_2 = 0.5$. Values are normalized by the bulk gap, $\Delta_1 = \omega_D \sinh^{-1}(\frac{\lambda_2}{\lambda_1\lambda_2 - \lambda_{12}^2})$ or $\Delta_2 = \omega_D \sinh^{-1}(\frac{\lambda_1}{\lambda_1\lambda_2 - \lambda_{12}^2})$. The solid and broken lines correspond to the condensation energy for bands 1 and 2, respectively. Lines plotted by squares, by triangles, and by circles are for the $2M_1 + 2M_2$ electron system, for $(2M_1 + 1) + 2M_2$ electron system and for $(2M_1 + 2) + 2M_2$ electron system, respectively. (a) The condensation energy for the interband coupling parameter $\lambda_{12} = 0.01$. (b) The condensation energy for $\lambda_{12} = 0.1$.

the others. We can also see that the condensation energy is affected by the interband interaction $\lambda_{12}$. This is mentioned in our previous work [218].

## 5.3.8. *Parity gap*

The parity gap of band $n$ is defined as

$$\Delta_n^p = E_n(2M_1 + 1, 2M_2)$$
$$-\frac{1}{2}\{E_n(2M_1, 2M_2) + E_n(2M_1 + 2, 2M_2)\} \qquad (5.76)$$

which was introduced by Matveev and Larkin and characterizes the difference of even–odd ground-state energies [237].

We have also calculated the parity gaps shown in Fig. 5.11. The solid and broken lines correspond to the parity gap for bands 1 and 2, respectively. Lines plotted by triangles and by squares are for the interband coupling parameter $\lambda_{12} = 0.01$ and for $\lambda_{12} = 0.1$, respectively. Other parameters used in this calculation are the same as for the calculation of the condensation energy. Values are normalized by the bulk gap.

For the condensed band, we obtain qualitatively the same result as that in the case of the single-band calculations, i.e., there is a minimal point and a tendency toward 1 for $d \to 0$. The mean level spacing giving the minimal point is, however, much less than that for the case of calculations

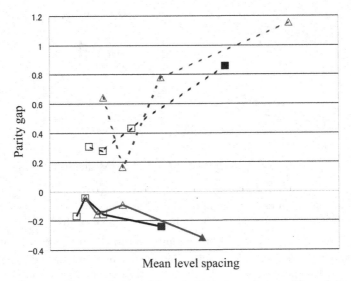

Fig. 5.11.   Parity gap. Parameters used in calculation are $\hbar\omega_D = 50$, and $\lambda_1 = \lambda_2 = 0.5$. The solid and broken lines correspond to the parity gap for bands 1 and 2, respectively. Lines plotted by triangles and by squares are for the interband coupling parameter $\lambda_{12} = 0.01$ and for $\lambda_{12} = 0.1$, respectively. Values are normalized by the bulk gap.

for the single-band. The parity gap is almost independent of the interband interaction $\lambda_{12}$. This is also mentioned in our previous work [218].

We have extended the Richardson's exact solution to the two-band system, and have derived a new coupled equation. To investigate the properties of the two-band superconductivity, we have solved the equation numerically and have determined the behavior of pair energy levels, the condensation energy, and the parity gap.

The band, whose mean level spacing is larger than that of the other band, degenerates and condenses faster. The behavior of the condensing band is qualitatively the same as that for the case of calculations for the single band. The co-existence of the normal band and the condensed one may be reflected in the opposite phases of the gaps of these bands. This phase character appears in every of the results of numerical calculations. Therefore, the phase of a gap is important to stabilize the two-band superconductivity.

We have also calculated the condensation energy and the parity gap for two-band superconductivity. The results suggest that the interband interaction $\lambda_{12}$ affects the condensation energy, but not the parity gap.

In summary, the expression of Richardson's exact solution for two-band superconductivity has been presented, by solving numerically a new coupled equation. Then, the behaviors of pair energy levels, the condensation energy, and the parity gap have been determined. The results for the condensed band is almost qualitatively the same as those for the single-band calculation, and the co-existence of the normal band and the condensed one may be originated from the opposite phases of the gaps of these bands.

## 5.4. Kondo Effect Coupled to Superconductivity

The Kondo effect has attracted a great interest, while considering the properties of semiconductor quantum dots. The Kondo effect can be understood as the magnetic exchange interaction between a localized impurity spin and free conduction electrons [242]. To minimize the exchange energy, the conduction electrons tend to screen the spin of the magnetic impurity, and the ensemble forms a spin singlet. In a quantum dot, some exotic properties of the Kondo effect have been observed [243, 244]. Recently, Sasaki *et al.* has found a significant Kondo effect in quantum dots with an even number of electrons [245]. The spacing of discrete levels in such quantum dots is comparable with the strength of the electron-electron Coulomb interaction. The Kondo effect in multilevel quantum dots has been investigated theoretically by several groups [246–248]. They have shown that the contribution from many levels enhances the Kondo effect in normal metals. There are some investigations on the Kondo effect in quantum dots revealing ferromagnetism [249], noncollinear magnetism [250], superconductivity [251] and so on [252, 253].

Properties of ultrasmall superconducting grains have been also theoretically investigated by many groups [228–230, 232–237, 254]. Black *et al.* have revealed the presence of a parity-dependent spectroscopic gap in the tunneling spectra of nanosize Al grains [228, 232]. For such ultrasmall superconducting grains, the bulk gap has been discussed in relation to physical properties such as the parity gap [237], condensation energy [233], and electron correlation [234] with the size dependence of the level spacing of samples [235]. In the previous works [218], we have also discussed physical properties such as condensation energy, parity gap, and electron correlation of two-gap superconductivity in relation to the size dependence and the effective pair scattering process. The possibility of new two-gap superconductivity has been also discussed by many groups [15, 138, 163, 173, 177, 179, 181, 184, 255, 256].

In a standard *s*-wave superconductor, the electrons form pairs with antialigned spins and are in a singlet state as well. When the superconductivity and the Kondo effect present simultaneously, the Kondo effect and the superconductivity are usually expected to be competing physical phenomena. The local magnetic moments from the impurities tend to align the spins of the electron pairs in the superconductor, which often results in a strongly reduced transition temperature. Buitelaar *et al.* have experimentally investigated the Kondo effect in a carbon nanotube quantum dot coupled to superconducting Au/Al leads [251]. They have found that the superconductivity of the leads does not destroy the Kondo correlations in the quantum dot at the Kondo temperature. A more subtle interplay has been proposed for exotic and not well-understood materials such as heavy-fermion superconductors, in which both effects might actually coexist [257].

In this paper, we investigate the Kondo effect and the superconductivity in ultrasmall grains by using a model which consists of the *sd* and reduced BCS Hamiltonians with the introduction of a pseudofermion. The mean-field approximation for the model is introduced, and we calculate physical properties of the critical level spacing and the condensation energy. These physical properties are discussed in relation to the coexistence of both the superconductivity and the Kondo regime. Finally, we derive the exact equation for the Kondo regime in a nanosystem and discuss the condensation energy from the viewpoint of the correlation energy.

### 5.4.1.  *Kondo regime coupled to superconductivity*

In nanosize superconducting grains, the quantum level spacing approaches the superconducting gap. It is necessary to treat the discretized energy levels of a small-sized system. For ultrasmall superconducting grains, we can consider the pairing-force Hamiltonian to describe the electronic structure of the system [240] and can determine the critical level spacing in the case where the superconducting gap function vanishes at a quantum level spacing [235]. In this section, we present a model for a system in the Kondo regime coupled to the superconductivity and discuss the physical properties such as critical level spacing and condensation energy by using the mean-field approximation in relation to the gap function, spin singlet-order as the Kondo effect, coexistence, and so on.

### 5.4.2.  *Model*

We consider a model coupled to the superconductivity for quantum dots to investigate the Kondo effect in normal metals, which can be expressed by the effective low-energy Hamiltonian obtained by the Schrieffer–Wolff transformation [258]:

$$H = H_0 + H_1 + H_2, \tag{5.77}$$

where

$$H_0 = \sum_{k,\sigma} \varepsilon_k a_{k\sigma}^\dagger a_{k\sigma} + \sum_\sigma E_\sigma d_\sigma^\dagger d_\sigma, \tag{5.78}$$

$$H_1 = J \sum_{k,k'} [S_+ a_{k'\downarrow}^\dagger a_{k\uparrow} + S_- a_{k'\uparrow}^\dagger a_{k\downarrow} + S_z (a_{k'\uparrow}^\dagger a_{k\uparrow} - a_{k'\downarrow}^\dagger a_{k\downarrow})], \tag{5.79}$$

$$H_2 = -g \sum_{k,k'} a_{k\uparrow}^\dagger a_{k\downarrow}^\dagger a_{k'\downarrow} a_{k'\uparrow}. \tag{5.80}$$

$a_{k\sigma}^\dagger$ $(a_{k\sigma})$ and $d_\sigma^\dagger$ $(d_\sigma)$ are the creation (annihilation) operator corresponding to conduction electrons and the effective magnetic particle as an impurity, respectively. In this study, we assume the magnetic particle is a fermion with $S = 1/2$ for simplicity. $E$ means an extraction energy given by $E_{\uparrow,\downarrow} = -E_0 \pm E_z$ included the Zeeman effect. The second term in Eq. (5.77) means the interaction between conduction electrons and the spin in a quantum dot. $S$ is the spin operator as $S_+ = d_\uparrow^\dagger d_\downarrow$, $S_- = d_\downarrow^\dagger d_\uparrow$, and $S_z = (d_\uparrow^\dagger d_\uparrow - d_\downarrow^\dagger d_\downarrow)/2$. The third term corresponds to the interaction between conduction electrons included in the pairing force Hamiltonian.

Here, we introduce a pseudofermion for the magnetic particle operator [259] as

$$\begin{aligned} d_\uparrow^\dagger = f_\downarrow, \, d_\uparrow = f_\downarrow^\dagger, \\ d_\downarrow^\dagger = -f_\uparrow, \, d_\downarrow = -f_\uparrow^\dagger. \end{aligned} \tag{5.81}$$

For this transformation, we have the condition

$$f_\uparrow^\dagger f_\uparrow + f_\downarrow^\dagger f_\downarrow = 1, \tag{5.82}$$

and we have $|\sigma\rangle = f_\sigma^\dagger |0\rangle$. The spin operator $S$ can be presented as $S_+ = f_\uparrow^\dagger f_\downarrow$, $S_- = f_\downarrow^\dagger f_\uparrow$, and $S_z = (f_\uparrow^\dagger f_\uparrow - f_\downarrow^\dagger f_\downarrow)/2$. The Hamiltonian can be rewritten as

$$H_0 = \sum_{k,\sigma} \tilde{\varepsilon}_k c_{k\sigma}^\dagger c_{k\sigma} + \sum_\sigma E f_\sigma^\dagger f_\sigma, \tag{5.83}$$

$$H_1 = J \sum_{k,k',\sigma,\sigma'} f_\sigma^\dagger f_{\sigma'} c_{k'\sigma'}^\dagger c_{k\sigma}, \tag{5.84}$$

$$H_2 = -g \sum_{k,k'} c_{k\uparrow}^\dagger c_{k\downarrow}^\dagger c_{k'\downarrow} c_{k'\uparrow}, \tag{5.85}$$

where $c_{k\sigma} = \sum_i U_{ik} a_{i\sigma}$ with $\tilde{\varepsilon}_k = \sum_{i,j} U_{ki}^\dagger [\varepsilon_i \delta_{ij} - J/2] U_{jk}$. For the sake of simplicity, we only focus on $E_z = 0$ without an external magnetic field: $E = E_0$.

### 5.4.3.    *Mean-field approximation*

In this section, we introduce the mean-field approximation for the present Hamiltonian (5.77). Eto *et al.* have presented the mean-field approximation for the Kondo effect in quantum dots [260].

In the mean-field approximation, we can introduce the spin-singlet-order parameter

$$\Xi = \frac{1}{\sqrt{2}} \sum_{k,\sigma} \langle f_\sigma^\dagger c_{k\sigma} \rangle. \tag{5.86}$$

This order parameter describes the spin couplings between the dot states and conduction electrons. The superconducting gap function can be expressed as

$$\Delta = \sum_k \langle c_{k\downarrow} c_{k\uparrow} \rangle. \tag{5.87}$$

Using these order parameters in Eqs. (5.84) and (5.85), we obtain the mean-field Hamiltonian

$$H_{\mathrm{MF}} = \sum_{k,\sigma} \tilde{\varepsilon}_k c_{k\sigma}^\dagger c_{k\sigma} + \sum_\sigma \tilde{E} f_\sigma^\dagger f_\sigma + \sqrt{2} J \sum_{k,\sigma} [\Xi f_\sigma c_{k\sigma}^\dagger + \Xi^* c_{k\sigma} f_\sigma^\dagger]$$

$$- g \sum_k [\Delta^* c_{k\downarrow} c_{k\uparrow} + \Delta c_{k\uparrow}^\dagger c_{k\downarrow}^\dagger]. \tag{5.88}$$

Constraint (5.82) is taken into account by the second term with a Lagrange multiplier $\lambda$. In this study, we assume a constant density of states with the energy region of the Debye energy, and the coupling constants can be expressed as $J = d\tilde{J}$ and $g = d\lambda$.

### 5.4.4.    *Critical level spacing in the Kondo effect*

By minimizing the expectation value of $H_{\mathrm{MF}}$ in Eq. (5.88), the order parameters can be determined self-consistently. First, we show the Kondo effect without the pairing force part ($g = 0$) in the framework of the mean-field approximation. Next, the Kondo effect in the presence of the superconductivity is discussed in relation to the critical level spacing and the condensation energy. Finally, we derive the exact equation for the Kondo effect in ultrasmall grains coupled to normal metals and discuss properties such as the condensation energy in relation to Richardson's exact equation for the superconductivity.

For ultrasmall superconducting grains, the critical level spacing $d_c^{\mathrm{BCS}}$ can be expressed as $d_c^{\mathrm{BCS}} = 4\omega_D e^\gamma \exp(-1/\lambda)$ for even number of electrons, where $\omega_D$ means the Debye energy. This result suggests that the gap function

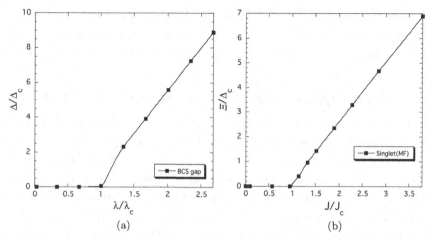

(a)                      (b)

Fig. 5.12. Gap function and spin singlet-order: (a) the gap function. The gap function vanishes in the region of $\lambda$ values less than $\lambda_c$. (b) Spin singlet-order parameter. In the case of $\tilde{J} < \tilde{J}_c$, the singlet-order vanishes. The system consists of eight energy levels and eight electrons with the level spacing $d = 1.0$ and $\omega_D = 1.0$.

of a nanosize system with the level spacing $d$ vanishes, when the coupling parameter $\lambda_c$ is less than the value $(\ln 4\omega_D/d + \gamma)^{-1}$. The bulk gap function $\Delta_c$ with $\lambda_c$ can be expressed as $\Delta_c = \omega_D \mathrm{sh}^{-1}(1/\lambda_c)$.

Figure 5.12(a) shows the gap function of a nanosize system in the framework of the standard BCS theory. We can find the region where the gap function vanishes, when the coupling becomes less than $\lambda_c$. This means the level spacing is larger than the gap function in this region.

Here, we drive the critical level spacing for only the Kondo regime ($\lambda = 0$). The equation determining the singlet-order parameter can be expressed as

$$\Xi = \sum_k \frac{\Xi(\xi_k - x)}{(\xi_k - x)^2 + \Xi^2}, \tag{5.89}$$

where $\xi_k = \tilde{\varepsilon}_k - \mu$, $x = [\tilde{\varepsilon}_k + \tilde{E} \pm \sqrt{(\tilde{\varepsilon}_k - \tilde{E})^2 + 4\Xi^2}]]/2$, and $\mu$ is the chemical potential. For the case of the critical level spacing, the solution shows that the spin singlet-order parameter vanishes. From Eq. (5.89), we can find the critical level spacing $d_c^{\mathrm{Kondo}}$ for the Kondo regime

$$d_c^{\mathrm{Kondo}} = 4\omega_D e^\gamma \exp\left[-\frac{1}{2\sqrt{2}\tilde{J}}\right]. \tag{5.90}$$

When the coupling parameter $\tilde{J}$ is smaller than $\tilde{J}_c = [2\sqrt{2}(\ln(4\omega_D/d) + \gamma)]^{-1}$, the spin singlet-order parameter vanishes.

Figure 5.12(b) presents the spin singlet-order parameter given by Eq. (5.86) in the case $g = 0$. In the region of $\tilde{J} < \tilde{J}_c$, the order parameter vanishes. This result suggests the critical level spacing in the Kondo effect.

### 5.4.5.  *Kondo effect coupled to the superconductivity*

Here, we consider a simple system which consists of 8 energy levels and eight electrons and investigate the critical level spacing and the condensation energy of the coupled system between the superconductivity and the Kondo regime in the framework of the mean-field approximation of Eq. (5.88).

Figure 5.13(a) shows the spin singlet-order parameter and the gap function for several cases. We can find the critical level spacings for the gap function and for the spin singlet-order parameter. When $\lambda < \lambda_c$ and $\tilde{J} > \tilde{J}_c$, we can find only the spin singlet-order parameters. In the region of $\lambda/\lambda_c$ from 1.4 to 1.7 with $\tilde{J}/\tilde{J}_c = 0.189$, we can find the coexistence of both the gap function and the spin singlet-order parameter. For $\lambda/\lambda_c$ larger than 1.7, only the gap function still exists, and the spin singlet-order parameter vanishes. At $\tilde{J}/\tilde{J}_c = 0.284$, we can find the coexistence in the region $\lambda/\lambda_c = 1.7$–2.3. These results suggest that strong local magnetic moments from the impurities make the transition temperature for superconductivity to be reduced. However, the weak couplings $\lambda$ of the superconductivity do not destroy the spin singlet-order parameter at all. These results are in good agreement with the experimental results [251]. We can find that there is the coexistence region for both the superconductivity and the Kondo regime.

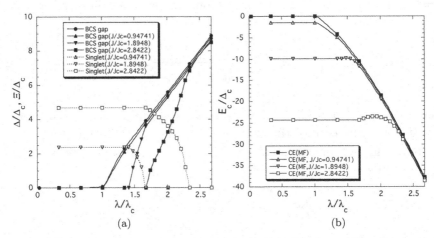

Fig. 5.13.  Physical properties in a coupled system: (a) Gap function and spin singlet-order parameter. (b) Condensation energy. $\tilde{J}/\tilde{J}_c = 0, 0.94741, 1.8948$, and $2.8422$. Other parameters are the same.

Figure 5.13(b) shows the condensation energy for several $\lambda$ and $\tilde{J}$ values. We have found that the condensation energy of the coupled system between the superconductivity and the Kondo regime becomes lower than that for the pure superconductivity. In the coexistence region, the highest value of the condensation energy appears in all cases.

### 5.4.6.  *Exact solution for the Kondo regime*

The standard BCS theory gives a good description of the phenomenon of superconductivity in large samples. However, when the size of a superconductor becomes small, the BCS theory fails. To investigate the physical properties such as the condensation energy, parity gap, etc., it is necessary to make a more accurate treatment. For the superconductivity in ultrasmall grains, the exact solution to the reduced BCS Hamiltonian presented by Richardson [240] has been applied to investigate the above-mentioned physical properties [236].

By using the wave function describing all pair electron excitations, we can derive the exact solution for the pairing force (reduced) Hamiltonian

$$2 - \sum_{k=1}^{N} \frac{\lambda}{\tilde{\varepsilon}_k - E_i} + \sum_{l=1,l\neq i}^{n} \frac{2\lambda}{E_l - E_i} = 0, \tag{5.91}$$

where $N$ and $n$ are the number of orbitals and the number of the occupied orbitals, respectively, and $E_i$ corresponds to the exact orbital. Figure 5.14 shows the condensation energy and the pairing energy level for the nanosize superconductivity. Note that the physical properties obtained in the mean-field approximation give a good description for the high density of states $(d \to \infty)$. We have found the different behavior of the condensation energy from that obtained in the mean-field approximation, as shown in Fig. 5.14(a). Figure 5.14(b) presents the qualitative behavior of the pairing energy level in the ground state. At $\lambda$ about 1.6, above two energy levels in Fig. 5.14(b) are completely paired. The pairing behavior has been already reported by many groups [182, 240].

Let us derive the exact equation for the Kondo regime in ultrasmall grains. We can consider the Hamiltonian $H = H_0 + H_1$ in Eq. (5.77). We introduce a creation operator describing all excited states at the spin singlet coupling between a conduction electron and a pseudofermion:

$$B_j^\dagger = \sum_{k,\sigma} \frac{c_{k\sigma}^\dagger f_\sigma}{\tilde{\varepsilon}_k - E_j}, \tag{5.92}$$

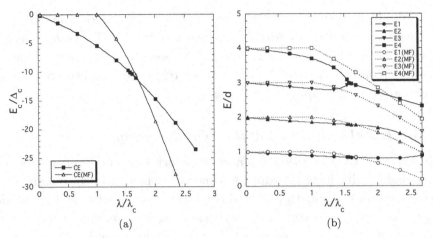

Fig. 5.14.   Exact solution for the superconductivity: (a) Condensation energy of the exact solution and that obtained in the mean-field approximation (b) Pairing energy level with energy level obtained in the mean-field approximation: Eight energy levels, eight electrons, $d = 1.0$, $\omega_D = 4.0$.

where $E_j$ means the exact eigenenergies in the Kondo regime. The exact eigenstate $|\Psi_n\rangle$ for the Kondo regime can be written as $|\Psi_n\rangle = \Pi_{\nu=1}^n B_\nu^\dagger |0\rangle$. Other electrons, which are not related to the spin singlet-order, contribute $E_{\text{single}} = \sum_{k=1}^n \tilde{\varepsilon}_k$ to the eigenenergy. The ground-state energy $E_{\text{GS}}$ can be written as $E_{\text{GS}} = \sum_{k=1}^n [E_k + \tilde{\varepsilon}_k]$.

By operating the Hamiltonian to the exact eigenstate, we obtain the condition

$$1 + \sum_{k=1}^N \frac{\tilde{J}}{\tilde{\varepsilon}_k - E_j} = 0. \tag{5.93}$$

This equation gives the exact solution for the Kondo regime. Note that the creation operator (5.92) might be a true boson one as compared with the case of the reduced BCS model.

Figure 5.15 shows the condensation energy of the exact solution in the Kondo regime with that obtained in the mean-field approximation. We can find the different behavior of the condensation energy from that obtained in the mean-field approximation. However, the behavior is similar to that in the case of the superconductivity in nanosize systems.

We have investigated properties of the Kondo regime coupled to the superconductivity in ultrasmall grains by using the mean-field approximation. In the framework of the mean-field approximation, we have found the critical level spacing for the Kondo regime. The result suggests that the

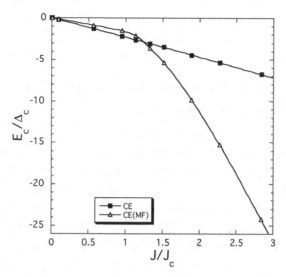

Fig. 5.15.   Condensation energy for the Kondo regime: all parameters used in the system are as follows: Eight energy levels, eight electrons, $d = 1.0$, and $\omega_D = 4.0$.

Kondo effect vanishes, when the level spacing becomes larger than the critical level spacing.

We have calculated physical properties of the critical level spacing and the condensation energy of the coupled system by using the mean-field approximation. From the results, we have found that strong local magnetic moments from the impurities make the transition temperature for superconductivity to be reduced. However, weak couplings $\lambda$ of the superconductivity do not destroy the spin singlet-order parameter at all. These results are in good agreement with the experimental results [251]. We have found that there is the coexistence region for both the superconductivity and the Kondo regime.

Finally, we have derived the exact equation for the Kondo regime in a nanosystem, which was not an easy task, and have discussed the condensation energy from the viewpoint of energy levels. The further study of the properties in the Kondo regime with the use of the exact equation will be presented elsewhere.

In summary, we have investigated the Kondo effect and the superconductivity in ultrasmall grains by using a model which involves the *sd* and reduced BCS Hamiltonians with the introduction of a pseudofermion. The mean-field approximation for the model have been introduced, and we have calculated physical properties of the critical level spacing and the condensation energy. These physical properties have been discussed in relation to the coexistence of both the superconductivity and the Kondo regime. Finally, we have derived the exact equation for the Kondo regime in a nanosystem and discuss the condensation energy from the viewpoint of energy levels.

## 5.5.    Interaction of Nanoscale Ferromagnetic Granules in London Superconductors

Recent experiments have fabricated structured arrays of ferromagnetic nanoparticles in proximity to a bulk superconductor. We consider the theory of interactions between two nanoscale ferromagnetic particles embedded in a superconductor. In the London limit approximation, we will show that the interactions between ferromagnetic particles can lead to either parallel or antiparallel spin alignment. The crossover between these is dependent on the ratio of the interparticle spacing and the London penetration depth. We show that a phase transition between spin orientations can occur, as the temperature is varied. Finally, we comment on the extension of these results to arrays of nanoparticles in different geometries. We show that a phase transition between spin orientations can occur under variation of the temperature. Finally, we comment on the extension of these results to arrays of nanoparticles in different geometries.

Magnetism and superconductivity are two competing collective ordered states in metals. In the case of ferromagnetism, the exchange interactions lead to the parallel alignment of electronic spins, while electron–phonon interactions in the BCS superconductivity lead to the spin singlet pairing of electrons. Clearly, these two types of order are generally mutually incompatible. In bulk systems, the frustration between the electron singlet pairing and the spin ordering is resolved by the FFLO (Fulde–Ferrell [261], Larkin–Ovchinnikov [262]) state. However, this has proved elusively experimentally, and few examples are known. In particular the FFLO state appears to be highly sensitive to disorder.

In recent years, however, there has been observed a great increase of the interest in the interactions between ferromagnetism and superconductivity in artificially structured systems. Advances in nanotechnology and microfabrication have made it possible to build hybrid structures containing both ferromagnetic and superconducting components which interact magnetically or via the proximity effect [263, 273]. Superconductor–ferromagnetic–superconductor planar structures have been found to show the $\pi$-junction Josephson behavior [265–267]. Ferromagnet–superconductor–ferromagnet spin valve structures have also been fabricated [268] with potential applications to spintronics. More complex types of structures have also been produced; for example, Moschalkov *et al.* fabricated arrays

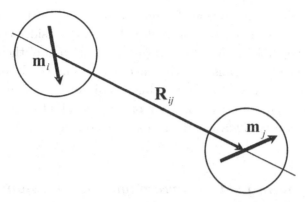

Fig. 5.16. Schematic diagram of a nanocomposite superconductor containing ferromagnetic granules.

of ferromagnetic nanoscale dots on superconducting substrates [269, 270]. A description of such structures based upon the Ginzburg–Landau theory was developed by Peeters [271].

In this section, we consider the interaction between magnetic nanoparticles embedded in a superconductor, as shown in Fig. 5.16. Theoretical studies of such systems can be carried out exactly in two limiting cases depending on the relative magnitudes of the London penetration depth $\lambda$, and the coherence length, $\xi$. If we consider the nanoparticles to be essentially point-like on the scale of both these characteristic lengths, then they correspond to effective point-like magnetic-moments of the form

$$\mathbf{M}(\mathbf{r}) = \sum_i \mathbf{m}_i \delta(\mathbf{r} - \mathbf{r}_i), \tag{5.94}$$

where the particle at $r_i$ has magnetic moment $m_i$. Interactions between these isolated moments arise directly from magnetic dipole–dipole forces modified by the screening of the bulk supercurrents. A second source of the interaction between the moments is the RKKY interaction modified by the presence of the BCS energy gap $\Delta$. It is clear that the range of the dipolar forces is determined by the penetration depth $\lambda$, while the usual oscillatory power-law RKKY interaction is truncated exponentially on a length scale of the order of $\xi_w$. Therefore, the dipolar forces dominate for superconductors in the London limit $r_0 \ll \lambda$, while the RKKY interactions are more important in the Pippard case $r_0 \gg l$, where $r_0^{-1} = \xi_w^{-1} + l^{-1}$, and $l$ is the mean free path [16]. Here, we consider the London limit and neglect RKKY interactions. Magnetic impurities interacting via RKKY interactions were considered by Larkin [272]. In the London limit, we first derive general expressions for the configuration of magnetic fields and the interaction energy of an ensemble of ferromagnetic granules. Then we consider the case of two interacting nanoparticles. It is found that, as a function of the temperature, an orientational phase transition can take place. The conditions for such a phase transition to occur are derived for a chain of ferromagnetic granules. Finally, we comment on the application of these results to the determination of the equilibrium configurations of more general lattices of ferromagnetic particles.

### 5.5.1.   *Magnetic field of ferromagnetic inclusions in a London superconductor*

We determine the magnetic configurations in superconducting nanocomposite systems by means of Maxwell's equations. The total current, $\mathbf{j}$, includes both normal and superconducting parts,

$$\mathbf{j} = \mathbf{j}_s + \mathbf{j}_n. \tag{5.95}$$

The role of the normal currents in a superconductor is negligible, since the superconductor exhibits weak magnetic characteristics in the normal state. However, a normal current will be present within the ferromagnetic inclusions. This normal-state current can be written in the traditional form

$$\mathbf{j}_n = \nabla \times \mathbf{M}, \tag{5.96}$$

where $\mathbf{M}$ is the magnetization of a material. The supercurrent obeys the usual London equation

$$\nabla \times \mathbf{j}_s = -\frac{n_s e^2}{m} \mathbf{B} \tag{5.97}$$

in SI units [68, 264], where $\mathbf{B}$ is the magnetic field, $n_s$ is the superfluid density, and $m$ and $e$ are the electron mass and charge, respectively.

Combining relations (5.95)–(5.97) with the Maxwell equation $\nabla \times \mathbf{B} = \mu_0 \mathbf{j}$, the magnetic field can be found in the form

$$\nabla \times (\nabla \times \mathbf{B}) + \lambda^{-2} \mathbf{B} = \mu_0 \nabla \times (\nabla \times \mathbf{M}), \tag{5.98}$$

where $\lambda$ is the London penetration depth of the field in the superconductor. Since $\operatorname{div} \mathbf{B} = 0$, Eq. (5.98) can be rewritten as

$$-\nabla^2 \mathbf{B} + \lambda^{-2} \mathbf{B} = \mu_0 \nabla \times (\nabla \times \mathbf{M}). \tag{5.99}$$

For the boundary conditions, we assume that the magnetic field in the superconductor vanishes far from the region of the ferromagnetic inclusions. According to these boundary conditions, the solution of Eq. (5.99) can be expressed as follows:

$$B(\mathbf{r}) = \frac{\mu_0}{4\pi} \int d^3 r' G\left(|\mathbf{r} - \mathbf{r}'|\right) \nabla \times (\nabla \times \mathbf{M}(\mathbf{r}')),\tag{5.100}$$

where the Green's function is given by

$$G\left(|\mathbf{r} - \mathbf{r}'|\right) = \frac{\exp\left(-|\mathbf{r} - \mathbf{r}'|/\lambda\right)}{|\mathbf{r} - \mathbf{r}'|}.\tag{5.101}$$

After the double integration by parts and some algebraic manipulations, this expression can be rewritten in the general form

$$B(\mathbf{r}) = \frac{\mu_0}{4\pi} \int d^3 r' \exp\left(-R/\lambda\right) \cdot \left\{ \left( \frac{3\mathbf{R}\,(\mathbf{R}\cdot\mathbf{M}(\mathbf{r}'))}{R^5} - \frac{\mathbf{M}(\mathbf{r}')}{R^3} \right) \right.$$
$$\left. \times \left(1 + \frac{R}{\lambda} + \frac{R^2}{\lambda^2}\right) - \frac{2\mathbf{R}\,(\mathbf{R}\cdot\mathbf{M}(\mathbf{r}'))}{\lambda^2 \cdot R^3} \right\},\tag{5.102}$$

where $\mathbf{R} = \mathbf{r} - \mathbf{r}'$, $R = |\mathbf{r} - \mathbf{r}'|$.

Expression (5.102) determines the magnetic field outside of the ferromagnetic inclusions. Since the magnetization of the system $\mathbf{M}(\mathbf{r})$ is defined by (5.96), we have in (5.101) that $\mathbf{M}(\mathbf{r}) = 0$ outside of the volume of ferromagnetic granules.

### 5.5.2. *Magnetic field of ferromagnetic quantum dots in a superconducting nanocomposite material*

Assuming that the sizes of the ferromagnetic inclusions are on the nanometer length scale, they will appear essentially point-like on the scale of the penetration depth $\lambda$. In this case, we can approximate the magnetization as a sum of point magnetic moments, as shown in Eq. (5.94). Using this approximation in the general expression Eq. (5.102), we find the result (see Appendix B):

$$B(\mathbf{r}) = \frac{\mu_0}{4\pi} \sum_i \exp\left(-R_i/\lambda\right) \cdot \left\{ \left( \frac{3\mathbf{R}_i\,(\mathbf{R}_i\cdot\mathbf{m}_i)}{R_i^5} - \frac{\mathbf{m}_i}{R_i^3} \right) \right.$$
$$\left. \times \left(1 + \frac{R_i}{\lambda} + \frac{R_i^2}{\lambda^2}\right) - \frac{2\mathbf{R}_i\,(\mathbf{R}_i\cdot\mathbf{m}_i)}{\lambda^2 \cdot R_i^3} \right\},\tag{5.103}$$

where $\mathbf{R}_i = \mathbf{r} - \mathbf{r}_i$ and $R_i = |\mathbf{r} - \mathbf{r}_i|$.

It is apparent from Eq. (5.103) that if the temperature of a superconductor approaches $T_c$, and the penetration depth $\lambda \to \infty$, then expression
(5.103) tends to the limit

$$B\left(r\right) = \frac{\mu_0}{4\pi} \sum_i \left( \frac{3\mathbf{R}_i\left(\mathbf{R}_i \cdot \mathbf{m}_i\right)}{R_i^5} - \frac{\mathbf{m}_i}{R_i^3} \right), \tag{5.104}$$

which describes the usual magnetic field of isolated dipoles in the normal-
state medium.

In the other limiting case, if the distance between granules is much more
than the depth penetration, we have

$$B\left(r\right) = \frac{\mu_0}{4\pi} \sum_i \frac{\exp(-R_i/\lambda)}{R_i \cdot \lambda^2} \cdot \left( \frac{\mathbf{R}_i\left(\mathbf{R}_i \cdot \mathbf{m}_i\right)}{R_i^2} - \mathbf{m}_i \right). \tag{5.105}$$

### 5.5.3.   *Interaction energy of quantum dots in a superconducting nanocomposite material*

To determine the collective states of the magnetic moments in superconducting nanocomposite materials in the London limit, we make use of the expression
for the free energy $F$ of the system [288]

$$F = \frac{1}{2\mu_0} \int d^3r \{B^2 + \lambda^2 \left(\nabla \times B\right)^2\}. \tag{5.106}$$

Integrating by parts and using the Gauss theorem, we transform Eq. (5.106)
to the following form:

$$F = \frac{1}{2\mu_0} \int d^3r\, B\left\{B + \lambda^2 \nabla \times \left(\nabla \times B\right)\right\} = -2\pi\lambda^2 \int d^3r\, B \cdot \nabla \times \left(\nabla \times M\right). \tag{5.107}$$

Integrating by parts and using the Gauss theorem again, we transform Eq.
(5.107) to the form

$$F = -2\pi\lambda^2 \int d^3r\, M \cdot \nabla \times \left(\nabla \times B\right). \tag{5.108}$$

Then, using the London equation (5.98) and omitting the magnetostatic
self-energy from consideration, we find the interaction energy of magnetic
moments. The obtained expression can be used to determine the collective

state of magnetization of an ensemble of granules:

$$U = \frac{\mu_0}{8\pi} \sum_i \sum_j \exp\left(-R_{ij}/\lambda\right) \cdot \left\{ \left( \frac{3\left(\mathbf{R_{ij}} \cdot \mathbf{m_j}\right)\left(\mathbf{R_{ij}} \cdot \mathbf{m_i}\right)}{R_{ij}^5} - \frac{\mathbf{m_i} \cdot \mathbf{m_j}}{R_{ij}^3} \right) \right.$$

$$\left. \times \left( 1 + \frac{R_{ji}}{\lambda} + \frac{R_{ij}^2}{\lambda^2} \right) - \frac{2\left(\mathbf{R_{ij}} \cdot \mathbf{m_j}\right)\left(\mathbf{R_{ij}} \cdot \mathbf{m_i}\right)}{\lambda^2 \cdot R_{ij}^3} \right\}, \tag{5.109}$$

where $\mathbf{R}_{ij} = \mathbf{r}_j - \mathbf{r}_i$, $R_{ij} = |\mathbf{r}_j - \mathbf{r}_i|$, $i \neq j$.

### 5.5.4. *Spin orientation phase transitions in a nanocomposite material with arrays of ferromagnetic quantum dots*

We begin by studying the magnetic configuration of an isolated pair of magnetic moments. The interaction energy of such a pair can be written as

$$U = -\frac{\mu_0}{4\pi} \exp\left(-R_{12}/\lambda\right) \cdot \left\{ \left( \frac{3\left(\mathbf{R_{12}} \cdot \mathbf{m_1}\right)\left(\mathbf{R_{12}} \cdot \mathbf{m_2}\right)}{R_{12}^5} - \frac{\mathbf{m_2} \cdot \mathbf{m_1}}{R_{12}^3} \right) \right.$$

$$\left. \times \left( 1 + \frac{R_{12}}{\lambda} + \frac{R_{12}^2}{\lambda^2} \right) - \frac{2\left(\mathbf{R_{12}} \cdot \mathbf{m_1}\right)\left(\mathbf{R_{12}} \cdot \mathbf{m_2}\right)}{\lambda^2 \cdot R_{12}^3} \right\}. \tag{5.110}$$

Let us introduce a coordinate system with the origin at the first magnetic moment $\mathbf{m}_1$ and the polar axis along the line connecting magnetic moments. In this coordinate system, the magnetic moments have the components $\mathbf{m}_i = m_i\left(\cos\varphi_i \sin\theta_i, \sin\varphi_i \sin\theta_i, \cos\theta_i\right)$, and their interaction energy (5.110) is written in the form

$$U = \frac{\mu_0}{4\pi} m_1 m_2 \cdot \frac{\exp(-R_{12}/\lambda)}{R_{12}^3} \cdot f\left(\theta_i, \varphi_i, R_{12}/\lambda\right), \tag{5.111}$$

where

$$f\left(\theta_i, \varphi_i, x\right) = -2\left(1 + x\right)\cos\theta_1 \cos\theta_2 + \left(1 + x + x^2\right)\sin\theta_1 \sin\theta_2 \cos\varphi \tag{5.112}$$

and $\varphi = \varphi_2 - \varphi_1$.

Differentiating the function $f\left(\theta_i, \varphi_i, x\right)$ with respect to angular variables and equating the results to zero, we find that there are four possible stable energy configurations:

$$\theta_1 = \theta_2 = 0,$$
$$\theta_1 = \theta_2 = \pi, \tag{5.113}$$
$$0 \leqslant \varphi < 2\pi,$$

$$\theta_1 = \theta_2 = \pi/2,$$
$$\varphi = \pi, \tag{5.114}$$

$$\theta_1 = \theta_2 = \pi/2,$$
$$\varphi = 0, \tag{5.115}$$

$$\theta_1 = 0, \quad \theta_2 = \pi,$$
$$0 \leqslant \varphi < 2\pi, \tag{5.116}$$

which are illustrated in Fig. 5.17.

The further analysis shows that configurations Eqs. (5.115) and (5.116) are saddle points, not energy minima. Evaluating the second derivatives of (5.112), we obtain the stability condition of configuration (5.113):

$$\left(\frac{R}{\lambda}\right)^2 - \frac{R}{\lambda} - 1 \leqslant 0. \tag{5.117}$$

This implies that the ferromagnetic ordering of the pair of magnetic moments is possible if

$$\frac{R}{\lambda} \leqslant \frac{1}{2}(1 + \sqrt{5}). \tag{5.118}$$

It turns out that if condition (5.118) is violated, then the alternative configuration (5.114) is a stable energy minimum. We can conclude that if the temperature changes, and the penetration depth parameter $\lambda(T)$ varies in such a way that condition (5.118) is not satisfied, then the ground-state orientation will change from (5.113) to (5.114).

This result is readily generalized to ordered arrays of ferromagnetic granules, and so we conclude that orientational phase transitions are possible in systems of quantum dots in a superconducting matrix. For example, it is clear that condition (5.118) can be applied to linear chains of quantum dots. On the other hand, the results for square or cubic lattices remain to be determined.

We have considered the interactions between nanoscale magnetic dots embedded in a bulk superconducting material. Our approach is valid for materials which are well described in the London limit $r_0 \ll \lambda$, since RKKY interactions are negligible. We have shown that, depending on the dimensionless parameter $R/\lambda$, different stable ground states occur. So, as

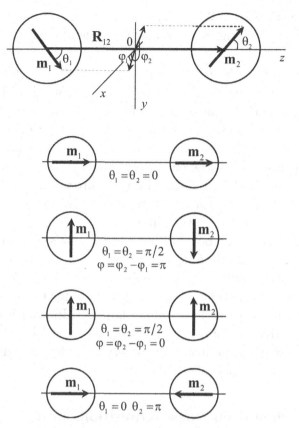

Fig. 5.17. Four energy saddle points of a pair of ferromagnetic quantum dots, as defined in Eq. (5.113). Of four, 1 and 2 correspond to the ground state, depending on condition Eq. (5.118). States 3 and 4 are never stable.

the temperature varies, orientational phase transitions will take place for periodic arrays of such quantum dots. Of course, our calculation does not include all types of interactions which define the orientation of magnetic moments in space. In particular, we neglect the energy of a magnetic anisotropy of granules which is determined by the shape of granules or the type of their crystal lattice. However, when the shape of granules is close to the spherical one and the lattice of a ferromagnet has the cubic symmetry, then Eq. (5.109) will be essentially exact.

In the experimental systems studied by V.V. Moshchalkov [270], a square of Pt/Co magnetic nanodots was deposited on the surface on the Pb superconductor, which is of type $I(k = 0.48)$. The dots where about $0.26\,\mu$m in diameter, and they were deposited on the grid with a spacing of $0.6\,\mu$m. For Pb, the penetration depth is $39\,$nm at low temperatures. So, this array was in the limit $R > \lambda$ and the dot–dot interaction would be expected to correspond to the antiferromagnetic alignment shown in the second state in Fig. 5.17. With increase in the temperature, the transition to the ferromagnetic alignment would occur according to Eq. (5.118) at $\lambda = 0.36R$, i.e., 219 nm.

According to the Casimir formula $\lambda(T) = \lambda(0)(1 - t^4)^{-1/2}$ with $t = T/T_c$, this would occur at $T = 7.14\,\mathrm{K}$, as compared with $T_c = 7.2\,\mathrm{K}$. Therefore, the experimental conditions for the transition to be observed are certainly feasible. Of course, for an exact comparison with theory in this case, our theory should be generalized to deal with magnetic particles near the surface rather than with those embedded in the bulk of a superconductor.

Of course, it would be interesting in the future to generalize our results to superconductors in the Pippard limit, where the RKKY interactions between quantum dots will dominate over dipolar forces [272].

Expression (5.110) can be used to study, by means of numerical methods, the magnet configurations and the orientational phase transitions in an ensemble of nanogranules. It is possible to determine the conditions of orientational transformations in the analytic form for the ordered structures (a chain of granules, plane and volume lattices). Inasmuch as the state of the magnetic subsystem of a specimen at phase transitions is changed, this phenomenon can be experimentally observed under changing a magnetic susceptibility in the region of low fields.

## 5.6.　Spin-orientation Phase Transitions in a Two-dimensional Lattice of Ferromagnetic Granules in a London-type Superconductor

In order to determine the collective state of magnetic moments in nanocomposite materials with the matrix made of a London-type superconductor, we use formula Eq. (5.103) for the energy of magnetic interaction.

First of all, we note that the realization of one or other magnetic configuration is defined by both the competition of diamagnetic effects from the side of the superconducting matrix and the magnetostatic interaction in the system of ferromagnetic granules. At lower temperatures eliminating the thermal disordering of the system, the magnetostatic interaction leads to a correlation of magnetic moments. As main conditions for the formation of magnetic configurations, we take the equivalence of all sites and the zero value of the net magnetic moment of the lattice. The planar lattice of granules by itself sets a preferred direction in space. Therefore, we will separate two configurations from the whole manifold of spatial orientations of magnetic moments. The first configuration is presented in Fig. 5.18. It is characterized by the orientation of the magnetization of granules which is orthogonal to the base plane. The alternation of the magnetization of

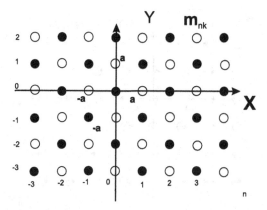

Fig. 5.18. Two-dimensional lattice of magnetic points. Magnetic moments (black color) are directed up or down (white color); $a$ is the lattice constant.

neighboring granules decreases, to a certain extent, the energy of magnetic interaction. In addition, such a distribution of magnetic moments favors a decrease of the amplitude of a magnetic field in the superconducting matrix, which is also advantageous from the energy viewpoint. Thus, the given configuration can be considered as a version of the magnetic order.

A configuration of the second type is shown in Fig. 5.19. It is characterized by the distribution of magnetic moments in the base plane of the lattice such that the magnetic moments are aligned as magnetic chains with alternating directions of the magnetization. Here, like the configuration presented in Fig. 5.18, the main requirement, the equivalence of the states of magnetic points, is satisfied. A similar distribution also decreases the energy of magnetostatic interaction, on the one hand, and favors a decrease in the amplitude of a magnetic field in a superconductor, on the other hand. At the same time, the planar orientation of magnetic moments has the basic distinction from the orthogonal one. For example, by means of a continuous deformation of the magnetization in the base plane, the configuration in Fig. 5.18 can be transferred into the structure shown in Fig. 5.19.

Such a system is characterized by a coherent rotation of magnetic moments by an angle $\pm\varphi$ relative to the principal direction of the lattice. In this case, there occurs both the modulation of the direction of moments at sites of the lattice and some increase in the energy of magnetic chains, but these processes are accompanied by the formation of magnetic vortices in cells, which promotes a decrease in the energy of magnetic interaction. The states of separate magnetic points in the lattice remain equivalent at the zero total magnetization.

Thus, the questions arise how the energy of the array of magnetic moments in the base plane shown in Fig. 5.20 depends on the angle $\pm\varphi$

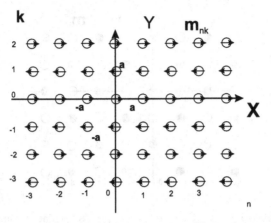

Fig. 5.19.   Two-dimensional lattice of magnetic points. Magnetic moments are aligned in the plane in the form of chains; $a$ is the lattice constant.

and to which value it is equal in the equilibrium state. To answer these questions, we consider relation Eq. (5.110) for the interaction energy and reduce it to a single sum by virtue of the fact that the states of magnetic points are equivalent. In this case, in order to calculate the energy of the lattice, it is sufficient to determine the energy of a single magnetic point, e.g., $m_{0,0}$, located at the origin of the coordinate system and then to multiply the result by the total number of magnetic points $N$. Relation Eq. (5.110) becomes significantly simpler:

$$U = -\frac{\mu_0}{4\pi}\frac{N}{2}\left\{3\left(1 - \frac{\partial}{\partial\alpha}\right) + \frac{\partial^2}{\partial\alpha^2}\right\}\sum_i^N \exp\left(-\alpha \cdot r_i/\delta\right)\frac{(\mathbf{r}_i \cdot \mathbf{m}_i)(\mathbf{r}_i \cdot \mathbf{m}_{0,0})}{r_i^5}$$

$$+ \frac{N}{2}\left\{1 - \frac{\partial}{\partial\alpha} + \frac{\partial^2}{\partial\alpha^2}\right\}\sum_i^N \exp\left(-\alpha \cdot r_i/\delta\right)\frac{\mathbf{m}_i \cdot \mathbf{m}_{0,0}}{r_i^3}. \tag{5.119}$$

By writing formula (5.119), we used the method of differentiation with respect to the parameter $\alpha$ which should be set equal to 1 after the calculations. The index "$i$" stands for the summation over all sites of the lattice. Performing the summation in relation Eq. (5.119), it is convenient to introduce the pair of indices $(n, k)$ defining the position of a site in the lattice (Fig. 5.20) instead of the running index of magnetic points "$i$". It is easy to see that the system represented in Fig. 5.20 *possesses* the translational invariance with a period of *2a* so that

$$\mathbf{m}_{n,k} = \mathbf{m}_{n+2l,k+2p},$$
$$l, p = \pm 1, \pm 2, \ldots. \tag{5.120}$$

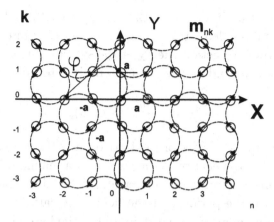

Fig. 5.20.   Part of the lattice with a modulated planar distribution of the magnetization; $a$ is the lattice constant. Circles stand for magnetic points, and the arrows on them indicate the directions of magnetic moments $m_{nk}$ in the base plane ($n, k$ — the spatial indices of magnetic points). The angle $\varphi$ defines the deviation of the moments of magnetic points from the principal direction of the lattice. The states of all points in the given configuration are equivalent. The net magnetic moment is zero. The separated circles denote schematically magnetic vortices. The dotted lines are tangents to the directions of magnetic moments at sites of the lattice.

The lattice has only four types of magnetic points differing from one another by a spatial orientation of magnetic moments. Their vector components depend on the angle $\varphi$ in the following manner:

$$\mathbf{m}_{n,k} = \mathbf{m}_{n+2l,k+2p}, \quad \mathbf{m}_{2l,2p} = \mathbf{m}_{0,0} = m \begin{pmatrix} \cos\varphi \\ \sin\varphi \\ 0 \end{pmatrix},$$

$$\mathbf{m}_{2l+1,2p} = \mathbf{m}_{1,0} = m \begin{pmatrix} \cos\varphi \\ -\sin\varphi \\ 0 \end{pmatrix},$$

$$\mathbf{m}_{2l,2p+1} = \mathbf{m}_{0,1} = m \begin{pmatrix} -\cos\varphi \\ \sin\varphi \\ 0 \end{pmatrix}, \tag{5.121}$$

$$\mathbf{m}_{2l+1,2p+1} = \mathbf{m}_{1,1} = m \begin{pmatrix} -\cos\varphi \\ -\sin\varphi \\ 0 \end{pmatrix},$$

$$l, p = \pm 1, \pm 2, \ldots,$$

where $m$ is the modulus of the magnetic moment of a site.

After the substitution of Eq. (5.121) in Eq. (5.119) and the summation over sites of the unbounded lattice, we get the following interesting result. It turned out that the interaction energy of the system of magnetic points (see Fig. 5.20) does not depend on the angle $\varphi$ and is determined by the relation

$$\frac{U_\parallel}{N} = \frac{\mu_0}{4\pi} \frac{m^2}{a^3} \cdot F\left(a/\delta\right), \tag{5.122}$$

where $N$ is the number of sites of the lattice, $m$ is the magnetic moment of a granule, $F\left(a/\delta\right)$ is the energy characteristic of a magnetic state which is a universal function of a single parameter and determines the dependence of the energy on both the period and the field penetration depth in the case where the magnetic moments are distributed in the base plane of the lattice. It can be represented in the form of a sum

$$F(a/\delta) = \frac{\mu_0}{4\pi}\left(-1 + \frac{\partial}{\partial\alpha} + \frac{\partial^2}{\partial\alpha^2}\right) \cdot \frac{1}{8}\sum_{l=1}^{\infty}\sum_{p=0}^{\infty} \frac{\exp(-2\alpha(a/\delta)\sqrt{l^2+p^2})}{(l^2+p^2)^{3/2}}$$

$$-\left(-1 + \frac{\partial}{\partial\alpha} + \frac{\partial^2}{\partial\alpha^2}\right) \cdot \frac{1}{4}\sum_{l=-\infty}^{\infty}\sum_{p=-\infty}^{\infty}$$

$$\times \frac{\exp(-\alpha(a/\delta)\sqrt{(2l+1)^2 + (2p+1)^2})}{((2l+1)^2 + (2p+1)^2)^{3/2}}$$

$$-\left(3 - 3\frac{\partial}{\partial\alpha} + \frac{\partial^2}{\partial\alpha^2}\right) \frac{1}{2}\sum_{l=-\infty}^{\infty}\sum_{p=-\infty}^{\infty}$$

$$\times \frac{((2l+1)^2 - (2p)^2) \cdot \exp(-\alpha(a/\delta)\sqrt{(2l+1)^2 + (2p)^2})}{((2l+1)^2 + (2p)^2)^{5/2}}. \tag{5.123}$$

Thus, there occurs the degeneration of the state in the parameter $\varphi$ in the presence of a tough correlation of the mutual orientations of moments of the ensemble of magnetic points. The energies of the configuration shown in Fig. 5.19 and the ensemble with a modulated distribution of the magnetization (see Fig. 5.20) coincide. In turn, the determination of the energy of magnetic interaction for the configuration possessing the orthogonal orientation of magnetic moments (Fig. 5.18) requires a smaller amount of calculations, because the first sum in formula Eq. (5.119) vanishes. The result of calculations can be represented in the form

$$\frac{U_\perp}{N} = \frac{\mu_0}{4\pi} \frac{m^2}{a^3} \cdot \Phi(a/\delta), \tag{5.124}$$

where $\Phi\left(a/\delta\right)$ is the energy characteristic of the magnetic state which a universal function of the single parameter and determines the dependence of the energy on both the period and the field penetration depth under the distribution of magnetic moments normally to the base plane of the lattice. This function can be represented in the form

$$\Phi\left(a/\delta\right) = \frac{\mu_0}{4\pi}\left(1 - \frac{\partial}{\partial\alpha} + \frac{\partial^2}{\partial\alpha^2}\right)$$

$$\times \frac{1}{2}\cdot\left\{\sum_{l=1}^{\infty}\sum_{p=0}^{\infty}\frac{4\cdot\exp(-\alpha\cdot(a/\delta)\cdot\sqrt{2\cdot l^2 + 2\cdot p^2})}{(2\cdot l^2 + 2\cdot p^2)^{3/2}}\right.$$

$$\left. - \sum_{l=-\infty}^{\infty}\sum_{p=-\infty}^{\infty}\frac{\exp(-\alpha\cdot(a/\delta)\cdot\sqrt{2\cdot(l+1/2)^2 + 2\cdot(p+1/2)^2})}{(2\cdot(l+1/2)^2 + 2\cdot(p+1/2)^2)^{3/2}}\right\}.$$

$$(5.125)$$

The calculation of the functions $F\left(a/\delta\right)$, $\Phi\left(a/\delta\right)$ on the basis of relations (5.123) and (5.125) is not a difficult task and can be realized with any mathematical software. The results are presented in the graphical form in Fig. 5.21.

In Fig. 5.21, we represent the plots of the energy characteristics of two different states of a magnetic lattice versus the ratio of the parameter of a cell and the penetration depth of the magnetic field, $a/\delta$. The limit $a/\delta \to 0$ corresponds to the transition of the matrix into the normal state. It is obvious that a lattice with planar orientation of magnetic moments (Figs. 5.19 and 5.20) possesses the lower energy in the normal state at $a/\delta = 0$. Therefore, the state with the perpendicular direction of moments (Fig. 5.18)

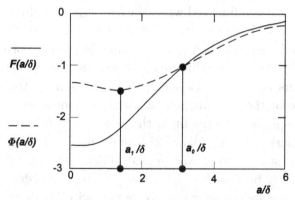

Fig. 5.21. Plots of the energy characteristics $F\left(a/\delta\right)$ and $\Phi\left(a/\delta\right)$ of states of the lattice with normal and planar orientations of magnetic moments, respectively. Values of $F(0)$ and $\Phi(0)$ correspond to the transition of the superconducting matrix into the normal state.

cannot be realized at all in the absence of a superconductor. As the temperature decreases and the penetration depth diminishes gradually, the parameter $a/\delta$ begins to grow. When this parameter attains the value $a_0/\delta \approx$ 3.3, the configuration with the orthogonal orientation of magnetic moments (Fig. 5.18) becomes more advantageous in energy, and the orientation phase transition occurs in the system. At a decrease in the temperature, a similar scenario of events completely corresponds to a reorientation of the magnetic moments of an isolated pair of magnetic points which was considered in the previous work [288].

In conclusion, we note that an analogous phase transition can be expected to occur in a planar lattice with rectangular cell. The only difference will consist in the elimination of the degeneration relative to the directions of magnetic moments in the base plane. Of course, the direct observation of a phase transition will be hampered, because the problem involves the magnetic lattice surrounded by a superconductor. However, a similar orientation transformation must happen also in a lattice applied on the surface of a massive superconductor, though values of the parameter $a_0/\delta$ will be different in this case.

## 5.7.   Energy Spectrum and Wave Function of Electrons in Hybrid Superconducting Nanowires

Recent advances in nanoscience have demonstrated that the fundamentally new physical phenomena can be found, when systems are reduced in size to dimensions, which become comparable to the fundamental microscopic length scales of the investigated material. Superconductivity is a macroscopic quantum phenomena, and, therefore, it is especially interesting to see how this quantum state is influenced when the samples are reduced to nanometer sizes. In such systems, new states of matter can be engineered that do not occur in bulk materials. A good example is the case of nanostructures composed of both superconducting and ferromagnetic metals. Here, the proximity effects couple the Cooper pair condensate to the spin polarized band structure of the ferromagnet, allowing the local coexistence of both pairing and magnetism. In the bulk, the possibility of such coexistence was examined by Larkin–Ovchinnikov [274] and Fulde–Ferrell [275], giving the so-called LOFF state. But it was proved difficult to find this state in bulk materials, possibly because its high sensitivity to a disorder.

In this section, we consider composite nanowires made from both superconducting and ferromagnetic metals. We consider the cylindrical geometry shown in Fig. 5.22, in which one metal forms the core of the nanowire and

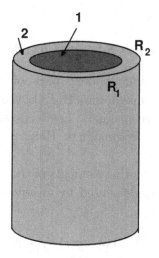

Fig. 5.22.   Nanowire.

the second forms an outer cylindrical sheath. Mesoscopic wires with a similar geometry have been examined by Mota and coworkers [276–280]. In the case of a ferromagnetic core, we can compare the spectrum with results for planar hybrid S/F nanostructures, as reviewed recently by Lyuksyutov and Pokrovsky [281] and by Buzdin [282].

Below, we will calculate the wave functions and the energy spectrum of electrons in hybrid superconducting nanowires. The spectrum of electrons spectrum of electrons reminds a fractal structure such as Hoftstadter's butterfly introduced in 1976 for the description of a behavior of electrons in a magnetic field [286].

### 5.7.1.   *Methodology*

We model the hybrid superconductor-ferromagnet nanowire system by extending the self-consistent theory of a type II vortex developed by Gygi and Schlüter [283] to the case of a ferromagnetic vortex core. We start with the effective Pauli Hamiltonian for the spin-polarized electronic states of a normal metal

$$\hat{H}^0_{\sigma\sigma'} = \frac{(\hat{p} + e\mathbf{A})^2}{2m_e}\delta_{\sigma\sigma'} + V_{\sigma\sigma'}(\mathbf{r}), \qquad (5.126)$$

where $-e$ is the electron charge, $\mathbf{A}(\mathbf{r})$ is the magnetic vector potential, and $V_\sigma(\mathbf{r})$ is a general spin-dependent single potential, where $\sigma$ is $\uparrow/\downarrow$. This potential can be assumed to be the sum of the ionic, Hartree, and exchange-correlation potentials of a self-consistent spin-polarized DFT calculation.

In this case, it has the form

$$V_\sigma(\mathbf{r}) = V_{ion}(\mathbf{r}) + V_H(\mathbf{r}) + V_{xc}(\mathbf{r}) + \mu_B(\mathbf{B} + \mathbf{B}_{xc})\sigma_{\sigma\sigma'}, \tag{5.127}$$

where $V_{xc}(\mathbf{r})$ is the spin-independent part of the exchange-correlation potential, $\mu_B$ is the Bohr magneton, $\mathbf{B}(\mathbf{r})$ is the physical local magnetic field, and $\mathbf{B}_{xc}$ is an effective magnetic field representing the exchange field of spin-polarization in the ferromagnet. Here, $\sigma_{\sigma\sigma'}$ is the vector of Pauli matrices.

To model the superconducting elements of the hybrid system, this single-particle Hamiltonian is supplemented by an effective attraction, which we take as the BCS contact term

$$V(\mathbf{r}, \mathbf{r}') = -g(\mathbf{r})\delta(\mathbf{r} - \mathbf{r}'), \tag{5.128}$$

where $g(\mathbf{r})$ is the local attractive potential strength at $\mathbf{r}$. This will be zero in the normal or ferromagnetic part of the nanowire and a constant $g$ in the superconducting parts. For simplicity, we neglect a retardation of the attraction in the rest of this paper.

The full effective Hamiltonian for our system is

$$\hat{H} = \int d^3r \left[ \sum_{\sigma\sigma'} \left( \hat{\psi}_\sigma^+(\mathbf{r})\hat{H}_{\sigma\sigma'}^0\hat{\psi}_{\sigma'}(\mathbf{r}) \right) - g(\mathbf{r})\hat{\psi}_\uparrow^+(\mathbf{r})\hat{\psi}_\downarrow^+(\mathbf{r})\hat{\psi}_\downarrow(\mathbf{r})\hat{\psi}_\uparrow(\mathbf{r}) \right]. \tag{5.129}$$

Here, $\hat{\psi}_\sigma^+(\mathbf{r})$ and $\hat{\psi}_\sigma(\mathbf{r})$ are the usual field operators for the electrons. In the Hartree–Fock–Gorkov approximation, this Hamiltonian is diagonalized by a spin-dependent Bogoliubov–Valatin transformation

$$\hat{\psi}_\sigma(\mathbf{r}) = \sum_n \left( u_{n\sigma}(\mathbf{r})\hat{\gamma}_n + v_{n\sigma}^*(\mathbf{r})\hat{\gamma}_n^+ \right), \tag{5.130}$$

$$\hat{\psi}_\sigma^+(\mathbf{r}) = \sum_n \left( u_{n\sigma}^*(\mathbf{r})\hat{\gamma}_n^+ + v_{n\sigma}(\mathbf{r})\hat{\gamma}_n \right). \tag{5.131}$$

The requirement that the quasiparticle creation and annihilation operators retain the fermion anticommutation laws,

$$\{\hat{\gamma}_n, \hat{\gamma}_{n'}^+\} = \delta_{nn'}, \tag{5.132}$$

implies that

$$\sum_{n\sigma} \left( u_{n\sigma}^*(\mathbf{r})u_{n\sigma'}(\mathbf{r}') + v_{n\sigma}(\mathbf{r})v_{n\sigma'}^*(\mathbf{r}') \right) = \delta(\mathbf{r} - \mathbf{r}')\delta_{\sigma\sigma'}. \tag{5.133}$$

The resulting set of Bogoliubov–de Gennes equations is

$$\begin{pmatrix} \hat{H}_1 + V_{\uparrow\uparrow} & V_{\uparrow\downarrow} & 0 & \Delta(\mathbf{r}) \\ V_{\downarrow\uparrow} & \hat{H}_1 + V_{\downarrow\downarrow} & -\Delta(\mathbf{r}) & 0 \\ 0 & -\Delta^*(\mathbf{r}) & -\hat{H}_1 - V_{\uparrow\uparrow} & -V_{\uparrow\downarrow} \\ \Delta^*(\mathbf{r}) & 0 & -V_{\downarrow\uparrow} & -\hat{H}_1 - V_{\downarrow\downarrow} \end{pmatrix} \begin{pmatrix} u_{n\uparrow\sigma} \\ u_{n\downarrow\sigma} \\ v_{n\uparrow\sigma} \\ v_{n\downarrow\sigma} \end{pmatrix}$$

$$= E_{n\sigma} \begin{pmatrix} u_{n\uparrow\sigma} \\ u_{n\downarrow\sigma} \\ v_{n\uparrow\sigma} \\ v_{n\downarrow\sigma} \end{pmatrix}, \tag{5.134}$$

where $\hat{H}_1 = (\hat{p} + e\mathbf{A})^2/2m_e - \mu$, and $\mu$ is the chemical potential.

The self-consistent pairing potential corresponds to a pure spin-singlet pairing state and is given by

$$\Delta(\mathbf{r}) = g\langle \hat{\psi}_\uparrow(\mathbf{r})\hat{\psi}_\downarrow(\mathbf{r})\rangle. \tag{5.135}$$

Consider the special case where the magnetization of the ferromagnet is in the same collinear direction, $\mathbf{B}_{xc}$, as the external field $\mathbf{B}$. Choosing it as the spin-quantization axis $\hat{z}$, we have $V_{\uparrow\downarrow} = V_{\downarrow\downarrow} = 0$. In this case, the full system of four matrix Bogoliubov–de Gennes equations is separated into a pair of $2 \times 2$ matrix equations

$$\begin{pmatrix} \hat{H}_1 + V_{\uparrow\uparrow} & \Delta(\mathbf{r}) \\ \Delta^*(\mathbf{r}) & -\hat{H}_1 - V_{\downarrow\downarrow} \end{pmatrix} \begin{pmatrix} u_{n\uparrow\sigma} \\ v_{n\downarrow\sigma} \end{pmatrix} = E_{n\sigma} \begin{pmatrix} u_{n\uparrow\sigma} \\ v_{n\downarrow\sigma} \end{pmatrix}, \tag{5.136}$$

$$\begin{pmatrix} \hat{H}_1 + V_{\downarrow\downarrow} & -\Delta(\mathbf{r}) \\ -\Delta^*(\mathbf{r}) & -\hat{H}_1 - V_{\uparrow\uparrow} \end{pmatrix} \begin{pmatrix} u_{n\downarrow\sigma} \\ v_{n\uparrow\sigma} \end{pmatrix} = E_{n\sigma} \begin{pmatrix} u_{n\downarrow\sigma} \\ v_{n\uparrow\sigma} \end{pmatrix}. \tag{5.137}$$

### 5.7.2. *Calculation of the spectrum*

We will study the Bogoliubov–de Gennes equations, by using a numerical method for their solution. The calculation algorithm is realized with the use of the most powerful FORTRAN language specially developed for mathematical calculations.

In view of the cylindrical symmetry of the system under study, we write the system of Bogoliubov–de Gennes equations in the form suitable for calculations. We will also use the fact that the amplitudes of the functions $u(r)$ and $v(r)$ tend to zero at the boundary point. The solutions of the equations are determined with the use of the Runge–Kutta method (Figs. 5.23 and 5.24).

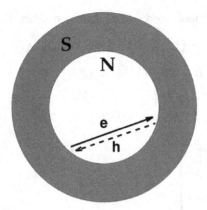

Fig. 5.23.   Billiard 1.

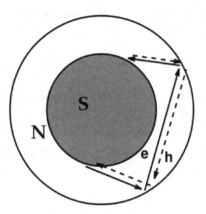

Fig. 5.24.   Billiard 2.

Taking the symmetry of our system (cylindrical) to account and choosing the gauge, in which the parameter $\Delta(r)$ is real, we can write the system of Bogoliubov–de Gennes equations in the form

$$\bar{u}(r,\theta,z) = u_{\mu n k_z}(r)e^{-i\mu\theta}e^{-ik_z z},$$

$$\bar{v}(r,\theta,z) = v_{\mu n k_z}(r)e^{i\mu\theta}e^{-ik_z z},$$

$$(5.138)$$

$$-\frac{\partial^2}{\partial r^2}u_{\mu n k_z}(r) - \frac{1}{r}\frac{\partial}{\partial r}u_{\mu n k_z}(r) + U(r)\cdot u_{\mu n k_z}(r) + \Delta(r)\cdot v_{\mu n k_z}(r) = 0,$$

$$-\frac{\partial^2}{\partial r^2}v_{\mu n k_z}(r) - \frac{1}{r}\frac{\partial}{\partial r}v_{\mu n k_z}(r) + U(r)\cdot v_{\mu n k_z}(r) - \Delta(r)\cdot u_{\mu n k_z}(r) = 0,$$

$$(5.139)$$

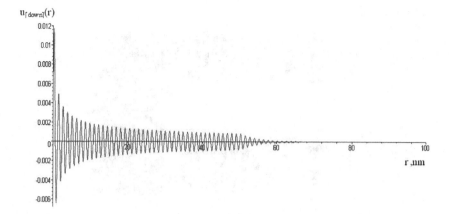

Fig. 5.25. Quasiparticle amplitude of the Andreev state.

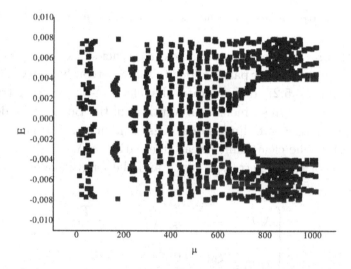

Fig. 5.26. Andreev spectrum as a function of the angular momentum for billiard 1.

where $U(r) = \frac{\mu^2}{r^2} - (E_f - \frac{k_z^2}{m_z} - E_{\mu n k_z})$. The plot of wave functions is given in Fig. 5.25 ($E_0 = 0.986669921875$ eV, $E_f = 1$ eV, $\Delta = 0.1$ eV, $k_z = 0$, $R_1 = 50$ nm, $R_2 = 100$ nm, $\mu = 1$). In Figs. 5.26 and 5.27, we present the results of calculations for the spectrum of Andreev states of the system.

The resulting spectrum as a function of the angular momentum is shown in Fig. 5.26 for billiard 1 and Fig. 5.27 for billiard 2. The spectra demonstrate the splitting caused by the boundary effect.

The Andreev billiard systems are the analog of a classical billiard. The Andreev billiard is interpreted as a ballistic motion with the Andreev reflection at the interface with a superconductor. The Andreev reflection is a

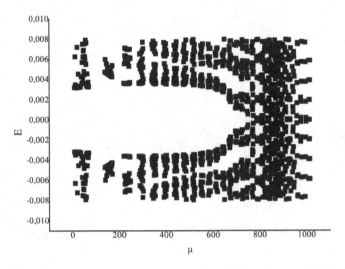

Fig. 5.27.    Andreev spectrum as a function of the angular momentum for billiard 2.

fundamental process that converts an electron incident on a superconductor into a hole, while a Cooper pair is added to the superconductivity condensate. Figures 5.26 and 5.27 demonstrate the electron-hole symmetry of the Andreev spectrum. One-dimensional solutions of the Bogoliubov–de Gennes equations pass into in two-dimensional ones, as $\mu$ increases. It is worth noting that we work in the clean limit (mean free path $l \gg \lambda, \xi$).

Estimates of the penetration depth and the length scale of order were performed in [285]:

$$\begin{cases} \lambda(T) = \dfrac{\lambda_0}{\sqrt{1-t^4}} \quad \text{where } t = \dfrac{T}{T_c}, \\[3mm] \xi(T) = \dfrac{\xi_0(1-0.25t)}{\sqrt{1-t}}, \end{cases} \tag{5.140}$$

where $\lambda_0 = 46\,\text{nm}$ and $\xi_0 = 74\,\text{nm}$, i.e., almost the characteristic lengths of lead at zero temperature in the clean limit. We have carried out also the calculations for MgB$_2$ nanowires with the parameters ($R_1 = 150\,\text{nm}$, $R_2 = 200\,\text{nm}$, $M_z = 0$, $k_z = 0$, $\Delta = 0.003\,\text{eV}$, $E_f = 1\,\text{eV}$). In Fig. 5.28 (NS, $R_1 = 150\,\text{nm}$, $R_2 = 200\,\text{nm}$, $M_z = 0$, $k_z = 0$, $\Delta = 0.003\,\text{eV}$, $E_f = 1\,\text{eV}$), we present the spectrum as a function of the angular momentum for MgB$_2$ nanowires (billiard 1), whereas Fig. 5.29 shows the analogous dependence in the case of billiard 2.

These spectra have visual similarity with the winds of a butterfly. They possess the property of self-similarity (fractality) and are one of the not numerous fractal structures discovered in quantum physics. Hoftstadter's

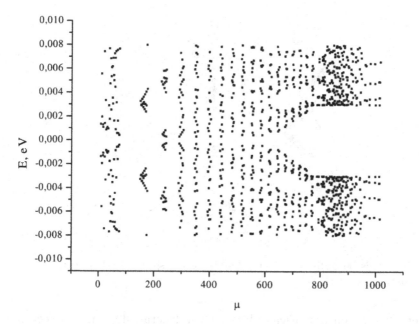

Fig. 5.28. Andreev spectrum for MgB$_2$ nanowires.

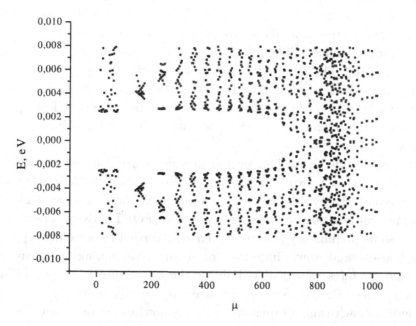

Fig. 5.29. Andreev spectrum for MgB$_2$ nanowires. (SN).

Fig. 5.30. This butterfly is a fractal describing the spectrum of electrons in a periodic structure with magnetic field.

butterfly is a mathematical object with fractal structure describing the behavior of electrons in a magnetic field. Under the action of a magnetic field, free electrons move along circumferences. The theory says that if the electrons are "locked" in a crystal atomic lattice, the trajectories of their motion will be much more complicated. An analogous situation is presented by the Andreev billiard. The fractal structures by themselves are rare in the quantum world. Therefore, Hoftstadter's butterfly was not discovered for a long time, which was partially related to the complexity of experiments. Then scientists from Manchester succeeded to observe an intricate pattern on graphene in a magnetic field representing Hoftstadter's butterfly [287] (Fig. 5.30).

We have studied hybrid nanowires, in which normal and superconducting regions are in close proximity, by using the Bogoliubov–de Gennes equations for superconductivity in a cylindrical nanowire. We have developed a method for numerical solutions of these equations in FORTRAN programme and present some preliminary results. We have succeeded to obtain the quantum energy levels and wave functions of a superconducting nanowire. The spectrum of states we calculated shows the interesting "Andreev Billiard" characteristics. Preliminary results are also obtained in the cases of a magnetic nanowire and a superconducting nanowire containing a vortex. The calculations for a superconducting nanowire can be extended to the cases of a superconductor-ferromagnet hybrid nanowire and a nanowire containing one or more superconducting flux quanta [284].

The obtained spectra of electrons remind Hoftstadter's butterfly [286].

## 5.8. Quantum Computer on Superconducting Qubits

### 5.8.1. *Principle of quantum computers*

Silicon microprocessors, being the main element of modern computers, have attained the limit of development. The miniaturization, i.e., the aspiration to place as much as possible components on a more and more smaller area of a chip, has approached the boundary of physical possibilities. Further, it will be impossible to conserve the stability of the operation of computers. Many researchers believe that silicon processors will begin to go into the past in at most five years, and the production of chips will be based on the other material — carbon nanotubes. It is worth noting that the computational processes are accompanied by the release of heat. R. Feynman said: "Any classical computation is a physical process running with the release of heat." As known, the calculations lead to an increase in the entropy and, hence, to the release of heat. The idea of the creation of quantum computers arose several decades ago, when it was proposed to reject the application of electric circuits in the processing of information and to pass to the use of quantum mechanics. The classical computers are processing the information only on the basis of ideas of one bit of information (it corresponds to the transition from state 0 to state 1 or conversely). The quantum computers can process the information, by basing on the ideas of a quantum bit (qubit) which allows one to realize simultaneously four logical operations ($0 + 0 = 0$, $0 + 1 = 1$, $1 + 0 = 1$, $1 + 1 = 2$).

Qubit is the abridged notation for quantum bit and means the unity of information coded in a quantum system which can be in the states $|0\rangle, |1\rangle$, and in any superposition of these states.

Let the state of a qubit be described by the state $|f\rangle$ which can be represented as a superposition of states $|0\rangle$ and $|1\rangle$:

$$|f\rangle = a|0\rangle + b|1\rangle,$$

where

$$a^2 + b^2 = 1. \tag{5.141}$$

Since the devices can register only classical quantities, the measurement of a quantum bit will give states $|0\rangle$ or $|1\rangle$ with probabilities determined by the squares of the coefficients $a$ and $b$.

Figure 5.31 shows the difference between an ordinary and a quantum qubit, here $a = b = 1/\sqrt{2}$.

The use of the principle of superposition allows one to increase the informational space exponentially with linear growth of the size of a physical

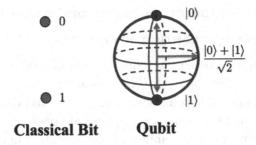

**Classical Bit**     **Qubit**

Fig. 5.31.   Ordinary qubit and quantum qubit.

system, because the register including $n$ qubits can be in a superposition of at once $2^n$ states.

In addition, quantum mechanics admits the existence of the so-called entangled states possessing no analogs in classical physics.

An ensemble of qubits is a collection of qubits in different but given states.

It is worth noting that, even on the level of mathematics, a quantum computer operates in a basically different way than that of a classical computer.

Input data are coded in "quantum cells of memory." In this case, the collection of qubits becomes a single quantum system. This system undergoes the sequence of elementary quantum operations. Quantum computations are a realization of the most astonishing idea to apply the principles of quantum mechanics to the world of computers. Ordinary computers, despite their complexity, use classical laws of mechanics. In the recent years, the theory of classical computations was developed on the basis of works by A. Turing. With the appearance of quantum computations, the new possibilities have arisen, and the situation is radically changed. Quantum methods can be successfully used in the solution of mathematical problems, though the time consumed for the solution of mathematical problems increases exponentially with the complexity of a problem.

For the theory of quantum calculations, the physical nature of qubits is not of crucial importance; the basically important point is that the system in the course of calculations obeys the laws of quantum mechanics.

### 5.8.2.  *Superconducting qubits*

The realization of a quantum computer requires the availability of systems with doubly degenerate ground state. For this reason, a great attention is paid to systems with two-level wells which can be fabricated by facilities of solid-state electronics. The modern technology allows one to produce circuits containing millions of transistors and Josephson junctions.

By observing the operation of electrical circuits at temperatures close to the absolute zero, the researchers found a new proof of the fact that the laws of quantum mechanics are suitable not only for the microscopic objects (atoms and electrons), but also for large electronic schemes which include superconducting bits (qubits).

Many years ago, the attempts to construct a Josephson computer on the basis of the Josephson tunnel logic hopelessly failed. The principal reason for the failure was the huge technological dispersion of parameters of tunnel junctions, which did not allow one to produce large microcircuits. A Josephson computer can be created only on the way having nothing in common with that based on semiconductors, namely it can be just a quantum computer.

The hope is related to two circumstances. First, the fabrication of superconducting qubits is quite possible in the framework of the up-to-date technology.

Second, the presence of a gap in the spectrum of excitations of a super-conductor allows one to expect the suppression of generation in a system.

Let us consider the first steps on the way of the construction of a superconducting quantum computer. First, we consider, in brief, the Josephson effect.

### 5.8.3. *The Josephson effect*

The Josephson effect is certainly one of the most interesting phenomena in superconductivity. The Josephson effect consists in the passage of a superconducting current through a thin dielectric layer separating two superconductors (the so-called Josephson junction).

It was predicted by an English physicist B. Josephson on the basis of superconductivity theory (1962, Nobel's prize in 1973) and discovered experimentally in 1963. Conduction electrons pass through a dielectric (a film of copper oxide $\sim 10$ Å in thickness) due to the tunneling effect [292,293].

In Fig. 5.32, we present a scheme of the Josephson tunneling between two superconductors. The Josephson stationary effect consists in that the superconducting current

$$J_c = J_0 \sin \varphi, \tag{5.142}$$

$$\frac{d\varphi}{dt} = \frac{2eV}{\hbar}, \tag{5.143}$$

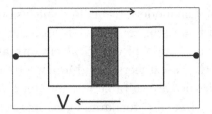

Fig. 5.32.   Scheme of the Josephson tunneling between two superconductors.

where $\varphi$ is the phase difference on the interface of superconductors, $V$ is the applied voltage, and $J_0$ is the critical current through the junction.

The Josephson effect indicates the existence of the electron ordering in superconductors, namely, the phase coherence: in the ground state, all electron pairs have the same phase $\varphi$ characterizing their wave function $\Psi_1 = \sqrt{n_{s1}}e^{i\varphi_1}$. According to quantum mechanics, the presence of a phase difference must cause a current through the junction. The discovery of such a current in experiment proves the existence of macroscopic phenomena in the nature which are directly determined by the phase of a wave function

$$\Psi = \sqrt{n}\, e^{i\psi}. \tag{5.144}$$

### 5.8.3.1.   *Current passing through two series-connected Josephson junctions*

The Josephson current in the scheme drawn in Fig. 5.33 can be easily determined with the help of the elementary Feynman approach to a Josephson junction as a two-level quantum-mechanical system [4]. By introducing an intermediate object, some island, with the wave function $\Psi_0$ into the two-level system characterized by the wave functions $\Psi_1 = \sqrt{n_{s1}/2}e^{i\varphi_1}$ and $\Psi_2 = \sqrt{n_{s2}/2}e^{i\varphi_2}$, we write the Schrödinger equation in the form

$$i\hbar\frac{d\Psi_1}{dt} = \frac{eV}{2}\Psi_1 + K\Psi_0, \tag{5.145}$$

$$i\hbar\frac{d\Psi_0}{dt} = K\Psi_1 + K\Psi_2 + E_0\Psi_0, \tag{5.146}$$

$$i\hbar\frac{d\Psi_2}{dt} = K\Psi_0 - \frac{eV}{2}\Psi_2. \tag{5.147}$$

The formula for a constant current running through two series-connected Josephson junctions at the zero external potential difference takes the form:

$$J = \hbar\frac{\partial\sqrt{n_{S1}}}{\partial t} = -\frac{K^2}{E_0}\sqrt{n_{S2}}\sin(\varphi_2 - \varphi_1). \tag{5.148}$$

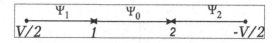

Fig. 5.33. Current passing through two series-connected Josephson junctions.

This relation can be rewritten as follows:

$$J_c = J_0 \sin(\varphi), \qquad (5.149)$$

where $\varphi = \varphi_2 - \varphi_1$ is the phase difference. The relation obtained is named the Josephson formula (the Josephson stationary effect) and determines the current of superconductive electron pairs due to the tunneling transition.

The Josephson coupling energy is an important parameter of the Josephson junction. From (5.142), we have

$$E = \int JV \, dt = \frac{hJ_0}{2} \int_0^\pi \sin \varphi \, d\varphi = -\frac{hJ_0}{2e} \cos \varphi. \qquad (5.150)$$

The Josephson effect is still one of the phenomena that make superconductors such a fascinating area of study. Despite the more than 40-year intense studies and numerous applications, the Josephson effect remains an important field of researches in connection with the use of small superconducting grains.

### 5.8.3.2. *SQUIDS*

The Josephson effect allowed one to construct superconducting interferometers, SQUIDs (superconducting quantum interference device), which contain parallel weak connections between superconductors [289]. In Fig. 5.34, we present a scheme with Josephson junctions.

The total current running from 1 to 2 is equal to

$$I = I_0 d \left[ \sin(\Delta \varphi_1) + \sin(\Delta \varphi_2) \right], \qquad (5.151)$$

where

$$\Delta \varphi_1 = \varphi_{2A} - \varphi_{1A},$$

$$\Delta \varphi_2 = \varphi_{2B} - \varphi_{1B}$$

are the phase differences on the first and the second Josephson junctions. There occurs a distinctive interference of the superconducting currents running through these connections.

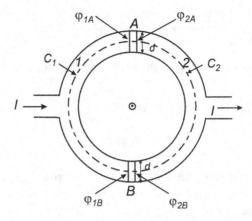

Fig. 5.34.   Basic diagram of a "quantum interferometer".

Inside the superconductor, the current is zero $\mathbf{j} \equiv 0$. We will use the following formula for the current $j$:

$$\mathbf{j} \sim h\nabla\varphi - \frac{2e}{c}\mathbf{A}. \tag{5.152}$$

We can write

$$\varphi_{1B} - \varphi_{1A} = \frac{2e}{hc}\int_{C_1}\mathbf{A}\cdot dl, \tag{5.153}$$

$$\varphi_{2B} - \varphi_{2A} = \frac{2e}{hc}\int_{C_2}\mathbf{A}\cdot dl. \tag{5.154}$$

Summing up these two equations, we get

$$\varphi_{1B} - \varphi_{2B} + \varphi_{2A} - \varphi_{1A} = \frac{2e}{hc}\oint\mathbf{A}\cdot dl = 2\pi\frac{\Phi}{\Phi_0}. \tag{5.155}$$

Thus, we have

$$\Delta\varphi_1 - \Delta\varphi_2 = 2\pi\frac{\Phi}{\Phi_0}, \tag{5.156}$$

where $\Phi$ is the total quantum flux.

The flux quantum is defined as

$$\Phi_0 = \frac{h}{2e}. \tag{5.157}$$

For a balance SQUD ring system, we can write

$$\Delta\varphi_1 = \varphi_0 + \pi\frac{\Phi}{\Phi_0},$$

$$\Delta\varphi_2 = \varphi_0 - \pi\frac{\Phi}{\Phi_0}.$$

(5.158)

The total current in the SQUD is

$$I = I_0 \sin(\Delta\varphi_1) + I_0 \sin(\Delta\varphi_2)$$

$$= I_0 \sin\left(\varphi_0 + \pi\frac{\Phi}{\Phi_0}\right) + I_0 \sin\left(\varphi_0 - \pi\frac{\Phi}{\Phi_0}\right)$$

$$= 2I_0 \sin(\varphi_0) \cos\left(\pi\frac{\Phi}{\Phi_0}\right) = I_{max}\left|\cos\left(\pi\frac{\Phi}{\Phi_0}\right)\right|$$

(5.159)

with $I_{max} = 2I_0 \sin(\varphi_0)$.

In this case, the critical current turns out to be periodically dependent on the flow of an external magnetic field, which allows one to use such a unit for the exact measurement of the magnetic field.

### 5.8.3.3.  *Flux qubit*

Let us consider the first steps on the way of the creation of a superconducting computer. The simplest superconducting system demonstrating the coherency is SQUID which is a superconducting ring including a Josephson junction at one point. The energy of this system contains two terms: Josephson transition energy $(\cos\Phi)$ and the energy related to the ring $L$:

$$H = -E_J \cos\left(2\pi\frac{\Phi}{\Phi_0}\right) + \frac{(\Phi - \Phi_x)^2}{2L}.$$

(5.160)

Here, $\Phi$ is the difference of superconductive phases at the junction. The superconductive phase in the ring is proportional to a magnetic flow applied to the ring (quantization of the magnetic flow). If $\Phi_x$ is equal exactly to a half of the magnetic flow, the potential of a SQUID becomes doubly degenerate.

Two minima of the well correspond to the currents in the ring passing in the clockwise ↻ and counterclockwise directions ↺, respectively [291].

A superposition of these states in SQUID was observed experimentally in works [291, 294]. These experiments demonstrated clearly the possibility to create a superposition of states in a system with a macroscopic number of particles. In the given case, the circular current including $10^{13}$ electrons was registered in a loop. The states participating in a superposition were macroscopically distinguishable, by differing from one another by the

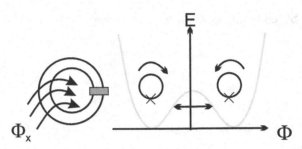

Fig. 5.35.   Scheme of the Josephson tunneling between two superconductors.

currents, whose difference was several microamperes. We indicate that, at the recent time, the important notion has been introduced in the course of studies of structures with Josephson junctions. It is the notion of macroscopic quantum coherence. In such systems, the Josephson energy can have two almost degenerate minima at values of the phase which are separated by a potential barrier (see Fig. 5.35). It is possible that the phase passes from one minimum to another one due to the quantum-mechanical tunneling, and the eigenstates of the system are superpositions of the states localized in the first and second minima.

The operation of a superconducting computer requires low temperatures which are needed, in particular, to suppress heat-induced excitations destroying the quantum mechanical state of a system.

It is worth noting that the best condition for the observation of these current states is defined by a size of the superconducting ring. If the ring size is taken to be 1 cm, there appear the effects of decoherence which will destroy the current states. But if the ring is taken much smaller (say, 5 $\mu$m), then it is possible to observe these states. These states were discovered in experiments with Rabi oscillations [294].

### 5.8.3.4.   *Charge qubit*

The second type of a superconducting bit can be realized in the "Cooper pair box" system which is characterized by two charge states: without excess Cooper pair $|0\rangle$ and with a single Cooper pair $|1\rangle$. "Cooper pair box" is a nanotransistor with Coulomb blockage with controlling voltage $V_g$, as shown in Fig. 5.36. A "Cooper pair box" is represented by an aluminum superconducting film of the order of 1 mm in size with the working temperature of about several milli-Kelvin, which is significantly lower than the superconducting temperature $T_c$. Its quantum state can be characterized by the number of Cooper pairs. According to the BCS theory, the ground-state of a superconductor is a superposition of states with different numbers

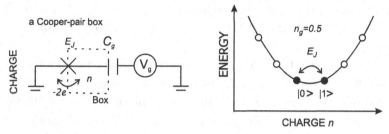

Fig. 5.36. Cooper-pair box [× denotes a Josephson junction (*JJ*)], $C_g$ and $V_g$ are the capacitor and the controlling voltage.

of Cooper pairs. The excited, i.e., unpaired states in an ordinary supercon-ductor are separated from the ground state by the energy gap. Therefore, as the number of electrons varies by 1, i.e., $N \to N \pm 1$, the ground-state energy must be changed by $\pm\delta$. The sign (plus or minus) depends on that the initial number of electrons $N$ is even or odd. The effects of parity of the number of electrons which are considered in this chapter in connection with mesoscopic superconductivity were successfully measured on nanotransistors with Coulomb blockade.

The Hamiltonian of such a system can be written as

$$H = E_c(n - n_g)^2 + E_j \cos \varphi, \qquad (5.161)$$

where $E_c$ and $E_j$ are the charging and Josephson energies, and $\varphi$ is the phase change. In the charging mode where $E_c \gg E_j$, only two lower charged states are of importance. The controlling voltage $V_g$ induces a charge in the box

$$n_g = C_g \frac{V_g}{2e}, \qquad (5.162)$$

where $2e$ is the charge of each Cooper pair, and $C_g$ is the gate capacitance. At $n_g = 1/2$, such a system operates as a two-level atomic system, in which the states $|0\rangle$ and $|1\rangle$ can be realized. The control is realized by a voltage $V_g$. On the basis of such a qubit, a system of two qubits was realized [291, 292], and the formation of entangled states was demonstrated. The effects of parity of the number of electrons which are considered in this chapter in connection with mesoscopic superconductivity were successfully measured on nanotransistors with Coulomb blockade.

### 5.8.3.5. *Phase qubit*

There exists one more possibility to realize SQUIDs with the use of high-temperature superconductors possessing the *d*-pairing (*n*-loop ones) [293].

The physics of $d$-pairing was considered in Chapter 2. Quantum processors on the base of these SQUIDs are developed at the Canadian company "D-wave".

We note also that, in addition to superconducting qubits, quantum computers use qubits possessing other physical properties. Scientists at the Yale University took a very fine aluminum plate in the fabrication of a quantum chip. A single qubit consists of one billion of aluminum atoms which behave themselves, nevertheless, as a single unit that can be in two energy states denoted as 0 and 1. Such quantum-mechanical states of a qubit cannot be long-term — their lifetime is about one microsecond. But it is sufficient for a chip to solve the so-called algorithm. We have considered the technologies of superconducting computers which represent a new type of quantum computers. These computers are based on the other mechanism obeying the laws of quantum mechanics. We recall that, till the recent time, the principle of devices was invariable, and the archetype of such devices is mechanical clocks. In such devices, all stages of their relative motion can be observed; therefore, it is quite simple to understand their structure. On the contrary, the principle of operation of quantum computers involves the specific features of quantum mechanics which are difficult to understand. Nevertheless, by possessing the quantum resources, we can solve the very difficult complicated problems. In particular, the most important potential field of application of quantum computers is the problem of the exact calculation of properties of quantum systems.

### 5.8.4. *Quantum computer*

Quantum computer is a computing unit using the quantum mechanical effects in its operation and realizing the execution of quantum algorithms. Quantum computers are working on the basis of quantum logic.

The idea of quantum computations was advanced independently and almost simultaneously by R. Feynman [297] at the beginning of the 1980s. In those works, it was proposed to use the operations with the states of a quantum system in computations.

As distinct from the classical systems, a quantum system can stays in the superposition state, being simultaneously in all possible physical states. In quantum computations, the elementary step is a separate operation over the $n$-qubit superposition corresponding to the principle of quantum parallelism. In a classical computer, the single ouput value for one input state is calculated, whereas a quantum computer calculates the output values for all input states.

In other words, the quantum computer can solve rapidly problems such that their solution with an ordinary classical computer would take hundreds of years.

Quantum computations are the basic innovation, since the invention of a microprocessor and quantum computers gives vast advantages as compared with the digital electronic computers on the basis of transistors. Quantum computers operate much faster and are more powerful than the modern ordinary computers. The former are else on the stage of development. However, several military agencies and national governments are financing the studies on quantum computations aimed at the development of quantum computers for business, civil population, and safety purposes.

The fabrication of a quantum computer as a real physical device is a fundamental problem of the physics of the 21st century. As of the beginning of 2020, only limited versions of a quantum computer are produced (the largest developed quantum registers have several tens of coupled qubits).

Of course, the huge number of very complex theoretical and practical problems should be solved on the way to the practical realization of quantum computers.

Till now, some versions of a construction of quantum computers are proposed on the basis of quantum points and superconducting junctions.

Moreover, several commercial companies have declared the creation of real processors on the basis of superconducting qubits used, for example, in a computer "Orion" (Fig. 5.37). It includes a microchip with 16 qubits. Each qubit is composed of niobium particles surrounded by solenoids.

In essence, "Orion" is an analog computer that solves the problems by the method of physical modeling. Such computer is suitable for the calculations

Fig. 5.37.   Photo of a 16-qubit processor "Orion".

within complicated models (financial, chemical, pharmaceutical, etc. ones) consuming much computer time.

Among the principal tasks in the creation of a quantum computer, it is worth to mention the use of as many as possible number of qubits and its design.

However, we should make a reservation about the existence of the problems related to the ambiguity of the interpretation of information. Therefore, the scientific society considers the declarations about the creation of quantum computers with prejudice.

### 5.8.4.1. *D-wave quantum computer*

In 2011, the Canadian firm D-wave created the first commercial quantum computer D-Wave-1 (DW-1) with 128 qubits and sold it for 10 mln dollars. It was purchased by the largest corporation Lockheed Martin of the military-industrial complex of the USA. The purpose, principle of operation, and architecture of DW-1 are based on the above-mentioned superconducting qubits. Unusual are the macroscopically large qubits in the register of this quantum computer that use the Josephson superconducting junctions. We now consider the operation of superconducting qubits in a quantum analog computer applying the physical mechanism of the phenomenon of superconductivity and the principles of functioning of Josephson junctions.

At super-low temperatures of one or several Kelvins, the electrons in a superconducting metal are coupled due to the electron–phonon interaction and form the so-called Cooper pairs that are bosons with the zero spin, zero momentum, and charge $-2e$. As a result of the interaction of bosons with one another, the correlations appear between them. Moreover, at a definite critical temperature, the phase transition into the state of superconductivity occurs. In this case, there appears a gap in the electron energy spectrum, namely, the forbidden zone $E$ in width. Above the gap, the one-electron energy levels are placed. The bosons not obeying the Pauli principle transit into a single quantum state with the energy lying below the gap bottom. Such correlated state of the gas of bosons that are in one quantum state is called the Bose-condensate. The Bose-condensate as a whole in the superconductor is described by a single complex wavefunction $\Psi = \sqrt{n}\, e^{i\psi}$ with a phase fixed for the entire condensate, i.e., it behaves itself as one quantum particle. The formation of such coherent state of the ensemble of many particles in a lengthy specimen with macroscopic sizes is one of the examples of phenomena of macroscopic quantum physics.

For niobium and aluminium that are most frequently used in superconductors, the critical temperatures of the superconducting phase transitions
$T_c$ are equal, respectively, to 9.3 and 1.2 K.

A Josephson qubit in the quantum computer DW-1 is a ring with a
characteristic size of about one micron. It is produced of a superconducting
metal, niobium, with two thin dielectric spacers. The ring is cooled down to
a temperature of several milli-Kelvin.

The quantum register of a computer including $n$ qubits is described by
the Hamiltonian

$$H = \sum_{i=1}^{n} h_i \sigma_{zi} + \sum_{i=1}^{n} \Delta_i \sigma_{xi} + \sum_{i,j} J_{ij} \sigma_{zi} \sigma_{zj}, \qquad (5.163)$$

where $h_i \sim (\Phi_i - \Phi_0/2)$ is a shift of the controlling magnetic flow through the
contour of the $i$th qubit from the central value, $\Delta_i$ is the tunneling energy
determining the frequency $\omega_{0i}$ of the transition between the ground and
excited levels of the $i$th qubit: $\omega_{0i} \sim \Delta_i$. The constant of interaction of the $i$th
and $j$th qubits $J_{ij}$ is proportional to the mutual inductance of these qubits.
The parameters of Hamiltonian (5.163) modeling a quantum computer can
be controlled and be changed selectively and directedly according to the
algorithm of solution of the problem.

As a result, we get a macroscopic magnetic qubit. As a computational
basis for it, we can use the spin or energy representation with regard for the
solved problem.

The quantum analog computer DW-1 has already successfully applied
to the solution of a number of problems of the optimization and pattern
recognition. We may assume its wide use in the control over the databases,
analysis and forecasting of financial risks, and other problems. The very form
of Hamiltonian (5.163) presents a possibility to solve a wide spectrum of
problems in the physics of ferromagnetism ($J_{ij} < 0$) and antiferromagnetism
($J_{ij} > 0$), as well as in the description of other physical systems whose
Hamiltonians can be efficiently reduced to the form (5.163).

The Josephson qubit is an experimentally realized object of the macroscopic quantum physics: its size is more than the size of an atomic qubit by
about 10,000 times!

The description of the realization of phenomena of the macroscopic
quantum physics by a quantum computer can be found, for example, in
[298, 299].

Quantum computers are so powerful, because we have qubits of the highest quality with the lowest level of errors. Such computers demonstrate the combination of the use of identical completely coupled qubits and the exact control.

The scientists are close to the great breakthrough in the design of large-scale quantum computers. However, the creation of highly efficient useful quantum computers is hampered by the problem of noises in them. In brief, this is a result of errors introduced by a quantum computer itself at the manipulation by qubits.

Noises become a heavier problem, as the number of qubits and the size of a computer system increase. This means that the mentioned problem sets a particular obstacle for the development of those types of large quantum computers which would be the basis for new revolutionary technologies in the future.

Potentially, the quantum computers can change many technologies, by allowing one to solve the problems unsolvable with available classical computers. But the former should be made noiseless to be reliable.

### 5.8.4.2. *Quantum advantage of Google*

Since the quantum computers will be able to make calculations that are too hard for ordinary computers, the humanity will attain the quantum advantage. This event promises us to enter the truly revolutionary period in all spheres of our life. For example, quantum computers will help us to produce fantastic composites for new transport means, to design a new generation of electronic devices, and to change the digital systems.

Recently, the company Google informed that its quantum processor Sycamore executed calculations for 3 min and 20 sec, whereas the classical supercomputer would make such work for 10 thou years.

The processor Sycamore was fabricated from aluminium, indium (very soft metal), and silicon with the use of the Josephson effect, when the superconduction current flows through the junctions of two superconductors. In order to reach the state with superconductivity, the processor with the qubits that are minimal units of information in a quantum computer was cooled down to a temperature close to the absolute zero (20 milli-Kelvin), which is equal approximately to minus 273 Celsius degrees.

With the help of the so-called attenuators (units that decrease the intensity of electromagnetic oscillations) and the additional filters, the supercooled Sycamore was connected with an ordinary electronics operating at room temperature. To read-off the information from the quantum computer became possible due to analog–digital transducers. The whole system supports the quantum state of qubits and was able to prove the "randomness of numbers that were created by a generator of random numbers" for about 3 min.

But the declaration about the advantage of quantum computers was answered at once by the experts of the IBM company which is the main competitor of Google in the field of quantum computers. They elucidated that the calculations executed with the quantum processor Google Sycamore bear only a technical character, and the IBM supercomputer Summit can make the same for only two-three days.

Apparently, the single laboratory experiment with the realization of a very specific processing procedure for a random sample outside any practical application gives no basis for the assertion that the quantum computers "exceed" the classical ones. In fact, the quantum computers will not "dominate", but should operate in a tandem with ordinary computers, since each type of computers has the own unique advantages.

It is expected that the growth of investments to quantum technologies and the increase in the demand for highly efficient computations for the analysis of data will ensure the fast development in this field in the nearest years. In addition, the appearance of local quantum computers for commercial applications will create the advantageous perspectives for the market of corporate quantum computations.

At the present time, a new quantum Internet is developed.

The quantum Internet that will join quantum computers able to solve the problems with incredible complexity, by ensuring the rapid information flow and opening the completely new trends for scientific studies and the economic growth.

## 5.8.5. *Topological quantum computers*

We now discuss briefly the idea of a topological quantum computer presented in Fig. 5.38.

Topological quantum calculations tend to realize a more elastic qubit, by using the non-Abelian forms of the matter for the storage of a quantum information. The non-Abelian forms of the matter are quasiparticles such as Majorana fermions and anyons (non-Abelian quasiparticles that are neither bosons, nor fermions). The most promising topological developments in the field of quantum calculations are related to the non-Abelian braiding of chiral Majorana fermions at quantum points. The type of anyons which is necessary to create a universal quantum computer was not yet confirmed experimentally (only some preliminary signs are available). Till now, this model of quantum calculations bears a purely theoretical character.

Fig. 5.38.   Photo of a topological quantum computer.

We recall that topology is a branch of mathematics concerning those properties of a body that are invariable at its deformations such as the tension or flattening. The topological properties are most stable. In order to break them, the much higher energy must be applied to the body, than in other cases.

By intuition, the idea is clear: at any deformations of the body, its topology is invariant. A torus can be transformed into a cup, but the topology remains the same.

As distinct from the ordinary digital computers operating on the basis of transistors, where the data are coded in binary digits (bits), the quantum computers use quantum bits. Those qubits can exist in a lot of states simultaneously, which gives potential for the execution of a huge number of simultaneous calculations in parallel, which significantly reduces the time of searching for the answer.

Since we need a more profound comprehension of the nature of a qubit as the base of quantum computers, it is useful to separate three significant aspects in the operation of this element:

1. *Superposition of states.* A qubit has the wave nature, i.e., the smooth changes in its state are described by a wave function. If the maximum and minimum of a wave are described as 1 and 0, respectively, then a qubit stays permanently in these extreme and all intermediate states of the wave described by a phase, but with different probabilities.

2. *Interference of qubits.* The interaction in a set of waves from different sources generates the interference pattern. By adjusting the phases of

the sources of waves, such patterns can be made stable. In the quantum computer, namely such pattern with a beforehand known required form signalizes that the answer to a computational problem is obtained.

The essence of algorithms for quantum calculations consists in the following. By fine non-destructive actions onto qubits, their jointly consistent oscillations should be directed to the generation of a required interference pattern.

3. *Entanglement of qubits*. In order that the superpositions of a set of qubits in a quantum computer behave themselves as a unique coordinated system called a coherent one, the qubits should stay in the state of quantum entanglement with one another.

To attain a coherent state for even several individual qubits is a highly complicated technical problem, because the quantum entanglement of particles under laboratory conditions is extremely brittle with a short life. As a rule, the coherence of a system is at once broken by a lowest electromagnetic "noise" and any other random actions of the environment at least on one of the qubits. Since the number of qubits in a useful computer is hundreds and thousands, it is obvious that the expected time of the appearance of such computing units on the market moves further and further into the future, until some revolutionarily new device appears.

The researchers at the Delft Technical University under the support of the company Intel managed to construct a qubit that is stable to a highest degree and can form the states that are stable against the noises and are tangled with other qubits. This qubit was produced on the basis of Majorana fermion quasiparticles, and, on its basis, Aleksei Kitaev invented a "topological quantum computer", by advancing a completely new conception of a quantum computing unit stable to distortions and noises from the environment. A. Kitaev made conclusion that the quantum calculations of states can be executed on quantum braids that are intertwined from filaments-trajectories.

We now describe briefly a Majorana fermion and the ideas by Kitaev.

### 5.8.5.1. *Majorana fermion*

A Majorana fermion is a particle that is identical to the own antiparticle. In other words, a Majorana fermion is produced from a superposition of an electron and a hole.

Moreover, this superposition is formed at the boundaries of a semiconductor and a superconductor (Fig. 5.39).

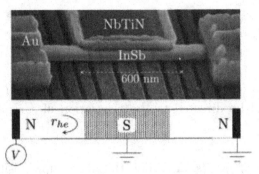

Fig. 5.39.   On the left, a InSb wire put on the NbTiN superconductor in a magnetic field is shown. On the ends of the wire and the chain, Majorana fermions are localized. Taken from [303].

The key specific feature of a Majorana fermion is in that it obeys a non-Abelian statistics. As a result, it can be used in quantum calculations.

We now give the basic positions of the Majorana theory.

From the mathematical viewpoint, the operator of annihilation of Majorana fermions $\gamma_i$ is defined as [300]

$$\gamma_i^\dagger = \gamma_i, \quad \{\gamma_i, \gamma_j\} = 2\delta_{ij}. \tag{5.164}$$

Generally speaking, any Dirac fermion $c_k$ (for the Dirac fermion we take the standard definition of a fermion operator) can be represented in terms of the Majorana operators

$$c = \frac{1}{\sqrt{2}} (\gamma_i + i\gamma_j),$$

$$c^\dagger = \frac{1}{\sqrt{2}} (\gamma_i - i\gamma_j),$$

$$\{c_i, c_j^\dagger\} = \delta_{ij},$$

$$\{c_i, c_j\} = 0.$$

$$\tag{5.165}$$

The key thing consists in the following. If the Majorana fermions are spatially separated, then the obtained Dirac fermion is also spatially nonlocal. In this case, its energy determined by the overlapping of the wave functions of spatially separated Majorana fermions is exponentially low [301]

$$H_n = \epsilon_n c_n^\dagger c_n, \quad \epsilon_n \propto e^{-L/\xi}, \tag{5.166}$$

where $L$ is the distance between Majorana fermions, and $\xi$ is a characteristic scale of the spatial coherence. We will consider that the Majorana fermions are well defined, if $L \gg \xi$. In this case, any local perturbations affect exponentially weakly the state determined by two Majorana fermions.

A Majorana fermion is a topological state protected by the electron–hole conjugation symmetry. This symmetry is analogous to the $C$ symmetry in the physics of elementary particles [300]:

$$\Xi H(k)\Xi = -H(-k), \quad \Xi^2 = -1, \tag{5.167}$$

if

$$H\psi_n = \epsilon_n\psi_n,$$

$$\Xi\psi_n = \psi_{-n}, \tag{5.168}$$

$$\epsilon_n = -\epsilon_{-n}.$$

The above presented relations mean that, for every state with energy $\epsilon > 0$, there exists a state with energy $-\epsilon > 0$. But if $\epsilon = 0$, then the operator of electron–hole symmetry transforms the state into itself:

$$\Xi\psi_{\mathrm{MF}} = \psi_{\mathrm{MF}}, \quad H\psi_{\mathrm{MF}} = 0. \tag{5.169}$$

The given state is a Majorana fermion.

In a realization of Majorana fermions crucial is the electron–hole conjugation symmetry. This symmetry arises in the Bogolyubov–de Gennes equations describing the superconductivity [300]. In this case, the operator of electron–hole conjugation is set as $\Xi = \sigma_y\tau_y K$, where the Pauli matrix $\sigma_i$ acts in the space of spins, $\tau_i$ acts in the electron–hole basis and $K$ is the operator of complex conjugation. However, the presence of only the superconductivity is not sufficient for a realization of spatially separated Majorana fermions. For this purpose, it is necessary to remove all types of generation.

As was mentioned above, a Majorana fermion is a topological state. To realize it, we need a nontrivial topology of the zone structure. It is known that the strong spin–orbit interaction in many materials leads to a nontrivial topology. So, it is logical to consider such materials. For the realization of spatially separated fermions the following conditions should be satisfied:

superconductivity + magnetic field + nontrivial topology.

The simplest way to the joining of all three conditions is given by the triplet superconduction parameter of order. It corresponds to the pairing of electrons with identical spins. If we produce a one-dimensional chain of superconductors with the triplet pairing, Majorana fermions will be localized on its ends. In the real superconductors, the spin degree of freedom is always present. In addition, the triplet parameter of order is sufficiently rare in the Nature. One of the candidates is strontium ruthenate $Sr_2RuO_4$.

### 5.8.5.2.  *Kitaev's superconducting chain*

Consider briefly the Kitaev one-dimensional model describing a spinless $p$-wave superconductor. Its Hamiltonian was given in [302, 303]:

$$H = -\mu \sum_{j=1}^{L} c_j^\dagger c_j - \sum_{j=1}^{L-1} (t_p c_j^\dagger c_{j+1} + \Delta c_j^\dagger c_{j+1}^\dagger + \text{h.c.}), \qquad (5.170)$$

where h.c. means the Hermitian conjugation, $\mu$ is the chemical potential $c_j$ is the electron annihilation operator at node $j$, and $L$ is the chain length. The tunneling $t_p$ and the superconductor gap $\Delta = |\Delta| e^{i\alpha}$ are the same for all nodes. For the sake of simplicity without any loss of generality, we take $\Delta = |\Delta|$. The Kitaev system can be realized experimentally by means of the contact of a nanowire with strong spin–orbit coupling (e.g., InSb and InAs nanowires, Fig. 5.39) with an $s$-wave superconductor in the Zeeman field [300]. For clarity, in this work we focus on the condition $\mu = 0$, $t_p = \Delta$ [301].

By concluding, we note that the quantum computers represent a technological trend that can change potentially the world of highly efficient calculations. This causes a huge interest of the scientific community and the significant volume of investments to their development.

At the present time, about ten different experimental models of quantum units are available. Each unit can include up to several tens of quantum objects. Such models are characterized by different levels of their development, but none of the models allows one to solve practical problems.

In the future, the quantum computers will be claimed in all fields of the human activity requiring great volumes of calculations, so that it will be impossible to operate without quantum technologies.

# Summary and Conclusions

In this book, we have concentrated our attention on the most urgent problems of superconductivity such as the nature of high-temperature superconductivity, mechanisms and symmetry of pairing, two-gap superconductivity in magnesium diborades, mesoscopic superconductivity and the problems of room-temperature superconductivity.

We considered some questions concerning the application of quantum field theory to the problems of superconductivity. Quantum field theory supplies the original and powerful means for the solution of certain problems of superconductivity. In the field of superconductivity we meet the problem-maximum — it consists in the creation of room-temperature superconductors. We consider this problem in our book, give some recommendations on the search for these superconductors and analyze the possibility of the fabrication of artificial materials possessing the property of superconductivity at room temperature.

The problem of high-temperature superconductivity in hydrides at high pressures is considered.

We also have touched the questions of the application of superconducting qubits to quantum computers. It is shown that the description of superconducting qubits is based on the laws of quantum mechanics.

We give a short review of the modern state of the problem of physical realization of quantum computers. The ideas of a topological quantum computer using qubits on the basis of Majorana fermion quasiparticles are considered.

It is worth noting that HTSC is investigated more than two decades with great efforts, but the whole pattern of the phenomenon is not else available. We are sure that the comprehension of HTSC will be attained when our knowledge of HTSC reaches the critical level which will be sufficient to understand the huge number of experimental data from a single viewpoint. The study is in progress. In particular, we mention the recent discovery of Cu-less superconductors with a high critical temperature which contain layers of FeAs.

The physical properties and electron models of the high-temperature Fe-based superconductors are analyzed.

The theoretical and experimental investigations of superconductivity and their practical applications at the present time form the wide, interesting and uncommon branch of science.

# Appendix A

## A.1. Two-particle Green Function for Multiband Superconductors

In this appendix, we will solve the problem on determination of two-particle Green function for multiband superconductors. In Chapter 4, we considered the multiband superconductivity using a two-particle Green function.

We will calculate a two-particle Green function for many-band superconductors.

The Hamiltonian describing the systems of interacting electrons and phonons of a crystal is written in the following form:

$$
H = \sum_{k,a,\nu} \varepsilon_k^\nu a_{k\sigma}^{+\nu} a_{k\sigma}^\nu + \frac{1}{N} \sum_{q,k_1,\sigma_1,\nu_1,k_2,\sigma_2,\nu_2} V_q a_{k_1\sigma_1}^{+\nu_1} a_{k_2\sigma_2}^{\nu_2} a_{k_2+q,\sigma_2}^{\nu_2} a_{k_1-q,\sigma_1}^{\nu_1}
$$

$$
+ \frac{1}{\sqrt{N}} \sum_{k,\sigma,\nu,q,s} \chi_q^s a_{k,\sigma}^{+\nu} a_{k-q,\sigma}^\nu Q_q^s + \sum_{q,s} \Omega_{qs} b_q^{+s} b_q^s, \tag{A.1}
$$

where $a_{k\sigma}^{+\nu}$ is the creation operator for the electron with the momentum $k$ in band $\nu$, $a_{k\sigma}^\nu$ is the annihilation operator for the electron with the momentum $k$ in band $\nu$, $b_q^{+s}$, $b_q^s$ are creation and annihilation operators for the phonons with the momentum $q$, respectively, $\varepsilon_k^\nu$ is the energy of the electron with the impulse $k$, $\Omega_{qs}$ is the energy of the phonons, $Q_q^s = b_q^{+s} + b_q^s$, $s$ is the number of phonon branch, $V_{-q} = V_q, X_q^{*s} = X_q^s$ are the Fourier components of the Coulomb interactions of electron and their coupling constant with the lattice phonons, respectively. We can see that both constants are independent on spin index of an electron. Let us introduce new operators for electron and phonon systems by rules

$$
a_k^\nu = e^S A_{k,\nu} e^{-S}, \quad b_q^s = e^S \beta_{s,q} e^{-S} \tag{A.2}
$$

where $S$ is the anti-Hermitian operator $(S^+ = -S)$.

In our unitary transformation, it will be more convenient to rewrite the term $H_{\text{e-ph}}$ in the following form:

$$H_{\text{e-ph}} = \sum_{k,q,s,v,\mu} \chi^{v,\mu}_{s,q} a^{+v}_{k} a^{\mu}_{k-q} Q^{s}_{q} \quad (\chi^{v,\mu}_{s,q} = \chi^{s}_{q} \delta_{v,\mu}), \tag{A.3}$$

where we united two indices and so $v = (v, \sigma)$ and $\mu = (\mu, \sigma')$ are the complex indices which characterize the number of crystals electron zone and the spin of electron. The Hamiltonian describing the system of interacting electrons and phonons of the crystal after transformation by a unitary operator is written in the following form [139–143]:

$$H = \sum_{k,v} \left( \varepsilon^{v}_{k} - \frac{1}{N} \sum_{s,q} \frac{|\chi^{s}_{q}|^{2}}{\Omega_{s,q}} \right) A^{+}_{k,v} A_{k,v}$$

$$+ \frac{1}{2N} \sum_{q,k,v,k',v'} \left( V_{q} - 2 \sum_{s} \frac{|\chi^{s}_{q}|^{2}}{\Omega_{s,q}} \right) A^{+}_{k,v} A^{+}_{k',v'} A_{k'-q,v'} A_{k-q,v}$$

$$+ \sum_{s,q} \Omega_{s,q} \beta^{+}_{s,q} \beta_{s,q} + (\text{higher order term}). \tag{A.4}$$

The unitary transformation gives rise to renormalization of the electron energy (first term) and renormalization of the Fourier component of the Coulomb electron–electron interaction. To calculate the density of electron states, we have to study the Green function for the case with approximation $t' \to t - 0$. The two-particle Green function can be written as follows:

$$G_{2} \left( \begin{matrix} k_{2}, v_{2}; k, v \\ k + q, v; k_{2} - q, v_{2} \end{matrix} \middle| t - t' \right)$$

$$= \langle -iT A_{k+q,v}(t) A_{k_{2}-q,v_{2}}(t) A^{+}_{k_{2},v_{2}}(t-0) A^{+}_{k,v}(t') \rangle (t' \to t - 0). \tag{A.5}$$

The equation for the Green function can be found from the motion of equation:

$$i\frac{\partial}{\partial t} \langle -iT A_{k,v}(t) A^{+}_{k,v}(t') \rangle$$

$$= \delta(t - t') + \left\langle -iT \left( \frac{\partial}{\partial t} A_{k,v}(t) \right) A^{+}_{k,v}(t') \right\rangle, \tag{A.6}$$

where

$$i\frac{\partial A_{k,v}}{\partial t} = [A_{k,v}.H] = \langle -iTA_{k,v}(t)A_{k,v}^+(t')\rangle = \langle -iTA_{k,v}(t)A_{k,v}^+(t')\rangle$$

$$= \tilde{\varepsilon}_k^v A_{k,v} + \frac{1}{N}\sum_{q,k_2,v_2}\tilde{V}_q A_{k_2,v_2}^+ A_{k_2+q,v_2}A_{k-q,v}$$

$$= \tilde{\varepsilon}_{k,v}A_{k,v} - \frac{1}{N}\sum_{q,k_2,v_2}\tilde{V}_q A_{k+q,v}A_{k_2-q,v_2}A_{k_2,v_2}^+, \tag{A.7}$$

where

$$\tilde{\varepsilon}_k^v = \varepsilon_k^v - \frac{1}{N}\sum_{s,q}\frac{|\chi_q^s|^2}{\Omega_{s,q}}, \tag{A.8}$$

$$\tilde{\varepsilon}_{k,v} = \tilde{\varepsilon}_k^v + \sum_{v_2}\left(\tilde{V}_{q=0} - n\frac{1}{N}\sum_q\tilde{V}_q\right). \tag{A.9}$$

Therefore, the equation for the Green function (A.6) is transformed after inserting in it (A.7) to the following form:

$$i\frac{\partial}{\partial t}\langle -iTA_{k,v}(t)A_{k,v}^+(t')\rangle$$

$$= \delta(t-t') + \tilde{\varepsilon}_{k,v}\langle -iTA_{k,v}(t)A_{k,v}^+(t')\rangle$$

$$- \frac{1}{N}\sum_{q,k_2,v_2}\tilde{V}_q\langle -iTA_{k+q,v}(t)A_{k_2-q,v_2}(t)A_{k_2,v_2}^+(t-0)A_{k,v}^+(t')\rangle. \tag{A.10}$$

Such two-particle Green function satisfies the equation of the Bethe–Salpeter type (we not split this function into two one-particle Green functions of the Gorkov type). The solution of this equation according to the Bogoliubov–Tyablikov method gives rise to the following expression for the Fourier component of the two-particle Green function:

$$G_2\left(\begin{array}{c}k_2,v;k_1,\mu\\k_1+q,\mu;k_2-q,v\end{array}\middle|\omega\right) \sim \frac{f(k_1,\mu;k_2,v;\omega)\sum_{\sigma,\sigma'}\varphi(\mu,v;\sigma,\sigma')}{1-VK(k_1,\mu;k_2,v;\omega)}, \tag{A.11}$$

where

$$K(k_1,\mu;k_2,v;\omega) = \frac{1}{N}\sum_q\frac{1-n_{k_1+q}^\mu-n_{k_2+q}^v}{\omega-\varepsilon_{k_1+q,\mu}-\varepsilon_{k_2+q,v}}, \tag{A.12}$$

where $V$ is the Fourier of the Coulomb interactions of electron, $\varphi\left(\mu, \nu; \sigma, \sigma'\right)$ are some functions depending on the frequency $\omega$, and momenta $k_1$ and $k_2$ of interacting electrons, $f\left(k_1, \mu; k_2, \nu; \omega\right)$ are expressed in terms of functions, $n^\mu_{k_1+q}$, $\varepsilon_{k_1+q,\mu}$ are numbers of filling and energy of electron, respectively,

$$\varphi\left(\mu, \nu; \sigma, \sigma'\right) = \delta_{\sigma\sigma}\delta_{\sigma'\sigma'} - \delta_{\mu\nu}\delta_{\sigma\sigma'}\delta_{\sigma\sigma'}, \qquad (A.13)$$

and $\sigma$, $\sigma'$ are the spins of the first $(\sigma)$ and second $(\sigma')$ electrons. The superconducting gap is given by the zeros of the denominator of (A.11) [140, 143]. Let us study the particular case supposing $k_1 = k_2 = k_0 + k$ and $\varepsilon_{k0,\mu} = \varepsilon_\mu$ corresponds to the extremum of the zone. Then, expanding the energy by momentum $k \pm q$ in series up to terms of second order, we can obtain:

$$\varepsilon_{k\pm q,\mu} = \varepsilon_\mu + \frac{(k \pm q)^2}{2m_\mu} = \varepsilon_f + \Delta_\mu + \frac{(k \pm q)^2}{2m_\mu}, \qquad (\mu = 1, 2, \ldots), \qquad (A.14)$$

where $m_\mu$ is the effective mass of electron in the $\mu$-energy band of the crystal, $\Delta\mu$ is a parameter which indicates the position of the extremum of $\mu$ band relative to the Fermi level, $\varepsilon_{k\pm q,\mu}$ is the energy of electrons. The sum in the denominator (A.11) is reduced to the following expression (we consider only one zone):

$$K\left(k, k', \omega\right) = \frac{1}{N} \sum_q \frac{1 - n_{k+q} - n_{k-q}}{\omega - \varepsilon_{k+q} - \varepsilon_{k-q}}$$

$$= 2N\left(\varepsilon_f\right) \int_0^\Delta \frac{d\varepsilon}{\omega - 2\left(\varepsilon_f + \Delta_1\right) - 2E - 2\varepsilon}$$

$$= 2N\left(\varepsilon_f\right)\left(-\ln\left|1 - \frac{\Delta}{a}\right|\right), \qquad (A.15)$$

where such assignments were used: $\varepsilon = q^2/2m$; $E = k^2/2m$; $a = \omega - 2(\varepsilon_f + \Delta_1) - 2E$; $m_1^* = m_1/m$ and $m$ are the reduced effective mass of electron in the crystal energy zone and mass of free electron, respectively,

$$N\left(\varepsilon_f\right) = \sqrt{2}\pi\, m_1^* \sqrt{m_1^* \varepsilon}\left(1 - n_{k+q} - n_{k-q}\right)\big|_{\varepsilon_f}, \qquad (A.16)$$

$$n_k = \left[\exp\left(\frac{\varepsilon_k - \varepsilon_f}{T}\right) + 1\right]^{-1}. \qquad (A.17)$$

The equation for superconducting gap has the following form:

$$1 - VN\left(\varepsilon_f\right)\left(-\ln\left|1 - \frac{\Delta}{a}\right|\right) = 0. \qquad (A.18)$$

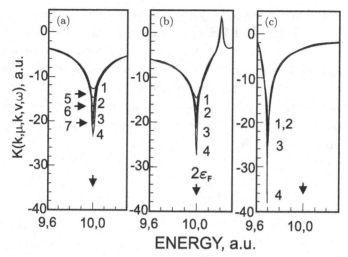

Fig. A.1. The effect of temperature on the $K(k,\mu;k,\nu;\omega)$ function ($E = k^2/2m = 0$) for different structures of the energy zones: (a) $\Delta_1 = -1$, $m_1^* = 1$; (b) $\Delta_2 = 0.2$, $m_2^* = -2$; (c) $\Delta_1 = -1$, $m_1^* = 1$, $\Delta_2 = 0.2$, $m_2^* = -2$; curves 4, $T = 2\,\mathrm{K}$; curves 3, $T = 10\,\mathrm{K}$; curves 2, $T = 50\,\mathrm{K}$; curves 1, $T = 100\,\mathrm{K}$; arrows 5–7 correspond to different $1/V$ values (arrow 5, $V = -0.07$; arrow 6, $V = -0.06$; arrow 7, $V = -0.05$). All energy values are taken as arbitrary values, i.e., $V = V/M$, where $M$ is a scale factor ($M = 1\,\mathrm{eV}$, for convenience; $m_i^* = \frac{m_i}{m}$, $m$ is the free electron mass).

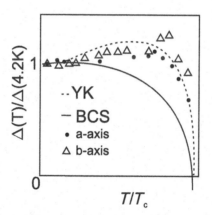

Fig. A.2. The temperature dependence of the superconducting gap: $\cdots$ theory, — BCS; • a-axis, △ b-axis-experiment.

The results of some numerical calculations are illustrated in Fig. A.1. We will study theoretically the situation close to the $Ba_2Sr_2CaCu_2O_8$ crystal because it has a high $T_c$. The results of some numerical calculations of the temperature dependence of superconductivity are illustrated in Fig. A.2. For illustration, the experimental dependences of the superconducting gap obtained for $Ba_2Sr_2CaCu_2O_8$ crystal are presented in Fig. A.2 together with theoretical curves of BCS type and our calculation. The obtained dependences are not similar to standard BCS ones. It clearly seen that our $r$ calculation predicting the maximum in the $T$ dependence of the superconducting gap gives the curve which is closer to the experimental results. In our approach, we take into account that all electron bands (the numerical calculations show that only bands located near the Fermi are important) contribute to the superconductivity.

# Appendix B

## B.1. A Solution Method for the Ferromagnetic Granules in the London Superconductors

In this appendix, we will solve the problem on determination of magnetic configurations in superconducting nanocomposite systems by means of Maxwell equation, which in a static case has the following form:

$$\nabla \times \mathbf{B} = \mu_0 j, \qquad (B.1)$$

where B is a magnetic field and $\mathbf{j}$ is a specific electric current.

The value $\mathbf{j}$ includes normal and superconducting parts. The role of normal currents in superconductors in the process of formation of magnetic fields is negligible since a superconductor substance exhibits weak magnetic characteristics in the normal state. However, ferromagnetic inclusions enter into the superconducting nanocomposites and the consideration of a normal current for these inclusions is necessarily required. Therefore, it may be supposed that

$$\mathbf{j} = \mathbf{j}_{sp} + \mathbf{j}_n. \qquad (B.2)$$

The current of superconductivity can be locally connected with the vector potential of system, which allows us to write:

$$\nabla \times \mathbf{j}_s = -\frac{n_s e^2}{m} \mathbf{B}. \qquad (B.3)$$

A specific current of charge carriers in the normal state can be written in the traditional form:

$$\mathbf{j}_n = \nabla \times \mathbf{M}, \qquad (B.4)$$

where M is a magnetization of material.

As a result, the term in (B.2) for the normal current is characterized only for the region filled by ferromagnetic material.

Combining relations (B.2), (B.3), (B.4) with the Maxwell equation $\nabla \times \mathbf{B} = \mu_0 j$ the magnetic field can be found in the form:

$$\nabla \times (\nabla \times \mathbf{B}) + \lambda^{-2}\mathbf{B} = \mu_0 \nabla \times (\nabla \times \mathbf{M}), \qquad (B.5)$$

where $\lambda$ is the London penetration depth of field in the superconductor. Since $\nabla \mathbf{B} = 0$, using relations (B.1), (B.3), (B.4), the equation for magnetic field can be presented in the form:

$$\nabla \times (\nabla \times \mathbf{B}) + \lambda^{-2}\mathbf{B} = \mu_0 \nabla \times (\nabla \times \mathbf{M}), \qquad (B.6)$$

where $\delta = \sqrt{c^2\Lambda/4\pi}$ is a London depth of field penetration in the superconductor. If the condition $\nabla \mathbf{B} = 0$ is satisfied without ferromagnetic inclusions, then Eq. (B.5) can be transformed to

$$\nabla \times (\nabla \times \mathbf{B}) + \lambda^{-2}\mathbf{B} = \mu_0 \nabla \times (\nabla \times \mathbf{M}). \qquad (B.7)$$

For boundary conditions, we assume that the magnetic field vanishes outside of the localization region of ferromagnetic inclusions. According to these boundary conditions, the solution of Eq. (B.7) can be presented as follows:

$$\mathbf{B}(\mathbf{r}) = \frac{\mu_0}{4\pi} \int d^3r' G\left(|\mathbf{r} - \mathbf{r}'|\right) \nabla \times (\nabla \times \mathbf{M}(\mathbf{r}')), \qquad (B.8)$$

where the Green's function is given by

$$G\left(|\mathbf{r} - \mathbf{r}'|\right) = \frac{\exp\left(-|\mathbf{r} - \mathbf{r}'|/\lambda\right)}{|\mathbf{r} - \mathbf{r}'|}.$$

Sizes of quantum dots, the field of which to be studied, are much less than the penetration depth $\lambda$. Henceforward, we consider magnetic moments as dotty magnetic dipole ones, supposing that

$$\mathbf{M}(\mathbf{r}) = \sum_i \mathbf{m}_i \delta(\mathbf{r} - \mathbf{r}_i), \qquad (B.9)$$

where $\delta(\mathbf{r} - \mathbf{r}_i)$ is delta-function of Dirac, $\mathbf{m}_i$ and is a magnetic moment of the $i$th granule.

In this case, the integrand in (B.7) holds the space derivatives from singular values. Let us transform (B.9) in such a way as to differentiate the kernel of integrand. For this purpose, we present the expression for the magnetic field as $\mathbf{H}(\mathbf{r}) = \int dv' \left\{ \left[\frac{\partial}{\partial \mathbf{r}'} \operatorname{rot} \mathbf{M}(\mathbf{r}')\right] + \left[\operatorname{rot} \mathbf{M}(\mathbf{r}') \frac{\partial}{\partial \mathbf{r}'}\right] \right\} \cdot G\left(|\mathbf{r} - \mathbf{r}'|\right)$, and, using the Gauss theorem, we rewrite the relation (B.9) as

$$\mathbf{H}(\mathbf{r}) = \int dv' \left\{ \left[ \frac{\partial}{\partial \mathbf{r}'} \operatorname{rot} \mathbf{M}(\mathbf{r}') \right] + \left[ \operatorname{rot} \mathbf{M}(\mathbf{r}') \frac{\partial}{\partial \mathbf{r}'} \right] \right\} \cdot G\left( |\mathbf{r} - \mathbf{r}'| \right).$$

(B.10)

The integration in the first term over the infinite surface vanishes. Then, integrating by parts in the second term of (B.10), we obtain

$$\mathbf{H}(\mathbf{r}) = \oint \left[ d\mathbf{S}' \cdot \operatorname{rot} \mathbf{M}(\mathbf{r}') \right] \cdot G\left( |\mathbf{r} - \mathbf{r}'| \right)$$
$$+ \int dv' \left[ \operatorname{rot} \mathbf{M}(\mathbf{r}') \frac{\partial}{\partial \mathbf{r}'} G\left( |\mathbf{r} - \mathbf{r}'| \right) \right],$$

(B.11)

where $\nabla = \mathbf{e}_x \frac{\partial}{\partial x} + \mathbf{e}_y \frac{\partial}{\partial y} + \mathbf{e}_z \frac{\partial}{\partial z}$, $\Delta = \frac{\partial^2}{\partial x^2} + \frac{\partial^2}{\partial y^2} + \frac{\partial^2}{\partial z^2} -$ is the Laplace operator.

In Eq. (B.11), the differentiation operators act no longer upon generalized functions but on the kernel of the integral operator.

Taking into account that the relation

$$\Delta G - \delta^{-2} G = -4\pi \delta \left( \mathbf{r} - \mathbf{r}' \right),$$

(B.12)

holds for the Green function $G\left( |\mathbf{r} - \mathbf{r}'| \right)$, Eq. (B.11) is transformed to

$$\mathbf{H}(\mathbf{r}) = \int dv' \exp(-R/\delta) \cdot \left\{ \left( \frac{3\mathbf{R}\left( \mathbf{R} \cdot \mathbf{M}(\mathbf{r}') \right)}{R^5} - \frac{\mathbf{M}(\mathbf{r}')}{R^3} \right) \cdot \left( 1 + \frac{R}{\delta} + \frac{R^2}{\delta^2} \right) \right.$$
$$\left. - \frac{2\mathbf{R}\left( \mathbf{R} \cdot \mathbf{M}(\mathbf{r}') \right)}{\delta^2 \cdot R^3} \right\},$$

$$\mathbf{R} = \mathbf{r} - \mathbf{r}', \quad R = |\mathbf{r} - \mathbf{r}'|.$$

(B.13)

The expression (B.13) determines the magnetic field out of ferromagnetic inclusions. Since the magnetization of system $\mathbf{M}(\mathbf{r})$ is defined by (B.8), it is considered in (B.13) that, out of the volume of granules, $\int dv' \mathbf{M}(\mathbf{r}') \delta(\mathbf{r} - \mathbf{r}') = 0$.

After integrating (B.13), we get the expression for magnetic field in the form of superposition of parts of quantum dots:

$$\mathbf{H}(\mathbf{r}) = \sum_i \exp(-R_i/\delta) \cdot \left\{ \left( \frac{3\mathbf{R}_i(\mathbf{R}_i \cdot \mathbf{m}_i)}{R_i^5} - \frac{\mathbf{m}_i}{R_i^3} \right) \cdot \left( 1 + \frac{R_i}{\delta} + \frac{R_i^2}{\delta^2} \right) \right.$$
$$\left. - \frac{2\mathbf{R}_i(\mathbf{R}_i \cdot \mathbf{m}_i)}{\delta^2 \cdot R_i^3} \right\},$$

$$\mathbf{R}_i = \mathbf{r} - \mathbf{r}_i, \quad R_i = |\mathbf{r} - \mathbf{r}_i|.$$

(B.14)

It has been apparent, when the temperature of superconductor and the depth penetration of magnetic field $\lambda \to \infty$ increase in the region adjacent to the ensemble of granules, then expression (B.14) tends to the limit

$$B\left(r\right) = \frac{\mu_0}{4\pi} \sum_i \left( \frac{3\mathbf{R}_i \left(\mathbf{R}_i \cdot \mathbf{m}_i\right)}{R_i^5} - \frac{\mathbf{m}_i}{R_i^3} \right), \qquad (B.15)$$

which describes the magnetic field of the system of dotty dipoles in the normal state medium.

In the another limiting case, if the distance between granules is much more than the depth penetration, we have

$$B\left(r\right) = \frac{\mu_0}{4\pi} \sum_i \frac{\exp(-R_i/\lambda)}{R_i \cdot \lambda^2} \cdot \left( \frac{\mathbf{R}_i \left(\mathbf{R}_i \cdot \mathbf{m}_i\right)}{R_i^2} - \mathbf{m}_i \right). \qquad (B.16)$$

# Bibliography

[1] Parks, R.D. (ed.) (1969). *Superconductivity* (Marcel-Dekker, New York).

[2] Bednorz, J.G. and Müller, K.A. (1986). Possible high-$T_c$ superconductivity in the Ba–La–Cu–O System, *Zeit. Phys. B* **64**, 1, pp. 189–193.

[3] Schrieffer, J.R. (1983). *Theory of superconductivity* (Benjamin, New York).

[4] Feynman, R.P. (1982). *Statistical mechanics* (Addison-Wesley, Redwood City).

[5] Anderson, P.W.(1987). The resonating valence bond state in LaCuO and superconductivity, *Science* **4**, pp. 1196–1198.

[6] Anderson, P.W. (1959). Theory of dirty superconductivity, *J. Phys. Chem. Solids* **11**, pp. 26–30.

[7] London, F. and London, H. (1935). The electromagnetic equations of the supraconductor, *Proc. Roy. Soc. (London) A* **147**, p. 71.

[8] Ryder, L.H. (1996). Quantum field theory, 2nd edn. (Cambridge University Press, Cambridge).

[9] Ginzburg, V.L. and Landau, L.D. (1950). On the theory of superconductivity, *Zh. Eksp. Theor. Fiz.* **20**, p. 11064.

[10] Clark, K. *et al.* (2010). Superconductivity in just four pairs of $(BETS)_2GaCl_4$ molecules, *Nat. Nanotechnol.* **5**, p.261.

[11] Tinkam, M. (1975). *Introduction to superconductivity* (McGraw-Hill, New York).

[12] Nambu, Y. (1960). Quasi-particles and gauge invariance in the theory of superconductivity, *Phys. Rev.* **117**, p. 648.

[13] Takahashi, Y. (1976). *Field theory for the solid state physics* (Baifuhkan, in Japanese).

[14] Nagao, H., Nishino,H., Shigeta Y. and Yamaguchi, K. (2000). Theoretical studies on anomalous phases of photodoped systems in two-band model, *J. Chem. Phys.* **113**, pp. 11237–1244.

[15] Kondo, J. (2002). Theory of multiband superconductivity, *J. Phys. Soc. Jpn.* **71**, p. 1353.

[16] Fetter, L.A. and Walecka, J.D. (1987). *Quantum theory of many-particle systems* (McGraw-Hill).

[17] Abrikosov, A.A., Gorkov, L.P. and Dzyaloshinski, I.E. (1975). *Methods of quantum field theory in statistical physics* (Dover Books).

[18] Aono, S. (2003). Molecular orbital theory: spinor representation, *J. Phys. Soc, Jpn.* **72**, pp. 3097–3105.

[19] Kimura, M., Kawabe, H., Nishikawa, K. and Aono, S. (1986). EPR observations of trivalent titanium in orthophosphate single crystals, *J. Chem. Phys.* **85**, pp. 1–6.

[20] Salem, L. and Longuet-Higgins, H.C. (1959). The alternation of bond lengths in long conjugated chain molecules, *Proc. Roy. Soc. London A* **255**, p. 172.

[21] Peierls, R.E. (1955). *Quantum theory of solids* (Oxford University Press, Oxford).

[22] Kimura, M., Kawabe, H., Nakajima, A., Nishikawa, K. and Aono, S. (1988). Possibility of superconducting and other phases in organic high polymers of polyacene carbon skeletons, *Bull. Chem. Soc. Jpn.* **61**, pp. 4239–4224.

[23] Kubo R. and Tomita, J. (1954). A general theory of magnetic resonance absorption, *Phys. Soc. Japan* **9**, pp. 888–919.

[24] Doi, Y. and Aono, S. (1974). Theory of line shape in nuclear magnetic resonance, *Prog. Theoret. Phys.* **51**, pp. 1019–1029.

[25] Hebel, L.C. and Slichter, C.P. (1959). Nuclear spin relaxation in normal and superconducting aluminum, *Phys. Rev.* **113**, pp. 1504–1519.

[26] Fibich, M. (1965). Phonon effects on nuclear spin relaxation in superconductors, *Phys. Rev. Lett.* **14**, pp. 561–564.

[27] Maki, K. (1967). *Gapless superconductors in superconductivity* (ed.) Parks, R.D. (Maxel Dekker).

[28] Uchino, K., Maeda, A. and Terasaki, I. (1959). *Nature of the high temperature superconductor* (Baifuukann, in Japanese).

[29] Aono, S. (1998). Superconductivity as symmetry breaking classical treatment, *Bull. Chem. Soc. Jpn.* **71**, pp. 49–56.

[30] Scalapino, D.J. (1995). The case for $d_{x^2-y^2}$ pairing in the cuprate superconductors, *Phys. Rep.* **250**, pp. 329–360.

[31] Gorkov, L.P. (1959). Microscopic derivation of the Ginzburg-Landau equations in the theory of superconductivity, *Sov. Phys. JETP* **9**, pp. 1364–1375.

[32] Negele, J.W. and Orland, H. (1987). *Quamtum many-particle systems* (Addison-Wesley, Reading, MA).

[33] Kawabe, H., Nishikawa, K. and Aono, S. (1994). Path integral approach to correlation energy, *Int. J. Quantum Chem.* **51**, pp. 265–283.

[34] Kruchinin, S. (2016). The quantum-field approach to superconductivity theory, *Rev. Theoret. Sci.* **4**, 2, pp. 117–144.

[35] Feynman, R.P. (1948). Space-time approach to non-relativistic quantum mechanics, *Rev. Mod. Phys.* **20**, pp. 367–387.

[36] Hubbard, J. (1959). Calculation of partition functions, *Phys. Rev. Lett.* **3**, pp. 77–78.

[37] Das, A. (2006). *Field theory: a path integral approach* (World Scientific, Singapore).

[38] DeWitt, B. (1984) *Supermanifolds* (Cambridge University Press, Cambridge).

[39] Ferraz, A. and Nguyen, A.V. (1995). Supersymmetry and electron-hole excitations in semiconductors, *Phys. Rev. B* **51**, pp. 10548–10555.

[40] Yang, C.N. (1962). Concept of off-diagonal long-range order and the quantum phases of liquid He and of superconductors, *Rev. Mod. Phys.* **34**, pp. 694–704.

[41] Dunne, L.J., Murrell, J.N. and Brändas, E.J. (1997). Off-diagonal long-range order, eta- pairing, and Friedel oscillations in high $T_c$ cuprate superconductors and the ground state of the extended Hubbard model, *Int. J. Quant. Chem.* **63**, pp. 675–684.

[42] Ginzburg, V.L., Kirzhnitz, D.A. (1982) (eds). *High-temperature superconductivity* (Consultants Bureau, New York).

[43] Fröhlich H.(1954). On the theory of superconductivity, *Proc. Roy. Soc. (London) A* **223**, pp. 296–305.

[44] Bogolyubov, N.N. (1958). A new method in the theory of superconductivity, *Zh. Eksp.Teor. Fiz.* **34**, pp. 58–65.

[45] Josephson, B.D. (1974). The discovery of tunneling superconductivity, *Rev. Mod. Phys.* **1074**, 46, pp. 251-254.

[46] Maroushkine, A. (2004). *Room-temperature superconductivity* (Cambridge Int. Sci. Publ., Cambridge).

[47] Davydov, A.S. (1990). *High-temperature superconductivity* (Naukova Dumka, Kiev).

[48] Kruchinin, S.P. (2014). Physics of high-$T_c$ superconductors, *Rev. Theoret. Sci.* **2**, pp. 124–145.

[49] Davydov, A.S. and Kruchinin, S.P. (1991). Interlayer effects in the newest high-T superconductors, *Physics C* **179**, pp. 461–468.

[50] Davydov, A.S. and Kruchinin, S.P. (1991). Dependence of the superconducting transition temperature on the number of cuprate layers in a unit cell of high-temperature superconductors, *Sov. J. Low. Temp. Phys.* **17**, 10, pp. 634–635.

[51] Timusk, T. and Statt, B. (1999). The pseudogap in high-temperature superconductors: an experimental survey, *Rep. Prog. Phys.* **62**, pp. 61–122.

[52] Plakida, N.M. (1995). *High-$T_c$ superconductivity: experiment and theory* (Springer), p. 306.

[53] Brusov, P. (1999). *Mechanisms of high temperature superconductivity* (Rostov State University Publishing), p. 685.

[54] Abrikosov, A.A. (1998). *Fundamentals of the theory of metals* (North-Holland, Amsterdam), p. 380

[55] Akhiezer, A.I., Pomeranchuk, I.Ya. (1959). On the interaction between conductivity electrons in ferromagnetics, *Zh. Eksp. Teor. Fiz.* **36**, p. 819.

[56] Akhiezer, A.I., Bar'yakhtar, V.G. and Peletminskii, S.V. (1968). *Spin waves* (North-Holland, Amsterdam), p. 369.

[57] Little, W.A. (1964). Possibility of synthesizing on organic superconductivity, *Phys. Rev.* **134**, p. 1416.

[58] Alexandrov, A.S., Ranninger, J. (1981). Bipolaronic superconductivity, *Phys. Rev.* **23**, 4, pp. 1796–1801.

[59] Pekar, S.I. (1951). *Studies on the electron theory of crystals* (Gostekhizdat, Moscow), p. 168.

[60] Lakhno, V. (2020). Translational-invariant bipolarons and superconductivity, *Condens. Matter* **5**, 2, p.30.

[61] Brusov, P. (1999). *Mechanisms of high temperature superconductivity* (Rostov State University Publishing, Rostov), p. 685.

[62] Ausloos, M. and Kruchinin, S. (eds). (1999). *Symmetry and pairing in super-conductors. Proceedings of the NATO ARW* (Kluwer Academic Publishers, Dordrecht), p. 410.

[63] Annett, J. and Kruchinin, S.(eds.) (2002). New trends in superconductivity. *Proceedings of a NATO ARW* (Kluwer Academic Publishers, Dordrecht), p. 435.

[64] Kruchinin, S. (eds.) (2003). Modern problems of superconductivity, Proceedings of Yalta conference, *Mod. Phys. Lett. B* **17**, 10–12, pp. 393–724.

[65] Sigrist, M. and Rice, T.M. (1995). Unusual paramagnetic phenomena in granular high-temperature superconductors — a consequence of d-wave pairing, *Rev. Mod Phys.* **67**, 2, pp. 503–513.

[66] Laughlin, R.B. (1998). Magnetic induction of $d_{x^2+y^2} + id_{xy}$ order in high-$T_c$ superconductors, *Phys. Rev. Lett.* **80**, p. 5188.

[67] Scalapino, D.J. (1999). The case for d pairing in the cuprate superconductors, *Phys. Rep.* **250**, pp. 331–370.

[68] Annet, J.F. (2003). *Superconductivity, superfluids and condensates* (University of Bristol, Oxford University Press), p. 183.

[69] Volovik, G.E., Gorkov, L.P. (1985). Superconducting classes in heavy-fermion systems, *Sov. Phys. JETP* **61**, p. 843.

[70] Mineev, V.P. and Samokhin, K.V. (1998). *Introduction in the theory of unconventional superconductivity* (Moscow Fiz.-Tekh. Inst. Press, Moscow), p. 141.

[71] Van Harligen, D.J. (1995). Phase -sensitive test of the symmetry of pairing state in the high-temperature superconductors- evidence for D symmetry, *Rev. Mod. Phys.* **67**, 2, pp. 515–535.

[72] Tsuei, C.C. and Kirtley, J.R. (2000). Pairing symmetry in cuprate superconductors, *Rev. Mod. Phys.* **72**, 4, pp. 969–1015.

[73] Pines, D. (1991). Spin fluctuations and high temperature superconductivity the antiferromagnetically correlated oxides:YBaCuO, LaSrCuO, *Physica C* **185–189**, pp. 120–129.

[74] Monthoux, P., Balatsky, A.V. Pines, D. (1991). Toward a theory of high-temperature superconductivity in the antiferromagnetically correlated cuprates oxides, *Phys. Rev. Lett.* **67**, 24, p. 3448.

[75] Balatsky, A.V., Vekhter, I.V. and Zhu, J.X. (2008). Impurity-induced state in convenventional and unconventional superconductors, *Rev. Mod. Phys.* **78**, pp. 373–433.

[76] Izyumov, Yu.A. (1999). Spin-fluctuation mechanism in high-$T_c$ superconductivity and symmetry of order parameter, *Usp. Phys.* **169**, 3, p. 215.

[77] Morya, T., Ueda, K. (2000). Spin fluctuations and high-temperature superconductivity, *Adv. Phys.* **49**, 5, pp. 556–605.

[78] Levi, B.G. (1993). In high-$T_c$ superconductors, is d-wave the new wave, *Phys. Today* **5**, pp. 17–20.

[79] Manske, D. (2004). *Theory of unconventional superconductors* (Springer, Heidelberg).

[80] Maple, M.B. (1986). New type superconductivity in f-electron system, *Physics Today* **4**, p. 72.

[81] Miyake, K. and Schmitt-Rink, S. (1986). Spin fluctuation mediated even party pairing in heavy fermion superconductors, *Phys. Rev. B* **34**, 9, pp. 6554–6556.

[82] Kruchinin, S.P. (1995). Functional integral of antiferromagnetic spin fluctuations in high-temperature superconductors, *Mod. Phys. Let. B* **9**, pp. 205–215.

[83] Kruchinin, S.P. and Patapis, S.K. (1996). Thermodynamics of d-wave pairing in cuprate superconductors, *J. Low Temp. Phys.* **105**, pp. 717–721.

[84] Kruchinin, S.P. and Patapis, S.K.(1997). Specific heat of antiferromagnetic spin fluctuations in cuprate superconductors, *Physica C* **282–285**, pp. 1397–1398.

[85] Kruchinin, S.P. (2000). The pseudogap of antiferromagnetic spin fluctuations in cuprate superconductors, *Physica C* **341–348**, pp. 901–902.

[86] Kruchinin, S.P. (2005). Condensation energy for spin fluctuation mechanism of pairing in high-Tc superconductors, *Proceedings a NATO Advanced Research Workshop "New challenges in superconductivity: Experimental advances and emerging theories"*, ed. Ashkenazi, J. (Kluwer Academic Publishers, Dordrecht), pp. 145–150.

[87] Popov, Y.N. (1987). *Functional integrals and collective excitations* (Cambridge University Press, Cambridge), p. 216.

[88] Conwall, J., Jackiw, R. and Tomhoulis, E. (1974). Effective action for composite operators, *Phys. Rev. D* **10**, 8, pp. 2428–2444.

[89] Weinberg, St. (1993). Effective action and renormalization group flow of anisotropic superconductors, Preprint UTTG 18, p. 1–17

[90] Shen, Z.X., Dessan, D.S., Wells, B.O., King, D.M., Spicer, W.E., Arko, A.J., Marshall, D., Lombarolo, L.W., Kapitulnik, A., Dickinson, P., Doniach, S., Carko, J.Di., Loesser, A.G. and Park, C.H. (1993). Anomalously large gap anisotropy in the a–b plane of $Bi_2Sr_2CaCu_2O_{8+\gamma}$, *Phys. Rev. Lett.* **70**, p. 1553.

[91] Moler, M.B., Barr, D.I., Urbach, J.S., Hardy, W.H. and Kapitulnik, A. (1994). Magnetic field dependence of the density of states of YBaCuO as determined from the specific heat, *Phys. Rev. Lett.* **73**, p. 2744.

[92] Momono, N., Ido, M., Nakano,T. and Oda, M. (1994). Low-temperature electronic specific heat in LaSrCuZnO, *Physica C* **235–240**, pp. 1739–1740.

[93] Loram, J.W., Mirza, K.A., Cooper, J.R. and Tallon, J.L. (1998). Specific heat evidence on the normal state pseudogap, *J. Phys. Chem. Solid* **59**, 10–12, pp. 2091–2094.

[94] Revaz, B., Genond, J.Y., Junod, A., Naumdier, K., Erb, A. and Walker, E. (1998). d-wave scaling relations in the mixed-state specific heat $YBa_2Cu_3O_7$, *Phys. Rev. Lett.* **80**, 15, pp. 3364–3367.

[95] Hasselbach, K., Kirtley, J.R. and Flouquet, J. (1994). Symmetry of the gap in superconducting UReSi, *Phys. Rev. B* **47**, pp. 509–512.

[96] Muhrer, G. and Schachinger, E. (2000). Free energy of a d-wave superconductor in the nearly antiferromagnetic Fermi liquid model, *J. Low. Temp. Phys.* **88–120**, 1–2, pp. 65–88.

[97] Dorbolo, S., Ausloos, M. and Houssa, M. (1998). Electronic specific heat of superconductors with van Hove singularities: effects of a magnetic field and thermal fluctuations, *Phys. Rev. B* **57**, pp. 5401–5411.

[98] Dobrosavljevic-Crujic, L. and Miranovic, P. (2003). Low temperature specific heat in anisotropic superconductors, *Physica C* **397**, pp. 117–122.

[99] Bogolyubov, Jr. N.N. and Kruchinin, S.P. (2003). Modern approach to the calculation of the correlation function for superconductivity model, *Mod. Phys. Lett. B* **17**, 10–12, pp. 709–724.

[100] Kamihara, Y., Watanabe, T., Hirano, M. and Hosono, H. (2008). Iron-based layered superconductor La[$O_{1-x}F_x$]FeAs ($x$ = 0.05-0.12) with $T_c$ = 26 K, *J. Am. Chem. Soc.* **130**, pp. 3296–3297.

[101] Song, C.L., Wang, Y.-L., Cheng, P., Jiang, Y.-P., Li, W., Zhang, T., Li, Z., He, K., Wang, L., Jia, J.-F., Hung, H.-H., Wu, C., Ma, X., Chen, X. and Xue, Q.-K. (2011). Direct observation of nodes and twofold symmetry in FeSe superconductor, *Science* **332**, 6036, p. 1410.

[102] Lee, J.J., Schmitt, F.T., Moore, R.G., Johnston, S., Cui, Y.-T., Li, W., Yi, M., Liu, Z.K., Hashimoto, M., Zhang, Y., Lu, D.H., Devereaux, T.P., Lee, D.-H. and Shen, Z.-X. (2014). Interfacial mode coupling as the origin of the enhancement of $T_c$ in FeSe films on $SrTiO_{3x}$, *Nature* **515**, p. 245.

[103] Sadovskii, M.V. (2008). High-temperature superconductivity in layered iron compounds, *Phys. Usp.* **51**, 12, pp. 1201–1227.

[104] Izyumov, Yu.A. and Kurmaev, E.Z. (2008). FeAs systems: a new class of high-temperature superconductors, *Phys. Usp.* **51**, 12, pp. 1261–1286.

[105] Graser, S., Maier, T.A., Hirschfeld, P.J. and Scalapino, D.J. (2009). Near-degeneracy of several pairing channels in multiorbital models for the Fe pnictides, *New J. Phys.* **11**, p. 025016.

[106] Shirage, P. *et al.* (2010). Synthesis of ErFeAsO-based superconductors by the hydrogen doping method, *Europhys. Lett.* **92**, pp. 57011–57015.

[107] Lebègue, S. (2007). Electronic structure and properties of the Fermi surface of the superconductor LaOFeP, *Phys. Rev. B* **75**, p. 035110.

[108] Rotter, M., Tegel, M. and Johrendt, D. (2008). Superconductivity at 38 K in the Iron Arsenide $Ba_{1-x}K_x$Fe$_2$As$_2$, *Phys. Rev. Lett.* **101**, p. 107006.

[109] Chen, G.F., Li, Z., Li, G., Hu, W.Z., Dong, J., Zhou, J., Zhang, X.-D., Zheng, P., Wang, N.-L. and Luo, J.-L. (2008). Superconductivity in hole-doped ($Sr_{1-x}K_x$)Fe$_2$As$_2$, *Chin. Phys. Lett.* **25**, p. 9.

[110] Tapp, J.H., Tang, Z., Lv, B., Sasmal, K., Lorenz, B., Chu, P.C. W. and Guloy, A.M. (2008). LiFeAs: An intrinsic FeAs-based superconductor with $T_c$ = 18 K, *Phys. Rev. B* **78**, p.060505.

[111] Gooch, M., Lv, B., Lorenz, B., Guloy, A.M. and Chu, C.W. (2009). Evidence of quantum criticality in the phase diagram of $K_xSr_{1-x}$Fe$_2$As$_2$ from measurements of transport and thermoelectricity, *Phys. Rev. B* **79**, p. 104504.

[112] Mizuguchi, Y., Tomioka, F., Tsuda, S., Yamaguchi, T. and Takano, Y. (2008). Superconductivity at 27 K in tetragonal FeSe under high pressure, *Appl. Phys. Lett.* **93**, p. 152505.

[113] Garbarino, G., Sow, A., Lejay, P., Sulpice, A., Toulemonde, P., Mezouar, M. and Núñez-Regueiro, M. (2009). High-temperature superconductivity ($T_c$ onset at 34 K) in the high-pressure orthorhombic phase of FeSe, *Europhys. Lett.* **86**, p. 27001.

[114] Gor'kov, L.P. (2016). Superconducting transition temperature: Interacting Fermi gas and phonon mechanisms in the nonadiabatic regime, *Phys. Rev. B* **93**, p. 054517.

[115] Gor'kov, L.P. (2016). Peculiarities of superconductivity in the single-layer FeSe/SrTiO$_3$ interface, *Phys. Rev. B* **93**, p. 060507.

[116] Mizuguchi, Y., Hara, Y., Deguchi, K., Tsuda, S., Yamaguchi, T. and Takeda, K., Kotegawa, H., Tou, H., and Takano, Y. (2010). Anion height dependence of $T_c$ for the Fe-based superconductor, *Supercond. Sci. Technol.* **23**, 5, p. 054013

[117] Kruchinin, S.P., Zolotovsky, A. and Kim, H.T. (2013). Band structure of new ReFe AsO superconductors, *J. Mod. Phys.* **4**, pp. 608-611.

[118] Lu, D. H., Yi, M., Mo, S.-K., Erickson, A.S., Analytis, J., Chu, J.-H., Singh, D.J., Hussain, Z., Geballe, T.H., Fisher, I.R. and Shen, Z.-X. (2008). Electronic structure of the iron-based superconductor LaOFeP, *Nature* **455**, p. 81.

[119] Liu, C., Kondo, T., Tillman, M.E., Gordon, R., Samolyuk, G.D., Lee, Y., Martin, C., McChesney, J.L., Bud'ko, S., Tanatar, M.A., Rotenberg, E., Canfield, P.C., Prozorov, R., Harmon, B.N. and Kaminski, A. (2008). Fermi surface and strong coupling superconductivity in single crystal NdFeAsO$_{1-x}$F$_x$, Preprint, arXiv:0806.2147.

[120] Coldea, A.I., Fletcher, J.D., Carrington, A., Analytis, J.G., Bangura, A.F., Chu, J.H., Erickson, A.S., Fisher, I.R., Hussey, N.E. and McDonald, R.D. (2008). Fermi surface of superconducting LaFePO determined from quantum oscillations, *Phys. Rev. Lett.* **101**, p. 216402.

[121] Hanaguri, T.T., Niitaka, S., Kuroki, K. and Takagi, H. (2010). Unconventional s-wave superconductivity in Fe(Se, Te), *Science*, **328**, pp. 474–476.

[122] De la Cruz, C., Huang, Q., Lynn, J.W., Li, J., Ratcliff, W., Zarestky, J.L., Mook, H.A., Chen, G.F., Luo, J.L. and Wang, N.L. (2008). Pengcheng dai magnetic order close to superconductivity in the iron-based layered LaO$_{1-x}$F$_x$FeAs systems, *Nature* **453**, p. 899.

[123] Dong, J., Zhang, H.J., Xu, G., Li, Z., Li, G., Hu, W.Z., Wu, D., Chen, G.F., Dai, X., Luo, J.L., Fang, Z. and Wang, N.L. (2008). Competing orders and spin-density-wave instability in La(O$_{1-x}$F$_x$)FeAs, *Europhys. Lett.* **83**, 2, p. 27006.

[124] Basov, D.N. and Chubukov, A.V. (2011). Manifesto for a higher $T_c$, *Nature Phys.* **7**, 4, pp. 272–276.

[125] Barzykin, V. and Gor'kov, L.P. (2008). On superconducting and magnetic properties of iron oxypnictides, *JETP Lett.* **88**, p.131.

[126] Mazin, I.I., Singh, D.J., Johannes, M.D. and Du, M.H. (2008). Unconventional superconductivity with a sign reversal in the order parameter of LaFeAsO$_x$F$_x$, *Phys. Rev. Lett.* **101**, p. 057003.

[127] Choi, H.-Y., Yun, J.H., Bang, Y. and Lee, H.C. (2009). Model for the inverse isotope effect of FeAs-based superconductors in the $\pi$-phase-shifted pairing state, Preprint, arXiv:0904.1864v3.

[128] Ding, L., He, C., Dong, J.K., Wu, T., Liu, R.H., Chen, X.H. and Li, S.Y. (2008). Specific heat of the iron-based high-$T_c$ superconductor SmO$_{1-x}$F$_x$FeA, *Phys. Rev. B* **77**, p. 180510.

[129] Wang, F., Zhai, H., Ran, Y., Vishwanath, A. and Lee, D.-H. (2009). Funtional renormalization group study of the pairing symmetry and pairing mechanism of the FeAs based high temperature superconductors, Preprint, arXiv:0804.4332.

[130] Mazin, I.I., Singh, D.J., Johannes, M.D., and Du, M.H. (2008). Unconventional superconductivity with a sign reversal in the order parameter of LaFeAsO$_{1-x}$F$_x$, *Phys. Rev. Lett.* **101**, p. 057003.

[131] Kuroki, K., Onari, S., Arita, R., Usui, H., Tanaka, Y., Kontani, H. and Aoki, H. (2008). Unconventional pairing originating from the disconnected Fermi surfaces of superconducting LaFeAsO, *Phys. Rev. Lett.* **101**, p. 087004.

[132] Bulut, N., Scalapino, D.J. and Scalettar, R.T. (1992). Nodeless d-wave pairing in a two-layer Hubbard model, *Phys. Rev. B* **45**, p. 5577.

[133] Okazaki, K., Ito, Y., Ota, Y., Kotani, Y., Shimojima, T., Kiss, T., Watanabe, S., Chen, C.-T., Niitaka, S., Hanaguri, T., Takagi, H., Chainani, A. and Shin, S. (2012). Evidence for a cos(4$\varphi$) modulation of the superconducting energy gap of optimally doped FeTe$_{0.6}$Se$_{0.4}$ single crystals using laser angle-resolved photoemission spectroscopy, *Phys. Rev. Lett.* **109**, p.237011.

[134] Raghu, S., Qi, X.-L, Liu, C.-X., Scalapino, D. and Zhang, S.-C. (2008). Minimal two-band model of the superconducting iron oxypnictides, *Phys. Rev. B* **77**, p. 220503.

[135] Kuroki, K., Onari, S., Arita, R., Usui, H., Tanaka, Y., Kontani, H. and Aoki, H. (2009). Unconventional pairing originating from the disconnected Fermi surfaces of superconducting LaFeAsO$_{1-x}$F$_x$, *Phys. Rev. Lett.* **102**, p. 109902.

[136] Seo, K., Bernevig, B.A. and Hu, J. (2008). Pairing symmetry in a two-orbital exchange coupling model of oxypnictides, *Phys. Rev. Lett.* **101**, p. 206404.

[137] Nagao, H., Kruchinin, S.P., Yaremko, A.M. and Yamaguchi, K. (2002). Multiband superconductivity, *Int. J. Mod. Phys. B* **16**, pp. 3419–3428.

[138] Kruchinin, S.P. and Nagao, H. (2005) Two-gap superconductivity in MgB, *Phys. Particles Nucl.* **36**, Suppl. 1, pp. 127–130.

[139] Kruchinin, S.P. and Yaremko, A.M. (1998). Many zone effect in cuprate superconductors, *Supercond. Sci. Technol.* **11**, pp. 4–8.

[140] Nagao, H., Kawabe, H., Kruchinin, S.P., Manske, D. and Yamaguchi, K. (2003). Theoretical studies on many band effects superconductivity by using renormalization group approach, *Mod. Phys. Lett. B* **17**, 10–12, pp. 423–431.

[141] Nagao, H., Kawabe, H., Kruchinin, S.P., Manske, D. and Yamaguchi, K. (2003). Superconductivity in two-band model by renormalization group approach, *Int. J. Mod. Phys. B* **17**, p. 3373–3376.

[142] Yaremko, A.M., Mozdor, E.V. and Kruchinin, S.P. (1996). Coupled states of electron-phonon system, superconductivity and HTSC of crystals, *Int. J. Mod. Phys. B* **10**, pp. 2665–2674.

[143] Kruchinin, S.P. and Nagao, H. (2006). *Multi-gap superconductivity in MgB$_2$, Proceedings a NATO Advanced Research Workshop "Symmetry and heterogeneity in High-T$_c$ superconductors"*, Bianconi, A. (ed.) (Kluwer Academic Publishers, Dorderecht), pp. 43–53.

[144] Nagao, H., Yaremko, A., Kruchinin, S.P., Yamaguchi, K. (2002). *Many band effects in superconductivity, Proceedings of a NATO Advanced Research*

*Workshop "New trends in superconductivity"*, Annett, J., and Kruchinin, S.P. (eds) (Kluwer Academic Publishers, Dordrecht), pp. 155–167.

[145] Kruchinin, S. (2016). Multiband superconductivity, *Rev. Theoret. Phys.* **4**, 2, pp. 165–178.

[146] Nakazawa, Y. and Kruchinin, S. (2018). Experimental and theoretical aspects of thermodynamic properties of quasi-1D and quasi-2D organic conductors and superconductors, *Inter. J. Mod. Phys. B* **30**, 13, p. 1042008.

[147] Kruchinin, S., Zolotovsky, A., Yamashita, S. and Nakazawa, Y. (2016). Thermodynamics of the d-wave pairing in organic superconductors, *Int. J. Mod. Phys. B* **30**, 13, p. 1042020.

[148] Bardeen, J., Cooper, L.N. and Schrieffer, J. R. (1957). Microscopic theory of superconductivity, *Phys. Rev.* **108**, pp. 1175.

[149] Anderson, P.W. and Zou, Z. (1988). "Normal" tunneling and "normal" transport: diagnostics for the resonating-valence-bond state, *Phys. Rev. Lett.* **60**, p. 132.

[150] Kampf, A.P. (1994). Magnetic correlations in high temperature superconductivity, *Phys. Rep.* **249**, p. 219.

[151] Emery, V.J. (1987). Theory of high-$T_c$ superconductivity in oxides, *Phys. Rev. Lett.* **58**, p. 2794.

[152] Hirsch, J.E. (1985). Attractive interaction and pairing in fermion systems with strong on-site repulsion, *Phys. Rev. Lett.* **25**, 1317.

[153] Zhang, F.C. and Rice, T.M. (1988). Effective Hamiltonian for the superconducting Cu oxides, *Phys. Rev. B* **37**, p. 3759.

[154] Fradkin, E. (1991). *Field theories of condensed matter systems* (Addison-Wesley).

[155] Nagaosa, N. and Lee, P.A. (1990). Normal-state properties of the uniform resonating-valence-bond state, *Phys. Rev. Lett.* **64**, p. 2450.

[156] Baskaran, G., Zou, Z. and Anderson, P. W. (1987). The resonating valence bond state and high-$T_c$ superconductivity-a mean field theory, *Solid State Commun.* **63**, p. 973.

[157] Fukuyama, H. and Yoshida, K. (1987). Critical temperature of superconductivity caused by strong correlation, *Jpn. J. Appl. Phys.* **26**, pp. 371–373.

[158] Yamamoto, S., Yamaguchi, K. and Nasu, K. (1990). Ab initio molecular-orbital study on electron correlation effects in $CuO_6$ clusters relating to high-$T_c$ superconductivity, *Phys. Rev. B* **42**, pp. 266–272.

[159] Nagao, H., Kitagawa, Y., Kawakami, T., Yoshimoto, T., Saito, H. and Yamaguchi, K. (2001). Theoretical studies on field-induced superconductivity in molecular crystals, *Int. J. Quantum Chem.* **85**, pp. 608–618.

[160] Kimura, M., Kawabe, H., Nakajima, A., Nishikawa, K. and Aono, S. (1988). Possibility of superconducting and other phases in organic high polymers polyacene carbon skeletons, *Bull. Chem. Soc. Jpn.* **61**, p. 4239.

[161] Torrance, J.B., Bezinge, A., Nazzal, A.I. and Parkin, S.S. (1989). Disappearance of high temperature superconductivity induced by high carrier concentrations, *Physica C* **162–164**, p. 291.

[162] Ord, T., Kristoffel, N. and Rago, K. (2003). $MgB_2$ superconductivity properties in a two-gap model, *Mod. Phys. Let. B* **17**, pp. 667–673.

[163] Suhl, H., Matthias, B.T. Walker, R. (1959). Bardeen–Cooper–Schrieffer theory of superconductivity in the case of overlapping bands, *Phys. Rev. Lett.* **3**, p. 552.

[164] Moskalenko, V.A. (1959). Superconductivity metals with overlapping energetic bands, *Fiz. Metalloved.* **8**, p. 503.

[165] Antipov, E. and Abakumov, A. (2008). Structural design of superconductors based on complex copper oxides, *Phys. Usp.* **51**, p. 180.

[166] Müller, K.A. (2006). The search for new high temperature superconductors, *Supercond. Sci. Technol.* **58**, pp. 1–3.

[167] Bennemann, K.H., Ketterson, J.B. (2008). *Superconductivity* (Springer, Heidelberg), **1**.

[168] Nagamatsu, J., Nakamura, N., Muranaka, T., Zentani, Y. and Akimitsu, J. (2001). Superconductivity at 39 K in magneszium dieboride, *Nature* **410**, p. 63.

[169] Canfield, J.P.C., Bud'ko, S.L. and Finnemore, D.K.(2003). An overview of the basic physical properties of $MgB_2$, *Physica C* **385**, 1–2, pp. 1–3.

[170] Binnig, G., Baratoff, A., Hoening, Bednorz, J.G. (1980). Two-band superconductivity in Nb-doped $SrTiO_3$, *Phys. Rev. Lett.* **45**, p. 1352.

[171] An, J.M. and Picket, W.E. (2001). Superconductivity of $MgB_2$: covalent bonds driven metallic, *Phys. Rev. Lett.* **86**, p. 4366.

[172] Kortus, J., Mazin, I.I., Belashenko, K.D., Antropov. V.P. and Boyer,I.L. (2001). Superconductivity of metallic boron in $MgB_2$, *Phys. Rev. Lett.* **86**, p. 4656.

[173] Kondo, J. (1963). Superconductivity in transition metals, *Prog. Theor. Phys.* **29**, p. 1.

[174] Kristoffel, N., Ord, T. and Rago, K. (2003). $MgB_2$ two-gap superconductivity with intra- and interband couplings, *Europhys. Lett.*, **61**, pp. 109–115.

[175] Szabo, P., Samuely, P., Klein, T., Marcus, J., Fruchart, D., Miraglia, S., Marcenat, C. and Jansen, A. (2001). Evidence for two superconducting energy gaps in $MgB_2$ by point-contact spectroscopy, *Phys. Rev. Lett.* **87**, p. 137005.

[176] McMillan, W.L. (1968). Transition temperature of strong-coupled superconductors, *Phys. Rev.* **167**, p. 331.

[177] Konsin, P., Kristoffel, N. and Örd, T. (1988). The interband interaction as a possible cause of high-temperature superconductivity, *Phys. Lett. A* **129**, pp. 339–342.

[178] Combescot, R. and Leyronas, X. (1995). Coupling between planes and chains in $YBa_2Cu_3O_7$: a possible solution for the order parameter controversy, *Phys. Rev. Lett.* **75**, pp. 3732–3735.

[179] Konsin, P. and Sorkin, B. (1998). Electric field effects in high-$T_c$ cuprates, *Phys, Rev. B* **58**, pp. 5795–5802.

[180] Nagao, H., Nishino, M., Shigeta, Y., Yoshioka, Y. and Yamaguchi, K. (2000). Theoretical studies on superconducting and other phases: Triplet superconductivity, ferromagnetism, and ferromagnetic metal, *J. Chem. Phys.* **113**, pp. 721–732.

[181] Kondo, J. (2001). Superconductivity of the two-dimensional Hubbard models with a small U, *J. Phys. Soc. Jpn.* **70**, p. 808.

[182] Nagao, H., Mitani, M., Nishino, M., Shigeta, Y., Yoshioka, Y. and Yamaguchi, K. (1999). Theoretical studies on anomalous phases in molecular systems with external field: Possibility of photo-induced superconductivity, *Int. J. Quantum Chem.* **75**, pp. 549–561.

[183] Nagao, H., Nishino, M., Mitani, M., Yoshioka, Y. and Yamaguchi, K. (1997). Possibilities of charge- and/or spin-mediated superconductors and photo-induced superconductors in the intermediate region of metal-insulator transitions, *Int. J. Quantum Chem.* **65**, pp. 947–957.

[184] Nagao, H., Mitani, M., Nishino, M., Shigeta, Y., Yoshioka, Y. and Yamaguchi, K. (1988). Possibility of charge-mediated superconductors in the intermediate region of metal-insulator transitions, *Int. J. Quantum Chem.* **70**, pp. 1075–1084.

[185] Bogoluibov, N.N. and Bogoluibov, Jr., N.N. (1986). *An introduction to quantum statistical mechanics* (Nauka, Moscow) (translated in English 1994).

[186] Bogoliubov, N.N. and Shirkov, D.V. (1959). *Introduction to the theory of quantized field* (Interscience Publishers Inc., New York).

[187] Mattuck, R.D. (1976). *A guide to Feynman diagrams in the many-body problem* 2nd edn. (McGraw-Hill, New York).

[188] Kimura, M., Kawabe, H., Nishikawa, K. and Aono, S. (1988). Possibility of superconducting and other phases in organic high polymers of polyacene carbon skeletons II. Screened electron–electron interaction, *Bull. Chem. Soc. Jpn.* **61**, pp. 4245–4252.

[189] Kimura, M., Kawabe, H., Nishikawa, K. and Aono, S.(1986). Superconducting and other phases in organic high polymers of polyacenic carbon skeletons. I. The method of sum of divergent perturbation series, *J. Chem. Phys.* **85**, pp. 3090–3096.

[190] Kimura, M., Kawabe, H., Nishikawa, K. and Aono, S.(1985). Superconducting and other phases in organic high polymers of polyacenic carbon skeletons. II. The mean field method, *J. Chem. Phys.* **85**, pp. 3097–3100.

[191] Kimura, M., Kawabe, H., Nakajima, A., Nishikawa, K. and Aono, S. (1988). Possibility of superconducting and other phases in organic high polymers of polyacene carbon skeletons. II. Screened electron–electron interaction. *Bull. Chem. Soc. Jpn.* **61**, pp. 4239–4244.

[192] Bogoluibov, N.N. and Tyablikov, S. (1959). Retarded and advanced Green functions in statistical physics, *Dokl. Acad. Sci. USSR* **126**, p. 53, *Sov. Phys. Dokl.* **4**, p. 589.

[193] *Highlights of the year, Physics* **12**, 145, https://physics.aps.org/articles/v12/145.

[194] Somayazulu, M. *et al.* (2019). Evidence for superconductivity above 260 K in lanthanum superhydride at megabar pressures, *Phys. Rev. Lett.* **122**, p. 027001.

[195] Drozdov, A.P., Kong, P.P., Minkov, V.S., Besedin, S.P., Kuzovnikov, M.A., Mozaffari, S., Balicas, L., Balakirev, F., Graf, D., Prakapenka, V.B., Greenberg, E., Knyazev, D.A., Tkacz, M. and Eremets M.I. (2018). Superconductivity at 250 K in lanthanum hydride under high pressures, Preprint, arXiv:1812.01561.

[196] Peng, F. *et al.* (2007). Hydrogen clathrate structures in rare earth hydrides at high pressures: possible route to room-temperature superconductivity, *Phys. Rev. Lett.* **119**, p.107001.

[197] Drozdov, A.P., Eremets, M.I., Troyan, I.A., Ksenofontov, V. and Shylin, S.I. (2015). Conventional superconductivity at 203 kelvin at high pressures in the sulfur hydride system, *Nature*, p. 525.

[198] Flores-Livas, J.A., Boeri, L., Sanna, A., Profeta, G., Arita, R. and Eremets, M. (2020). A perspective on conventional high-temperature superconductors at high pressure: Methods and materials, *Phys. Rep.* **856**, 4, pp. 1–78.

[199] Ginzburg, V.L. (1999). What problems of physics and astrophysics seem now to be especially important and interesting (thirty years later, already on the verge of XXI century)? *Phys. Usp.* **42**, 4, p. 353.

[200] Snider, E., Dasenbrock-Gammon, N., McBride, R., Debessai, M., Vindana, H., Vencatasamy, K., Lawler, K., Salamat, A. and Dias, R.P. (2020). Room-temperature superconductivity in a carbonaceous sulfur hydride, *Nature* **586**, pp. 373–377.

[201] Ashcroft, N. (1968). Metallic hydrogen: A high-temperature superconductor? *Phys. Rev. Lett.* **21**, pp. 1748–1749.

[202] Bardeen, J., Cooper, L.N. and Schrieffer, J.R. (1957). Theory of superconductivity, *Phys. Rev.* **108**, pp. 1175–1204.

[203] Gor'kov, L.P. (1958). On the energy spectrum of superconductors, *Sov. Phys. JETP* **7**, 3, p. 505.

[204] Abrikosov, A.A., Dzyaloshinskii, I. and Gor'kov, L.P. (1975). *Methods of quantum field theory in statistical physics* (Dover, New York).

[205] Fetter, A., Walecka, J.D. (1971). *Quantum theory of many-particle systems* (Dover, New York).

[206] Eliashberg, G.M. (1960). Interactions between electrons and lattice vibrations in a superconductor, *Sov. Phys. JETP* **38**, p. 696.

[207] Dynes, R. (1972). McMillan's equation and the $T_c$ of superconductors, *Solid State Commun.* **10**, 7, pp. 615–618.

[208] McMillan, W.L. (1968). Transition temperature of strong-coupled superconductors, *Phys. Rev.* **167**, pp. 331–344.

[209] Wen, X.-D. *et al.* (2011). Graphane sheets and crystals under pressure, *Proc. Natl. Acad. Sci. USA* **108**, p.6833.

[210] Allen, P.B. and Dynes, R.C. (1975). Transition temperature of strong-coupled superconductors reanalyzed, *Phys. Rev. B* **12**, p.905.

[211] Richarson, R.W. (1963). A restricted class of exact eigenstates of the pairing-force hamiltonian, *Phys. Lett* **3**, pp. 277–279.

[212] Matveev, K.A. and Larkin, A.I. (1997). Parity effect in ground state energies of ultrasmall superconducting grains, *Phys. Rev. Lett.* **78**, pp. 3749–3752.

[213] Ralph, D.C., Black, C.T. and Tinkham, M. (1997). Gate-voltage studies of discrete electronic staes in aluminum nanoparticles, *Phys. Rev. Lett.* **78**, p. 4087.

[214] Schechter, M., Imry, Y., Levinson, Y. and von Delft, J. (2001). Thermodynamic properties of a small superconducting grain, *Phys. Rev. B* **63**, pp. 214518–214534.

[215] Sierra, G. (2000). Conformal field theory and the exact solution of the BCS hamiltonian, *Nucl. Phys. B* **572**, pp. 517–527.

[216] Stenuit, G., Michotte, S., Govaerts, J. and Piraux, L.P. (2003). Vortex matter in lead nanowires, *Eur. Phys. J. B* **33**, pp. 103–110.

[217] Kasumov, A.Yu. *et al.* (1999). Supercurrents through single-walled carbon nanotubes, *Science* **284**, pp. 1508–1511.

[218] Nagao, H., Kawabe, H. and Kruchinin. S. (2006). Nanosize two-gap superconductivity. *Proceedings of a NATO ARW "Electron correlation in new materials and nanosystems"*, (Scharnberg, K. and Kruchinin S., eds.) (Springer), pp. 117–127.

[219] Kawabe, H., Nagao, H., Kruchinin, S. (2006). Exact solution of two-band superconductivity, *Proceedings of a NATO ARW "Electron correlation in new materials and nanosystem"*, (Scharnberg, K. and Kruchinin, S. eds.) (Springer), pp. 129–139.

[220] Nagao, H. and Kruchinin, S.P. (2008). Kondo effect coupled to superconductivity ultrasmall grains. *Proceedings NATO ARW "Electron transport in nanosystems"*, (Bonca, J. and Kruchinin, S. eds.) (Springer), pp. 105–115.

[221] Kruchinin, S. and Nagao, H. (2012). Nanoscale superconductivity, *Int. J. Mod. Phys. B* **26**, p. 1230013.

[222] Kruchinin, S.P., Kawabe, H., Nagao, H. and Nakazawa, Y. (2013). Condensation energy for two-gap superconducting state in nanoparticles, *J. Nanoparticles* **3**, pp. 1–6.

[223] Dzhezherya, Yu., Novak, I.Yu. and Kruchinin, S. (2010). Orientational phase transitions of lattice of magnetic dots embedded in a London type superconductors, *Supercond. Sci. Technol.* **23**, p.1–5.

[224] Kruchinin, S. (2015). Kondo effect in superconducting nanoparticles, *Quantum Matter* **4**, p. 373–377.

[225] Kruchinin, S. (2015). Richardson solution for superconductivity in ultrasmall grains, *Quantum Matter* **4**, p. 378–383.

[226] Kruchinin, S. (2016). Energy spectrum and wave function of electron in hybrid superconducting nanowires, *Int. J. Mod. Phys. B* **30**, 13, p. 1042008.

[227] Kruchinin, S.P., Dzezherya, Yu. and Annett, J.(2009). Interactions of nanoscale ferromagnetic granules in a London superconductors, *Supercond. Sci. Technol.* **19**, pp. 381–384.

[228] Ralph, D.C., Black, C.T. and Tinkham, M. (1995). Spectroscopic mesurements of discrete electronic states in single metal particles, *Phys. Rev. Lett.* **74**, p. 3241.

[229] Smith, R.A. and Ambegaokar, V. (1996). Effect of level statistics on superconductivity in ultrasmall metallic grains, *Phys. Rev. Lett.* **77**, p. 4962.

[230] von Delft, J., Zaikin, A.D., Golubev, D.S. and Tichy, W. (1996). Parity-affected superconductivity in ultrasmall metallic grains, *Phys. Rev. Lett.* **77**, pp. 3189–3192.

[231] Yamaji, K. and Shimoi, Y. (1994). Superconducting transition of the two-chain Hubbard model indicated by diagonalization calculations, *Physica C* **222**, pp. 349–360.

[232] Black, C.T., Ralph, D.C. and Tinkham, M. (1996). Spectroscopy of the superconducting dap in individual nanometer-scale aluminum particles, *Phys. Rev. Lett.* **76**, pp. 688–611.

[233] Braun, F. and von Delft, J. (1998). Fixed-N superconductivity: The crossover from the bulk to the few-electron limit, *Phys. Rev. Lett.* **81**, pp. 4712–4715.

[234] von Delft, J. and Braun, F. (1999). Superconductivity in ultrasmall metallic grains, *Phys. Rev. B* **59**, pp. 9527–9544.

[235] Gladilin, V.N., Fomin, V.M. and Devreese, J.T. (2002). Shape of nanosize superconducting grains: does it influence pairing characteristics, *Solid State Comm.* **121**, pp. 519–523.

[236] Jankó, B., Smith, A. and Ambegaokar, V. (1994). BCS superconductivity with fixed number parity, *Phys. Rev. B* **50**, pp. 1152–1161.

[237] Matveev, K.A. and Larkin, A.L. (1997). Parity effect in ground state energies of ultrasmall superconducting grains, *Phys. Rev. Lett.*, **78**, pp. 3749–3752.

[238] Richardson, R.W. and Sherman, N. (1964). Pairing models of Pb206, Pb204 and Pb202, *Nucl. Phys.* **523**, pp. 253–268.

[239] Richardson, R.W. (1977). Pairing in the limit of a large number of particles, *J. Math. Phys.* **6**, pp. 1802–1811.

[240] Richardson, R.W. (1966). Numerical study of the 8-32-particle eigenstate of the pairing hamiltonian, *Phys. Rev.* **141**, pp. 949–955.

[241] Sierra, G., Dukelsky, J., Dussel, G.G., von Delft, J. and Braun, F. (2000). Exact study of the effect of level statistics in ultrasmall superconducting grains, *Phys. Rev. B* **61**, pp. 11890–11893.

[242] Kouwenhoven, L. and Glazman, L. (2001). The revival of the Kondo effect, *Phys. World* **14**, 1, pp. 33–38.

[243] van der Wiel, G., De Franceschi, S., Fujisawa, T., Elzerman, J.M., Tarucha, S. and Kouwenhoven, L.P. (2000). The Kondo effect in the unitary limit, *Science* **289**, pp. 2105–2108.

[244] Inoshita, T. (1998). Kondo effect in quantum dots, *Science* **281**, pp. 526–527.

[245] Sasaki, S., De Franceschi, S., Elzerman, J.M., van der Wiel, W.G., Eto, M., Tarucha, S. and Kouwenhoven, L.P. (2000). Kondo effect in an integer-spin quantum dot, *Nature (London)* **405**, pp. 764–767.

[246] Inoshita, T., Shimizu, A., Kuramoto, Y. and Sakaki, H. (1993). Correlated electron transport through a quantum dot: The multiple-level effect, *Phys. Rev. B* **48**, pp. 114725–14728.

[247] Izumida, W., Sakai, O. and Shimizu, Y. (1998). Kondo effect in single quantum dot systems-study with numerical renormalization group method, *J. Phys. Soc. Jpn.* **67**, p. 2444.

[248] Yeyati, A.L., Flores, F. and Martin-Rodero, A.(1999). Transport in multilevel quantum dots: from the Kondo effect to the Coulomb blockade regime, *Phys. Rev. Lett.* **83**, pp. 600–603.

[249] Pasupathy, A.N., Bialczak, R.C., Martinek, J., Grose, J.E., Donev, L.A.K., McEuen, P.L. and Ralph, D.C. (2004). The Kondo effect in the presence of ferromagnetism, *Science* **306**, pp. 86–90.

[250] Matsubayashi, D. and Eto, M. (2007). Spin splitting and Kondo effect in quantum dots coupled to noncollinear ferromagnetic leads, *Phys. Rev. B* **75**, p. 165319.

[251] Buitelaar, M.R., Nussbaumer, T. and Schönenberger, C. (2002). Quantum dot in the Kondo regime coupled to superconductors, *Phys. Rev. Lett.* **89**, p. 256801.

[252] Ueda, A. and Eto, M. (2006). Resonant tunneling and Fano resonance in quantum dots with electron–phonon interaction, *Phys. Rev. B* **73**, pp. 235353–235365.

[253] Eto, M. and Nazarov, Y.V. (2000). Enhancement of Kondo effect in quantum dots with an even number of electrons, *Phys. Rev. Lett.* **85**, pp. 1306–1309.

[254] Reich, S., Leitus, G., Popovitz-Biro, R. and Schechter, M. (2003). Magnetization of small lead particles, *Phys. Rev. Lett.* **91**, p. 147001.

[255] Yamaji, K. and Shimoi, Y.(1994). Superconducting transition of the two-chain Hubbard model indicated by diagonalization calculations, *Physica C* **222**, pp. 349–360.

[256] Combescot, R. and Leyronas, X. (1995). Coupling between planes and chains in $YBa_2Cu_3O_7$: a possible solution for the order parameter controversy, *Phys. Rev. Lett.* **75**, pp. 3732–3735.

[257] Cox, D.L. and Maple, M.B. (1995). Electronic pairing in exotic superconductors, *Phys. Today* **48**, 2, p. 32.

[258] Hewson, A. C. (1993). *The Kondo problem to heavy fermion* (Cambridge University Press, Cambridge).

[259] Yoshimori, A. and Sakurai, A. (1970). Functional integral approach to the bound state due to the s-d exchange Interaction, *Suppl. Prog. Theor. Phys.* **46**, pp. 162–181.

[260] Eto, M. and Nazarov, Y.V. (2001). Mean field theory of the Kondo-effect in quantum dots with a even numbers of electrons, *Phys. Rev. B* **64**, p. 85322.

[261] Fulde, P. and Ferrell, A. (1964). Superconductivity in a strong spin-exchange field, *Phys. Rev.* **135**, pp. A550–A563.

[262] Larkin, A. and Ovchinnikov, Y. (1965). Inhomogeneous state of superconductors, *Sov. Phys. JETP* **20**, p. 762.

[263] Lyuksyutov, I.F. and Pokrovsky, V.L. (2005). Ferromagnet superconductor hybrids, *Adv. Phys.* **54** pp. 67–136.

[264] Kruchinin, S. (2017). *Problems and solutions in special relativity and electromagnetism* (World Scientific), p. 140.

[265] Ryazanov, V.V., Oboznov, V.A., Rusanov, A.V., Veretennikov, A.V., Golubov, A.A. and Aarts, J. (2001). Coupling of two superconductors through a ferromagnet: Evidence for a $\pi$-junction, *Phys. Rev. Lett.* **86**, pp. 2427–2430.

[266] Frolov, S.M., Van Harlingen, D.J., Oboznov, V.A., Bolginov, V.V. and Ryazanov, V.V. (2004). Measurement of the current-phase relation of superconductor/ferromagnet/superconductor Josephson junctions, *Phys. Rev. B* **70**, p. 144505.

[267] Bell, C., Loloee, R., Burnell, G. and Blamire, M.G. (2005). Characteristics of strong ferromagnetic Josephson junctions with epitaxial barriers, *Phys. Rev. B* **71**, p. 180501.

[268] Ruotolo, A., Pepe, G.P., Bell, C. Leung, C.W. and Blamire, M.G. (2005). Modulation of the dc Josephson current in pseudo-spin-valve Josephson multilayers, *Supercond. Sci. Technol.* **18**, pp. 921–926.

[269] Lange, M., Van Bael, M.J., Van Look, L., Raedts, S., Moschalkov, V.V. and Bruynseraede, Y. (2002). Nanostructured superconductor ferromagnet bilayers. in *New Trends in Superconductivity*, Annett, J.F. and Kruchinin, S. (eds.), *NATO Science Series II Mathematics Physics and Chemistry* Vol. **67**, pp. 365–373 (Kluwer Academic Publishers, Dordrecht).

[270] Aladyshkin, A., Yu., Silhanek, A.V., Gillijns, W. and Moshchalkov, V.V. (2009). Nucleation of superconductivity and vortex matter in superconductor-ferromagnet hybrids, *Supercond. Sci.Technol.* **22**, pp. 1–48

[271] Milosovec, M.V. and Peeters, F.M. (2005). Field-enhanced critical parameters in magnetically nanostructured superconductors, *Europhys. Lett.* **70**, pp. 670–676.

[272] Fetter, V.M. and Larkin, A.I. (2002). Spin glass versus superconductivity, *Phys. Rev. B* **66**, p. 64526.

[273] Andreev, A.F. (1998). Superfluidity, superconductivity, and magnetism in mesoscopics, *Usp. Phys.*, **168**, 6, pp. 656–663.

[274] Larkin, A. and Ovchinnikov, Y. (1965). *Nonuniform state of superconductors*, *Sov. Phys. JETP* **20**, 3, p. 762–770.

[275] Fulde, P. and Ferrell, A. (1964). Superconductivity in a strong spin-exchange field, *Phys. Rev.* **135**, p.A550.

[276] Mota, A.C., Visani, P., Polloni, A. and Aupke, K. (1994). Coherent phenomena in mesoscopic cylinders of Cu and Ag in proximity with a superconductor, *Physica B* **197**, pp. 95–100.

[277] Müller-Allinger, F.B. and Mota, A.C. (2000). Reentrance of the induced diamagnetism in gold-niobium proximity cylinders, *Phys. Rev. B* **62**, pp. 6120.

[278] Müller-Allinger, F.B. and Mota, A.C. (2000). Paramagnetic reentrant effect in high purity mesoscopic AgNb proximity structures, *Phys. Rev. Lett.* **84**, pp. 3161–3164.

[279] Müller-Allinger, F.B. and Mota, A.C. (2000). Magnetic response of clean NS proximity cylinders, *Physica B* **284–288**, p.683–684.

[280] Caroli. C. and Matricon, J. (1965). Excitations électroniques dans les supraconducteurs de 2ème espèce. II. Excitations de basse énergie, *Phys. Kondens. Mater.* **3**, pp. 380–401.

[281] Lyuksyutov, I.F. and Pokrovsky, V.L. (2005). Ferromagnet-superconductor hybrids, *Adv. Phys.* **54**, pp. 67–136.

[282] Buzdin, A.I. (2005). Proximity effects in superconductor-ferromagnet heterostructures, *Rev. Mod. Phys.* **77**, pp. 935–976.

[283] Gygi, F. and Schlüter, M. (1991). Self-consistent electronic structure of a vortex line in a type-II superconductor, *Phys. Rev. B* **43**, pp. 7609–7621.

[284] Wang, J. *et al.* (2010). Interplay between superconductivity and ferromagnetism in crystalline nanowires, *Nature Phys.* **6**, p.389.

[285] Shanenko, A.A., Croitoru, D., Vagov, A. and Peeters, M. (2008). *Proceedings of NATO ARW "physical properties of nanosystems"*, Bonca, J. and Kruchinin, S. (eds.) (Springer, Berlin), pp. 340–348.

[286] Hofstadter, D. (1976). Energy levels and wave functions of Bloch electrons in rational and irrational magnetic fields, *Phys. Rev. B* **14**, pp. 2239–2249.

[287] Kretinin, A.V., Cao, Y., Tu, J.S., Yu, G.L., Jalil, R., Novoselov, K.S., Haigh, S.J., Gholinia, A., Mishchenko, A., Lozada, M., Georgiou, T., Woods, C.R., Withers, F., Blake, P., Eda, G., Wirsig, A., Hucho, C., Watanabe, K., Taniguchi, T., Geim, A.K. and Gorbachev, R.V. (2014). Electronic properties of graphene encapsulated with different two-dimensional atomic crystals, *Nano Lett.* **14**, 6, pp. 3270–3276.

[288] Milosevic, M.V., Yampolskii, S.V. and Peeters, F.M. (2002). Magnetic pinning of vortices in a superconducting film: the (anti)vortex-magnetic dipole interaction energy in the London approximation, *Phys. Rev. B* **66**, p. 174519.

[289] de Gennes P.-G.(1966). *Superconductivity of metals and alloys* (Addision-Wesley, Advanced Book Programme).

[290] Van der Wal, C. (2000). Quantum superposition of macroscopic persistent-current states, *Science* **290**, p. 773.

[291] You, J.Q. and Nori, F. (2005). Superconducting circuits and quantum information, *Phys. Today* **12**, pp. 42–45.

[292] Nakamura, Y., Pashkin, Y.A. and Tsai, J.S. (1999). Quantum oscillations in two coupled charge qubits, *Nature* **398**, pp. 786–788.

[293] Makhlin, Y., Schon, G. and Shnirman, A. (2001). Quantum-state engineering with Josephson-junction devices, *Rev. Mod. Phys.* **73**, pp. 357–400.

[294] Chiorescu, I., Nakamura, Y., Harmans, C.J.P.M. and Mooij J.E. (2003). Coherent quantum dynamics of a superconducting flux qubit, *Science* **299**, pp. 1869–1871.

[295] Manin, Yu.I. (1980). *Computable and noncomputable* (Sov. Radio, Moscow, in Russian), p. 15.

[296] Wiesner, S. (1983). Conjugate coding, *SIGACT News* **15**, 1, pp. 78–88.

[297] Feynman, R.P. (1982). Simulating physics with computers, *Int. J. Theoret. Phys.* **21**, 6, pp. 467—488.

[298] Emelyanov, V.I. and Vladimirova, V.V. (2012). *The quantum physics. Bits and qubits* (Moscow State University), 176 p.

[299] Greenstein, G., Zajonc, A.G. (2006). *The quantum challenge: modern research on the foundations of quantum mechanics (physics and astronomy)* (Jones and Bartlett Publishers), p. 400.

[300] Qi, Xiao-Liang, Zhang, Shou-Cheng 2011. Topological insulators and super-conductors, *Rev. Mod. Phys.* **83**, 4, pp. 1057–1110.

[301] Kitaev, A.Yu. (2001). Unpaired Majorana fermions in quantum wires, *Phys.-Uspekhi.* **44**, 10S, p. 131.

[302] Zhang, Q. and Wu, B. (2018). Majorana modes in solid state systems and its dynamics, *Front. Phys.* **13**, 2, p. 137101.

[303] Mourik, V. *et al.* (2012). Signatures of Majorana Fermions in hybrid superconductor-semiconductor nanowire devices, *Science* **336**, 6084, p. 1003.

# Index